The Cambridge Handbook of Physics Formulas

The Cambridge Handbook of Physics Formulas is a quick-reference aid for students and professionals in the physical sciences and engineering. It contains more than 2 000 of the most useful formulas and equations found in undergraduate physics courses, covering mathematics, dynamics and mechanics, quantum physics, thermodynamics, solid state physics, electromagnetism, optics, and astrophysics. An exhaustive index allows the required formulas to be located swiftly and simply, and the unique tabular format crisply identifies all the variables involved.

The Cambridge Handbook of Physics Formulas comprehensively covers the major topics explored in undergraduate physics courses. It is designed to be a compact, portable, reference book suitable for everyday work, problem solving, or exam revision. All students and professionals in physics, applied mathematics, engineering, and other physical sciences will want to have this essential reference book within easy reach.

Graham Woan is Professor of Astrophysics in the School of Physics and Astronomy at the University of Glasgow. Prior to this he taught physics at the University of Cambridge where he also received his degree in Natural Sciences, specialising in physics, and his PhD, in radio astronomy. His research interests range widely with a special focus on low-frequency radio astronomy. His publications span journals as diverse as *Astronomy & Astrophysics*, *Geophysical Research Letters*, *Advances in Space Science*, the *Journal of Navigation* and *Emergency Prehospital Medicine*. He was co-developer of the revolutionary CURSOR radio positioning system, which uses existing broadcast transmitters to determine position, and he is the designer of the Glasgow Millennium Sundial.

The Cambridge Handbook of Physics Formulas

2003 Edition

GRAHAM WOAN
School of Physics & Astronomy
University of Glasgow

CAMBRIDGE
UNIVERSITY PRESS

CAMBRIDGE UNIVERSITY PRESS
Cambridge, New York, Melbourne, Madrid, Cape Town, Singapore,
São Paulo, Delhi, Dubai, Tokyo, Mexico City

Cambridge University Press
32 Avenue of the Americas, New York, NY 10013-2473, USA

www.cambridge.org
Information on this title: www.cambridge.org/9780521575072

First published 2000
10th printing 2010

Printed in the United States of America

A catalog record for this publication is available from the British Library.

Library of Congress Cataloging in Publication Data

Woan, Graham, 1963–
The Cambridge handbook of physics formulas / Graham Woan.
p. cm.
ISBN 0-521-57349-1 (hardback) – ISBN 0-521-57507-9 (pbk.)
1. Physics – Formulas. I. Title.
QC61.W67 1999
530′.02′12 – dc21 99-15228
CIP

ISBN 978-0-521-57349-8 Hardback
ISBN 978-0-521-57507-2 Paperback

Contents

Preface

In *A Brief History of Time*, Stephen Hawking relates that he was warned against including equations in the book because "each equation... would halve the sales." Despite this dire prediction there is, for a scientific audience, some attraction in doing the exact opposite.

The reader should not be misled by this exercise. Although the equations and formulas contained here underpin a good deal of physical science they are useless unless the reader *understands* them. Learning physics is not about remembering equations, it is about appreciating the natural structures they express. Although its format should help make some topics clearer, this book is not designed to teach new physics; there are many excellent textbooks to help with that. It is intended to be useful rather than pedagogically complete, so that students can use it for revision and for structuring their knowledge *once they understand the physics*. More advanced users will benefit from having a compact, internally consistent, source of equations that can quickly deliver the relationship they require in a format that avoids the need to sift through pages of rubric.

Some difficult decisions have had to be made to achieve this. First, to be short the book only includes ideas that can be expressed succinctly in equations, without resorting to lengthy explanation. A small number of important topics are therefore absent. For example, Liouville's theorem can be algebraically succinct ($\dot{\varrho} = 0$) but is meaningless unless $\dot{\varrho}$ is thoroughly (and carefully) explained. Anyone who already understands what $\dot{\varrho}$ represents will probably not need reminding that it equals zero. Second, empirical equations with numerical coefficients have been largely omitted, as have topics significantly more advanced than are found at undergraduate level. There are simply too many of these to be sensibly and confidently edited into a short handbook. Third, physical data are largely absent, although a periodic table, tables of physical constants, and data on the solar system are all included. Just a sighting of the marvellous (but dimensionally misnamed) *CRC Handbook of Chemistry and Physics* should be enough to convince the reader that a good science data book is thick.

Inevitably there is personal choice in what should or should not be included, and you may feel that an equation that meets the above criteria is missing. If this is the case, I would be delighted to hear from you so it can be considered for a subsequent edition. Contact details are at the end of this preface. Likewise, if you spot an error or an inconsistency then please let me know and I will post an erratum on the web page.

Acknowledgments This venture is founded on the generosity of colleagues in Glasgow and Cambridge whose inputs have strongly influenced the final product. The expertise of Dave Clarke, Declan Diver, Peter Duffett-Smith, Wolf-Gerrit Früh, Martin Hendry, Rico Ignace, David Ireland, John Simmons, and Harry Ward have been central to its production, as have the linguistic skills of Katie Lowe. I would also like to thank Richard Barrett, Matthew Cartmell, Steve Gull, Martin Hendry, Jim Hough, Darren McDonald, and Ken Riley who all agreed to field-test the book and gave invaluable feedback.

My greatest thanks though are to John Shakeshaft who, with remarkable knowledge and skill, worked through the entire manuscript more than once during its production and whose legendary red pen hovered over (or descended upon) every equation in the book. What errors remain are, of course, my own, but I take comfort from the fact that without John they would be much more numerous.

Contact information A website containing up-to-date information on this handbook and contact details can be found through the Cambridge University Press web pages at `us.cambridge.org` (North America) or `uk.cambridge.org` (United Kingdom), or directly at `radio.astro.gla.ac.uk/hbhome.html`.

Production notes This book was typeset by the author in LaTeX 2_ε using the CUP Times fonts. The software packages used were *WinEdt*, MiKTeX, *Mayura Draw*, *Gnuplot*, *Ghostscript*, *Ghostview*, and *Maple V*.

Comments on the 2003 edition I am grateful to all those who have suggested improvements, in particular Javier Hasbun, Martin Hendry, Wolfgang Jitschin, Joseph Katz, and Alan Watson. Although this edition contains only minor revisions to the original its production was also an opportunity to update the physical constants and periodic table entries and to reflect recent developments in cosmology.

How to use this book

The format is largely self-explanatory, but a few comments may be helpful. Although it is very tempting to flick through the pages to find what you are looking for, the best starting point is the index. I have tried to make this as extensive as possible, and many equations are indexed more than once. Equations are listed both with their equation number (in square brackets) and the page on which they can be found. The equations themselves are grouped into self-contained and boxed "panels" on the pages. Each panel represents a separate topic, and you will find descriptions of all the variables used at the right-hand side of the panel, usually adjacent to the first equation in which they are used. You should therefore not need to stray outside the panel to understand the notation. Both the panel as a whole and its individual entries may have footnotes, shown below the panel. Be aware of these, as they contain important additional information and conditions relevant to the topic.

Although the panels are self-contained they may use concepts defined elsewhere in the handbook. Often these are cross-referenced, but again the index will help you to locate them if necessary. Notations and definitions are uniform over subject areas unless stated otherwise.

Chapter 1 Units, constants, and conversions

1.1 Introduction

The determination of physical constants and the definition of the units with which they are measured is a specialised and, to many, hidden branch of science.

A quantity with dimensions is one whose value must be expressed relative to one or more standard units. In the spirit of the rest of the book, this section is based around the International System of units (SI). This system uses seven base units[1] (the number is somewhat arbitrary), such as the kilogram and the second, and defines their magnitudes in terms of physical laws or, in the case of the kilogram, an object called the "international prototype of the kilogram" kept in Paris. For convenience there are also a number of derived standards, such as the volt, which are defined as set combinations of the basic seven. Most of the physical observables we regard as being in some sense fundamental, such as the charge on an electron, are now known to a relative standard uncertainty,[2] u_r, of less than 10^{-7}. The least well determined is the Newtonian constant of gravitation, presently standing at a rather lamentable u_r of 1.5×10^{-3}, and the best is the Rydberg constant ($u_r = 7.6 \times 10^{-12}$). The dimensionless electron g-factor, representing twice the magnetic moment of an electron measured in Bohr magnetons, is now known to a relative uncertainty of only 4.1×10^{-12}.

No matter which base units are used, physical quantities are expressed as the product of a numerical value and a unit. These two components have more-or-less equal standing and can be manipulated by following the usual rules of algebra. So, if $1 \cdot \text{eV} = 160.218 \times 10^{-21} \cdot \text{J}$ then $1 \cdot \text{J} = [1/(160.218 \times 10^{-21})] \cdot \text{eV}$. A measurement of energy, U, with joule as the unit has a numerical value of U/J. The same measurement with electron volt as the unit has a numerical value of $U/\text{eV} = (U/\text{J}) \cdot (\text{J}/\text{eV})$ and so on.

[1]The **metre** is the length of the path travelled by light in vacuum during a time interval of 1/299 792 458 of a second. The **kilogram** is the unit of mass; it is equal to the mass of the international prototype of the kilogram. The **second** is the duration of 9 192 631 770 periods of the radiation corresponding to the transition between the two hyperfine levels of the ground state of the caesium 133 atom. The **ampere** is that constant current which, if maintained in two straight parallel conductors of infinite length, of negligible circular cross-section, and placed 1 metre apart in vacuum, would produce between these conductors a force equal to 2×10^{-7} newton per metre of length. The **kelvin**, unit of thermodynamic temperature, is the fraction 1/273.16 of the thermodynamic temperature of the triple point of water. The **mole** is the amount of substance of a system which contains as many elementary entities as there are atoms in 0.012 kilogram of carbon 12; its symbol is "mol." When the mole is used, the elementary entities must be specified and may be atoms, molecules, ions, electrons, other particles, or specified groups of such particles. The **candela** is the luminous intensity, in a given direction, of a source that emits monochromatic radiation of frequency 540×10^{12} hertz and that has a radiant intensity in that direction of 1/683 watt per steradian.

[2]The relative standard uncertainty in x is defined as the estimated standard deviation in x divided by the modulus of x ($x \neq 0$).

1.2 SI units

SI base units

physical quantity	*name*	*symbol*
length	metre[a]	m
mass	kilogram	kg
time interval	second	s
electric current	ampere	A
thermodynamic temperature	kelvin	K
amount of substance	mole	mol
luminous intensity	candela	cd

[a]Or "meter".

SI derived units

physical quantity	*name*	*symbol*	*equivalent units*
catalytic activity	katal	kat	$\mathrm{mol\,s^{-1}}$
electric capacitance	farad	F	$\mathrm{C\,V^{-1}}$
electric charge	coulomb	C	$\mathrm{A\,s}$
electric conductance	siemens	S	Ω^{-1}
electric potential difference	volt	V	$\mathrm{J\,C^{-1}}$
electric resistance	ohm	Ω	$\mathrm{V\,A^{-1}}$
energy, work, heat	joule	J	$\mathrm{N\,m}$
force	newton	N	$\mathrm{m\,kg\,s^{-2}}$
frequency	hertz	Hz	$\mathrm{s^{-1}}$
illuminance	lux	lx	$\mathrm{cd\,sr\,m^{-2}}$
inductance	henry	H	$\mathrm{V\,A^{-1}\,s}$
luminous flux	lumen	lm	$\mathrm{cd\,sr}$
magnetic flux	weber	Wb	$\mathrm{V\,s}$
magnetic flux density	tesla	T	$\mathrm{V\,s\,m^{-2}}$
plane angle	radian	rad	$\mathrm{m\,m^{-1}}$
power, radiant flux	watt	W	$\mathrm{J\,s^{-1}}$
pressure, stress	pascal	Pa	$\mathrm{N\,m^{-2}}$
radiation absorbed dose	gray	Gy	$\mathrm{J\,kg^{-1}}$
radiation dose equivalent[a]	sievert	Sv	$[\mathrm{J\,kg^{-1}}]$
radioactive activity	becquerel	Bq	$\mathrm{s^{-1}}$
solid angle	steradian	sr	$\mathrm{m^2\,m^{-2}}$
temperature[b]	degree Celsius	°C	K

[a]To distinguish it from the gray, units of $\mathrm{J\,kg^{-1}}$ should not be used for the sievert in practice.
[b]The Celsius temperature, T_C, is defined from the temperature in kelvin, T_K, by $T_\mathrm{C} = T_\mathrm{K} - 273.15$.

SI prefixes[a]

factor	*prefix*	*symbol*	*factor*	*prefix*	*symbol*
10^{24}	yotta	Y	10^{-24}	yocto	y
10^{21}	zetta	Z	10^{-21}	zepto	z
10^{18}	exa	E	10^{-18}	atto	a
10^{15}	peta	P	10^{-15}	femto	f
10^{12}	tera	T	10^{-12}	pico	p
10^{9}	giga	G	10^{-9}	nano	n
10^{6}	mega	M	10^{-6}	micro	μ
10^{3}	kilo	k	10^{-3}	milli	m
10^{2}	hecto	h	10^{-2}	centi	c
10^{1}	deca[b]	da	10^{-1}	deci	d

[a]The kilogram is the only SI unit with a prefix embedded in its name and symbol. For mass, the unit name "gram" and unit symbol "g" should be used with these prefixes, hence 10^{-6} kg can be written as 1 mg. Otherwise, any prefix can be applied to any SI unit.
[b]Or "deka".

Recognised non-SI units

physical quantity	*name*	*symbol*	*SI value*
area	barn	b	10^{-28} m^2
energy	electron volt	eV	$\simeq 1.602\,18 \times 10^{-19}$ J
length	ångström	Å	10^{-10} m
	fermi[a]	fm	10^{-15} m
	micron[a]	μm	10^{-6} m
plane angle	degree	°	$(\pi/180)$ rad
	arcminute	′	$(\pi/10\,800)$ rad
	arcsecond	″	$(\pi/648\,000)$ rad
pressure	bar	bar	10^5 N m^{-2}
time	minute	min	60 s
	hour	h	3 600 s
	day	d	86 400 s
mass	unified atomic mass unit	u	$\simeq 1.660\,54 \times 10^{-27}$ kg
	tonne[a,b]	t	10^3 kg
volume	litre[c]	l, L	10^{-3} m^3

[a]These are non-SI names for SI quantities.
[b]Or "metric ton."
[c]Or "liter". The symbol "l" should be avoided.

1.3 Physical constants

The following 1998 CODATA recommended values for the fundamental physical constants can also be found on the Web at `physics.nist.gov/constants`. Detailed background information is available in *Reviews of Modern Physics*, Vol. 72, No. 2, pp. 351–495, April 2000.

The digits in parentheses represent the 1σ uncertainty in the previous two quoted digits. For example, $G = (6.673 \pm 0.010) \times 10^{-11}\,\mathrm{m^3\,kg^{-1}\,s^{-2}}$. It is important to note that the uncertainties for many of the listed quantities are correlated, so that the uncertainty in any expression using them in combination cannot necessarily be computed from the data presented. Suitable covariance values are available in the above references.

Summary of physical constants

speed of light in vacuum[a]	c	2.997 924 58	$\times 10^{8}\,\mathrm{m\,s^{-1}}$
permeability of vacuum[b]	μ_0	4π	$\times 10^{-7}\,\mathrm{H\,m^{-1}}$
		$=12.566\,370\,614\ldots$	$\times 10^{-7}\,\mathrm{H\,m^{-1}}$
permittivity of vacuum	ϵ_0	$1/(\mu_0 c^2)$	$\mathrm{F\,m^{-1}}$
		$=8.854\,187\,817\ldots$	$\times 10^{-12}\,\mathrm{F\,m^{-1}}$
constant of gravitation[c]	G	6.673(10)	$\times 10^{-11}\,\mathrm{m^3\,kg^{-1}\,s^{-2}}$
Planck constant	h	6.626 068 76(52)	$\times 10^{-34}\,\mathrm{J\,s}$
$h/(2\pi)$	$\hbar$	1.054 571 596(82)	$\times 10^{-34}\,\mathrm{J\,s}$
elementary charge	e	1.602 176 462(63)	$\times 10^{-19}\,\mathrm{C}$
magnetic flux quantum, $h/(2e)$	Φ_0	2.067 833 636(81)	$\times 10^{-15}\,\mathrm{Wb}$
electron volt	eV	1.602 176 462(63)	$\times 10^{-19}\,\mathrm{J}$
electron mass	m_e	9.109 381 88(72)	$\times 10^{-31}\,\mathrm{kg}$
proton mass	m_p	1.672 621 58(13)	$\times 10^{-27}\,\mathrm{kg}$
proton/electron mass ratio	m_p/m_e	1 836.152 667 5(39)	
unified atomic mass unit	u	1.660 538 73(13)	$\times 10^{-27}\,\mathrm{kg}$
fine-structure constant, $\mu_0 c e^2/(2h)$	α	7.297 352 533(27)	$\times 10^{-3}$
inverse	$1/\alpha$	137.035 999 76(50)	
Rydberg constant, $m_e c \alpha^2/(2h)$	R_∞	1.097 373 156 854 9(83)	$\times 10^{7}\,\mathrm{m^{-1}}$
Avogadro constant	N_A	6.022 141 99(47)	$\times 10^{23}\,\mathrm{mol^{-1}}$
Faraday constant, $N_A e$	F	9.648 534 15(39)	$\times 10^{4}\,\mathrm{C\,mol^{-1}}$
molar gas constant	R	8.314 472(15)	$\mathrm{J\,mol^{-1}\,K^{-1}}$
Boltzmann constant, R/N_A	k	1.380 650 3(24)	$\times 10^{-23}\,\mathrm{J\,K^{-1}}$
Stefan–Boltzmann constant, $\pi^2 k^4/(60\hbar^3 c^2)$	σ	5.670 400(40)	$\times 10^{-8}\,\mathrm{W\,m^{-2}\,K^{-4}}$
Bohr magneton, $e\hbar/(2m_e)$	μ_B	9.274 008 99(37)	$\times 10^{-24}\,\mathrm{J\,T^{-1}}$

[a]By definition, the speed of light is exact.

[b]Also exact, by definition. Alternative units are $\mathrm{N\,A^{-2}}$.

[c]The standard acceleration due to gravity, g, is defined as exactly $9.806\,65\,\mathrm{m\,s^{-2}}$.

General constants

speed of light in vacuum	c	2.997 924 58	$\times 10^{8}$ m s^{-1}
permeability of vacuum	μ_0	4π	$\times 10^{-7}$ H m^{-1}
		=12.566 370 614...	$\times 10^{-7}$ H m^{-1}
permittivity of vacuum	ϵ_0	$1/(\mu_0 c^2)$	F m^{-1}
		=8.854 187 817...	$\times 10^{-12}$ F m^{-1}
impedance of free space	Z_0	$\mu_0 c$	Ω
		=376.730 313 461...	Ω
constant of gravitation	G	6.673(10)	$\times 10^{-11}$ m^3 kg^{-1} s^{-2}
Planck constant	h	6.626 068 76(52)	$\times 10^{-34}$ J s
in eV s		4.135 667 27(16)	$\times 10^{-15}$ eV s
$h/(2\pi)$	$\hbar$	1.054 571 596(82)	$\times 10^{-34}$ J s
in eV s		6.582 118 89(26)	$\times 10^{-16}$ eV s
Planck mass, $(\hbar c/G)^{1/2}$	m_{Pl}	2.176 7(16)	$\times 10^{-8}$ kg
Planck length, $\hbar/(m_{\mathrm{Pl}} c) = (\hbar G/c^3)^{1/2}$	l_{Pl}	1.616 0(12)	$\times 10^{-35}$ m
Planck time, $l_{\mathrm{Pl}}/c = (\hbar G/c^5)^{1/2}$	t_{Pl}	5.390 6(40)	$\times 10^{-44}$ s
elementary charge	e	1.602 176 462(63)	$\times 10^{-19}$ C
magnetic flux quantum, $h/(2e)$	Φ_0	2.067 833 636(81)	$\times 10^{-15}$ Wb
Josephson frequency/voltage ratio	$2e/h$	4.835 978 98(19)	$\times 10^{14}$ Hz V^{-1}
Bohr magneton, $e\hbar/(2m_e)$	μ_B	9.274 008 99(37)	$\times 10^{-24}$ J T^{-1}
in eV T^{-1}		5.788 381 749(43)	$\times 10^{-5}$ eV T^{-1}
μ_B/k		0.671 713 1(12)	K T^{-1}
nuclear magneton, $e\hbar/(2m_p)$	μ_N	5.050 783 17(20)	$\times 10^{-27}$ J T^{-1}
in eV T^{-1}		3.152 451 238(24)	$\times 10^{-8}$ eV T^{-1}
μ_N/k		3.658 263 8(64)	$\times 10^{-4}$ K T^{-1}
Zeeman splitting constant	$\mu_B/(hc)$	46.686 452 1(19)	m^{-1} T^{-1}

Atomic constants[a]

fine-structure constant, $\mu_0 c e^2/(2h)$	α	7.297 352 533(27)	$\times 10^{-3}$
inverse	$1/\alpha$	137.035 999 76(50)	
Rydberg constant, $m_e c \alpha^2/(2h)$	R_∞	1.097 373 156 854 9(83)	$\times 10^{7}$ m^{-1}
$R_\infty c$		3.289 841 960 368(25)	$\times 10^{15}$ Hz
$R_\infty hc$		2.179 871 90(17)	$\times 10^{-18}$ J
$R_\infty hc/e$		13.605 691 72(53)	eV
Bohr radius[b], $\alpha/(4\pi R_\infty)$	a_0	5.291 772 083(19)	$\times 10^{-11}$ m

[a]See also the Bohr model on page 95.
[b]Fixed nucleus.

Electron constants

electron mass	m_e	9.109 381 88(72)	$\times 10^{-31}$ kg
in MeV		0.510 998 902(21)	MeV
electron/proton mass ratio	m_e/m_p	5.446 170 232(12)	$\times 10^{-4}$
electron charge	$-e$	−1.602 176 462(63)	$\times 10^{-19}$ C
electron specific charge	$-e/m_e$	−1.758 820 174(71)	$\times 10^{11}$ C kg^{-1}
electron molar mass, $N_A m_e$	M_e	5.485 799 110(12)	$\times 10^{-7}$ kg mol^{-1}
Compton wavelength, $h/(m_e c)$	λ_C	2.426 310 215(18)	$\times 10^{-12}$ m
classical electron radius, $\alpha^2 a_0$	r_e	2.817 940 285(31)	$\times 10^{-15}$ m
Thomson cross section, $(8\pi/3)r_e^2$	σ_T	6.652 458 54(15)	$\times 10^{-29}$ m^2
electron magnetic moment	μ_e	−9.284 763 62(37)	$\times 10^{-24}$ J T^{-1}
in Bohr magnetons, μ_e/μ_B		−1.001 159 652 186 9(41)	
in nuclear magnetons, μ_e/μ_N		−1 838.281 966 0(39)	
electron gyromagnetic ratio, $2\lvert\mu_e\rvert/\hbar$	γ_e	1.760 859 794(71)	$\times 10^{11}$ s^{-1} T^{-1}
electron g-factor, $2\mu_e/\mu_B$	g_e	−2.002 319 304 3737(82)	

Proton constants

proton mass	m_p	1.672 621 58(13)	$\times 10^{-27}$ kg
in MeV		938.271 998(38)	MeV
proton/electron mass ratio	m_p/m_e	1 836.152 667 5(39)	
proton charge	e	1.602 176 462(63)	$\times 10^{-19}$ C
proton specific charge	e/m_p	9.578 834 08(38)	$\times 10^{7}$ C kg^{-1}
proton molar mass, $N_A m_p$	M_p	1.007 276 466 88(13)	$\times 10^{-3}$ kg mol^{-1}
proton Compton wavelength, $h/(m_p c)$	$\lambda_{C,p}$	1.321 409 847(10)	$\times 10^{-15}$ m
proton magnetic moment	μ_p	1.410 606 633(58)	$\times 10^{-26}$ J T^{-1}
in Bohr magnetons, μ_p/μ_B		1.521 032 203(15)	$\times 10^{-3}$
in nuclear magnetons, μ_p/μ_N		2.792 847 337(29)	
proton gyromagnetic ratio, $2\mu_p/\hbar$	γ_p	2.675 222 12(11)	$\times 10^{8}$ s^{-1} T^{-1}

Neutron constants

neutron mass	m_n	1.674 927 16(13)	$\times 10^{-27}$ kg
in MeV		939.565 330(38)	MeV
neutron/electron mass ratio	m_n/m_e	1 838.683 655 0(40)	
neutron/proton mass ratio	m_n/m_p	1.001 378 418 87(58)	
neutron molar mass, $N_A m_n$	M_n	1.008 664 915 78(55)	$\times 10^{-3}$ kg mol^{-1}
neutron Compton wavelength, $h/(m_n c)$	$\lambda_{C,n}$	1.319 590 898(10)	$\times 10^{-15}$ m
neutron magnetic moment	μ_n	−9.662 364 0(23)	$\times 10^{-27}$ J T^{-1}
in Bohr magnetons	μ_n/μ_B	−1.041 875 63(25)	$\times 10^{-3}$
in nuclear magnetons	μ_n/μ_N	−1.913 042 72(45)	
neutron gyromagnetic ratio, $2\lvert\mu_n\rvert/\hbar$	γ_n	1.832 471 88(44)	$\times 10^{8}$ s^{-1} T^{-1}

Muon and tau constants

muon mass	m_μ	1.883 531 09(16)	$\times 10^{-28}$ kg
in MeV		105.658 356 8(52)	MeV
tau mass	m_τ	3.167 88(52)	$\times 10^{-27}$ kg
in MeV		1.777 05(29)	$\times 10^{3}$ MeV
muon/electron mass ratio	m_μ/m_e	206.768 262(30)	
muon charge	$-e$	$-1.602\,176\,462(63)$	$\times 10^{-19}$ C
muon magnetic moment	μ_μ	$-4.490\,448\,13(22)$	$\times 10^{-26}$ J T^{-1}
in Bohr magnetons, μ_μ/μ_B		4.841 970 85(15)	$\times 10^{-3}$
in nuclear magnetons, μ_μ/μ_N		8.890 597 70(27)	
muon g-factor	g_μ	$-2.002\,331\,832\,0(13)$	

Bulk physical constants

Avogadro constant	N_A	6.022 141 99(47)	$\times 10^{23}$ mol^{-1}
atomic mass constant[a]	m_u	1.660 538 73(13)	$\times 10^{-27}$ kg
in MeV		931.494 013(37)	MeV
Faraday constant	F	9.648 534 15(39)	$\times 10^{4}$ C mol^{-1}
molar gas constant	R	8.314 472(15)	J mol^{-1} K^{-1}
Boltzmann constant, R/N_A	k	1.380 650 3(24)	$\times 10^{-23}$ J K^{-1}
in eV K^{-1}		8.617 342(15)	$\times 10^{-5}$ eV K^{-1}
molar volume (ideal gas at stp)[b]	V_m	22.413 996(39)	$\times 10^{-3}$ m^3 mol^{-1}
Stefan–Boltzmann constant, $\pi^2 k^4/(60\hbar^3 c^2)$	σ	5.670 400(40)	$\times 10^{-8}$ W m^{-2} K^{-4}
Wien's displacement law constant,[c] $b = \lambda_m T$	b	2.897 768 6(51)	$\times 10^{-3}$ m K

[a]= mass of ^{12}C/12. Alternative nomenclature for the unified atomic mass unit, u.
[b]Standard temperature and pressure (stp) are $T = 273.15$ K (0°C) and $P = 101\,325$ Pa (1 standard atmosphere).
[c]See also page 121.

Mathematical constants

pi (π)	3.141 592 653 589 793 238 462 643 383 279 ...
exponential constant (e)	2.718 281 828 459 045 235 360 287 471 352 ...
Catalan's constant	0.915 965 594 177 219 015 054 603 514 932 ...
Euler's constant[a] (γ)	0.577 215 664 901 532 860 606 512 090 082 ...
Feigenbaum's constant (α)	2.502 907 875 095 892 822 283 902 873 218 ...
Feigenbaum's constant (δ)	4.669 201 609 102 990 671 853 203 820 466 ...
Gibbs constant	1.851 937 051 982 466 170 361 053 370 157 ...
golden mean	1.618 033 988 749 894 848 204 586 834 370 ...
Madelung constant[b]	1.747 564 594 633 182 190 636 212 035 544 ...

[a]See also Equation (2.119).
[b]NaCl structure.

1.4 Converting between units

The following table lists common (and not so common) measures of physical quantities. The numerical values given are the SI equivalent of one unit measure of the non-SI unit. Hence 1 astronomical unit equals 149.597 9 × 10^9 m. Those entries identified with a "*" in the second column represent exact conversions; so 1 abampere equals exactly 10.0 A. Note that individual entries in this list are not recorded in the index, and that values are "international" unless otherwise stated.

There is a separate section on temperature conversions after this table.

unit name	*value in SI units*	
abampere	10.0*	A
abcoulomb	10.0*	C
abfarad	1.0*	$\times 10^{9}$ F
abhenry	1.0*	$\times 10^{-9}$ H
abmho	1.0*	$\times 10^{9}$ S
abohm	1.0*	$\times 10^{-9}$ Ω
abvolt	10.0*	$\times 10^{-9}$ V
acre	4.046 856	$\times 10^{3}$ m^2
amagat (at stp)	44.614 774	mol m^{-3}
ampere hour	3.6*	$\times 10^{3}$ C
ångström	100.0*	$\times 10^{-12}$ m
apostilb	1.0*	lm m^{-2}
arcminute	290.888 2	$\times 10^{-6}$ rad
arcsecond	4.848 137	$\times 10^{-6}$ rad
are	100.0*	m^2
astronomical unit	149.597 9	$\times 10^{9}$ m
atmosphere (standard)	101.325 0*	$\times 10^{3}$ Pa
atomic mass unit	1.660 540	$\times 10^{-27}$ kg
bar	100.0*	$\times 10^{3}$ Pa
barn	100.0*	$\times 10^{-30}$ m^2
baromil	750.1	$\times 10^{-6}$ m
barrel (UK)	163.659 2	$\times 10^{-3}$ m^3
barrel (US dry)	115.627 1	$\times 10^{-3}$ m^3
barrel (US liquid)	119.240 5	$\times 10^{-3}$ m^3
barrel (US oil)	158.987 3	$\times 10^{-3}$ m^3
baud	1.0*	s^{-1}
bayre	100.0*	$\times 10^{-3}$ Pa
biot	10.0	A
bolt (US)	36.576*	m
brewster	1.0*	$\times 10^{-12}$ m^2 N^{-1}
British thermal unit	1.055 056	$\times 10^{3}$ J
bushel (UK)	36.368 72	$\times 10^{-3}$ m^3
bushel (US)	35.239 07	$\times 10^{-3}$ m^3
butt (UK)	477.339 4	$\times 10^{-3}$ m^3
cable (US)	219.456*	m
calorie	4.186 8*	J

continued on next page ...

unit name	*value in SI units*	
candle power (spherical)	4π	lm
carat (metric)	200.0*	$\times 10^{-6}$ kg
cental	45.359 237	kg
centare	1.0*	m^2
centimetre of Hg (0 °C)	1.333 222	$\times 10^{3}$ Pa
centimetre of H_2O (4 °C)	98.060 616	Pa
chain (engineers')	30.48*	m
chain (US)	20.116 8*	m
Chu	1.899 101	$\times 10^{3}$ J
clusec	1.333 224	$\times 10^{-6}$ W
cord	3.624 556	m^3
cubit	457.2*	$\times 10^{-3}$ m
cumec	1.0*	$m^3\,s^{-1}$
cup (US)	236.588 2	$\times 10^{-6}\,m^3$
curie	37.0*	$\times 10^{9}$ Bq
darcy	986.923 3	$\times 10^{-15}\,m^2$
day	86.4*	$\times 10^{3}$ s
day (sidereal)	86.164 09	$\times 10^{3}$ s
debye	3.335 641	$\times 10^{-30}$ C m
degree (angle)	17.453 29	$\times 10^{-3}$ rad
denier	111.111 1	$\times 10^{-9}\,kg\,m^{-1}$
digit	19.05*	$\times 10^{-3}$ m
dioptre	1.0*	m^{-1}
Dobson unit	10.0*	$\times 10^{-6}$ m
dram (avoirdupois)	1.771 845	$\times 10^{-3}$ kg
dyne	10.0*	$\times 10^{-6}$ N
dyne centimetres	100.0*	$\times 10^{-9}$ J
electron volt	160.217 7	$\times 10^{-21}$ J
ell	1.143*	m
em	4.233 333	$\times 10^{-3}$ m
emu of capacitance	1.0*	$\times 10^{9}$ F
emu of current	10.0*	A
emu of electric potential	10.0*	$\times 10^{-9}$ V
emu of inductance	1.0*	$\times 10^{-9}$ H
emu of resistance	1.0*	$\times 10^{-9}\,\Omega$
Eötvös unit	1.0*	$\times 10^{-9}\,m\,s^{-2}\,m^{-1}$
esu of capacitance	1.112 650	$\times 10^{-12}$ F
esu of current	333.564 1	$\times 10^{-12}$ A
esu of electric potential	299.792 5	V
esu of inductance	898.755 2	$\times 10^{9}$ H
esu of resistance	898.755 2	$\times 10^{9}\,\Omega$
erg	100.0*	$\times 10^{-9}$ J
faraday	96.485 3	$\times 10^{3}$ C
fathom	1.828 804	m
fermi	1.0*	$\times 10^{-15}$ m
Finsen unit	10.0*	$\times 10^{-6}\,W\,m^{-2}$
firkin (UK)	40.914 81	$\times 10^{-3}\,m^3$

continued on next page ...

unit name	*value in SI units*	
firkin (US)	34.068 71	$\times 10^{-3}$ m^3
fluid ounce (UK)	28.413 08	$\times 10^{-6}$ m^3
fluid ounce (US)	29.573 53	$\times 10^{-6}$ m^3
foot	304.8*	$\times 10^{-3}$ m
foot (US survey)	304.800 6	$\times 10^{-3}$ m
foot of water (4 °C)	2.988 887	$\times 10^{3}$ Pa
footcandle	10.763 91	lx
footlambert	3.426 259	cd m^{-2}
footpoundal	42.140 11	$\times 10^{-3}$ J
footpounds (force)	1.355 818	J
fresnel	1.0*	$\times 10^{12}$ Hz
funal	1.0*	$\times 10^{3}$ N
furlong	201.168*	m
g (standard acceleration)	9.806 65*	m s^{-2}
gal	10.0*	$\times 10^{-3}$ m s^{-2}
gallon (UK)	4.546 09*	$\times 10^{-3}$ m^3
gallon (US liquid)	3.785 412	$\times 10^{-3}$ m^3
gamma	1.0*	$\times 10^{-9}$ T
gauss	100.0*	$\times 10^{-6}$ T
gilbert	795.774 7	$\times 10^{-3}$ A turn
gill (UK)	142.065 4	$\times 10^{-6}$ m^3
gill (US)	118.294 1	$\times 10^{-6}$ m^3
gon	$\pi/200$*	rad
grade	15.707 96	$\times 10^{-3}$ rad
grain	64.798 91*	$\times 10^{-6}$ kg
gram	1.0*	$\times 10^{-3}$ kg
gram-rad	100.0*	J kg^{-1}
gray	1.0*	J kg^{-1}
hand	101.6*	$\times 10^{-3}$ m
hartree	4.359 748	$\times 10^{-18}$ J
hectare	10.0*	$\times 10^{3}$ m^2
hefner	902	$\times 10^{-3}$ cd
hogshead	238.669 7	$\times 10^{-3}$ m^3
horsepower (boiler)	9.809 50	$\times 10^{3}$ W
horsepower (electric)	746*	W
horsepower (metric)	735.498 8	W
horsepower (UK)	745.699 9	W
hour	3.6*	$\times 10^{3}$ s
hour (sidereal)	3.590 170	$\times 10^{3}$ s
Hubble time	440	$\times 10^{15}$ s
Hubble distance	130	$\times 10^{24}$ m
hundredweight (UK long)	50.802 35	kg
hundredweight (US short)	45.359 24	kg
inch	25.4*	$\times 10^{-3}$ m
inch of mercury (0 °C)	3.386 389	$\times 10^{3}$ Pa
inch of water (4 °C)	249.074 0	Pa
jansky	10.0*	$\times 10^{-27}$ W m^{-2} Hz^{-1}

continued on next page ...

unit name	*value in SI units*	
jar	10/9*	$\times 10^{-9}$ F
kayser	100.0*	m^{-1}
kilocalorie	4.186 8*	$\times 10^{3}$ J
kilogram-force	9.806 65*	N
kilowatt hour	3.6*	$\times 10^{6}$ J
knot (international)	514.444 4	$\times 10^{-3}$ m s^{-1}
lambert	$10/\pi$*	$\times 10^{3}$ cd m^{-2}
langley	41.84*	$\times 10^{3}$ J m^{-2}
langmuir	133.322 4	$\times 10^{-6}$ Pa s
league (nautical, int.)	5.556*	$\times 10^{3}$ m
league (nautical, UK)	5.559 552	$\times 10^{3}$ m
league (statute)	4.828 032	$\times 10^{3}$ m
light year	9.460 73*	$\times 10^{15}$ m
ligne	2.256*	$\times 10^{-3}$ m
line	2.116 667	$\times 10^{-3}$ m
line (magnetic flux)	10.0*	$\times 10^{-9}$ Wb
link (engineers')	304.8*	$\times 10^{-3}$ m
link (US)	201.168 0	$\times 10^{-3}$ m
litre	1.0*	$\times 10^{-3}$ m^{3}
lumen (at 555 nm)	1.470 588	$\times 10^{-3}$ W
maxwell	10.0*	$\times 10^{-9}$ Wb
mho	1.0*	S
micron	1.0*	$\times 10^{-6}$ m
mil (length)	25.4*	$\times 10^{-6}$ m
mil (volume)	1.0*	$\times 10^{-6}$ m^{3}
mile (international)	1.609 344*	$\times 10^{3}$ m
mile (nautical, int.)	1.852*	$\times 10^{3}$ m
mile (nautical, UK)	1.853 184*	$\times 10^{3}$ m
mile per hour	447.04*	$\times 10^{-3}$ m s^{-1}
milliard	1.0*	$\times 10^{9}$ m^{3}
millibar	100.0*	Pa
millimetre of Hg (0 °C)	133.322 4	Pa
minim (UK)	59.193 90	$\times 10^{-9}$ m^{3}
minim (US)	61.611 51	$\times 10^{-9}$ m^{3}
minute (angle)	290.888 2	$\times 10^{-6}$ rad
minute	60.0*	s
minute (sidereal)	59.836 17	s
month (lunar)	2.551 444	$\times 10^{6}$ s
nit	1.0*	cd m^{-2}
noggin (UK)	142.065 4	$\times 10^{-6}$ m^{3}
oersted	$1000/(4\pi)$*	A m^{-1}
ounce (avoirdupois)	28.349 52	$\times 10^{-3}$ kg
ounce (UK fluid)	28.413 07	$\times 10^{-6}$ m^{3}
ounce (US fluid)	29.573 53	$\times 10^{-6}$ m^{3}
pace	762.0*	$\times 10^{-3}$ m
parsec	30.856 78	$\times 10^{15}$ m

continued on next page ...

unit name	*value in SI units*	
peck (UK)	9.092 18*	$\times10^{-3}$ m^3
peck (US)	8.809 768	$\times10^{-3}$ m^3
pennyweight (troy)	1.555 174	$\times10^{-3}$ kg
perch	5.029 2*	m
phot	10.0*	$\times10^{3}$ lx
pica (printers')	4.217 518	$\times10^{-3}$ m
pint (UK)	568.261 2	$\times10^{-6}$ m^3
pint (US dry)	550.610 5	$\times10^{-6}$ m^3
pint (US liquid)	473.176 5	$\times10^{-6}$ m^3
point (printers')	351.459 8*	$\times10^{-6}$ m
poise	100.0*	$\times10^{-3}$ Pa s
pole	5.029 2*	m
poncelot	980.665*	W
pottle	2.273 045	$\times10^{-3}$ m^3
pound (avoirdupois)	453.592 4	$\times10^{-3}$ kg
poundal	138.255 0	$\times10^{-3}$ N
pound-force	4.448 222	N
promaxwell	1.0*	Wb
psi	6.894 757	$\times10^{3}$ Pa
puncheon (UK)	317.974 6	$\times10^{-3}$ m^3
quad	1.055 056	$\times10^{18}$ J
quart (UK)	1.136 522	$\times10^{-3}$ m^3
quart (US dry)	1.101 221	$\times10^{-3}$ m^3
quart (US liquid)	946.352 9	$\times10^{-6}$ m^3
quintal (metric)	100.0*	kg
rad	10.0*	$\times10^{-3}$ Gy
rayleigh	$10/(4\pi)$	$\times10^{9}$ s^{-1} m^{-2} sr^{-1}
rem	10.0*	$\times10^{-3}$ Sv
REN	1/4 000*	S
reyn	689.5	$\times10^{3}$ Pa s
rhe	10.0*	Pa^{-1} s^{-1}
rod	5.029 2*	m
roentgen	258.0	$\times10^{-6}$ C kg^{-1}
rood (UK)	1.011 714	$\times10^{3}$ m^2
rope (UK)	6.096*	m
rutherford	1.0*	$\times10^{6}$ Bq
rydberg	2.179 874	$\times10^{-18}$ J
scruple	1.295 978	$\times10^{-3}$ kg
seam	290.949 8	$\times10^{-3}$ m^3
second (angle)	4.848 137	$\times10^{-6}$ rad
second (sidereal)	997.269 6	$\times10^{-3}$ s
shake	100.0*	$\times10^{-10}$ s
shed	100.0*	$\times10^{-54}$ m^2
slug	14.593 90	kg
square degree	$(\pi/180)^{2*}$	sr
statampere	333.564 1	$\times10^{-12}$ A
statcoulomb	333.564 1	$\times10^{-12}$ C

continued on next page ...

unit name	*value in SI units*	
statfarad	1.112 650	$\times 10^{-12}$ F
stathenry	898.755 2	$\times 10^{9}$ H
statmho	1.112 650	$\times 10^{-12}$ S
statohm	898.755 2	$\times 10^{9}$ Ω
statvolt	299.792 5	V
stere	1.0*	m^3
sthéne	1.0*	$\times 10^{3}$ N
stilb	10.0*	$\times 10^{3}$ cd m^{-2}
stokes	100.0*	$\times 10^{-6}$ m^2 s^{-1}
stone	6.350 293	kg
tablespoon (UK)	14.206 53	$\times 10^{-6}$ m^3
tablespoon (US)	14.786 76	$\times 10^{-6}$ m^3
teaspoon (UK)	4.735 513	$\times 10^{-6}$ m^3
teaspoon (US)	4.928 922	$\times 10^{-6}$ m^3
tex	1.0*	$\times 10^{-6}$ kg m^{-1}
therm (EEC)	105.506*	$\times 10^{6}$ J
therm (US)	105.480 4*	$\times 10^{6}$ J
thermie	4.185 407	$\times 10^{6}$ J
thou	25.4*	$\times 10^{-6}$ m
tog	100.0*	$\times 10^{-3}$ W^{-1} m^2 K
ton (of TNT)	4.184*	$\times 10^{9}$ J
ton (UK long)	1.016 047	$\times 10^{3}$ kg
ton (US short)	907.184 7	kg
tonne (metric ton)	1.0*	$\times 10^{3}$ kg
torr	133.322 4	Pa
townsend	1.0*	$\times 10^{-21}$ V m^2
troy dram	3.887 935	$\times 10^{-3}$ kg
troy ounce	31.103 48	$\times 10^{-3}$ kg
troy pound	373.241 7	$\times 10^{-3}$ kg
tun	954.678 9	$\times 10^{-3}$ m^3
XU	100.209	$\times 10^{-15}$ m
yard	914.4*	$\times 10^{-3}$ m
year (365 days)	31.536*	$\times 10^{6}$ s
year (sidereal)	31.558 15	$\times 10^{6}$ s
year (tropical)	31.556 93	$\times 10^{6}$ s

Temperature conversions

From degrees Celsius[a]	$T_K = T_C + 273.15$	(1.1)	T_K temperature in kelvin T_C temperature in °Celsius
From degrees Fahrenheit	$T_K = \frac{T_F - 32}{1.8} + 273.15$	(1.2)	T_F temperature in °Fahrenheit
From degrees Rankine	$T_K = \frac{T_R}{1.8}$	(1.3)	T_R temperature in °Rankine

[a]The term "centigrade" is not used in SI, to avoid confusion with "10^{-2} of a degree".

1.5 Dimensions

The following table lists the dimensions of common physical quantities, together with their conventional symbols and the SI units in which they are usually quoted. The dimensional basis used is length (L), mass (M), time (T), electric current (I), temperature (Θ), amount of substance (N), and luminous intensity (J).

physical quantity	*symbol*	*dimensions*	*SI units*
acceleration	a	$\mathrm{L\,T^{-2}}$	$\mathrm{m\,s^{-2}}$
action	S	$\mathrm{L^2\,M\,T^{-1}}$	$\mathrm{J\,s}$
angular momentum	$\boldsymbol{L}, \boldsymbol{J}$	$\mathrm{L^2\,M\,T^{-1}}$	$\mathrm{m^2\,kg\,s^{-1}}$
angular speed	ω	$\mathrm{T^{-1}}$	$\mathrm{rad\,s^{-1}}$
area	A, S	$\mathrm{L^2}$	$\mathrm{m^2}$
Avogadro constant	N_{A}	$\mathrm{N^{-1}}$	$\mathrm{mol^{-1}}$
bending moment	$\boldsymbol{G}_{\mathrm{b}}$	$\mathrm{L^2\,M\,T^{-2}}$	$\mathrm{N\,m}$
Bohr magneton	μ_{B}	$\mathrm{L^2\,I}$	$\mathrm{J\,T^{-1}}$
Boltzmann constant	k, k_{B}	$\mathrm{L^2\,M\,T^{-2}\,\Theta^{-1}}$	$\mathrm{J\,K^{-1}}$
bulk modulus	K	$\mathrm{L^{-1}\,M\,T^{-2}}$	Pa
capacitance	C	$\mathrm{L^{-2}\,M^{-1}\,T^4\,I^2}$	F
charge (electric)	q	$\mathrm{T\,I}$	C
charge density	ρ	$\mathrm{L^{-3}\,T\,I}$	$\mathrm{C\,m^{-3}}$
conductance	G	$\mathrm{L^{-2}\,M^{-1}\,T^3\,I^2}$	S
conductivity	σ	$\mathrm{L^{-3}\,M^{-1}\,T^3\,I^2}$	$\mathrm{S\,m^{-1}}$
couple	$\boldsymbol{G}, \boldsymbol{T}$	$\mathrm{L^2\,M\,T^{-2}}$	$\mathrm{N\,m}$
current	I, i	I	A
current density	$\boldsymbol{J}, \boldsymbol{j}$	$\mathrm{L^{-2}\,I}$	$\mathrm{A\,m^{-2}}$
density	ρ	$\mathrm{L^{-3}\,M}$	$\mathrm{kg\,m^{-3}}$
electric displacement	$\boldsymbol{D}$	$\mathrm{L^{-2}\,T\,I}$	$\mathrm{C\,m^{-2}}$
electric field strength	$\boldsymbol{E}$	$\mathrm{L\,M\,T^{-3}\,I^{-1}}$	$\mathrm{V\,m^{-1}}$
electric polarisability	α	$\mathrm{M^{-1}\,T^4\,I^2}$	$\mathrm{C\,m^2\,V^{-1}}$
electric polarisation	$\boldsymbol{P}$	$\mathrm{L^{-2}\,T\,I}$	$\mathrm{C\,m^{-2}}$
electric potential difference	V	$\mathrm{L^2\,M\,T^{-3}\,I^{-1}}$	V
energy	E, U	$\mathrm{L^2\,M\,T^{-2}}$	J
energy density	u	$\mathrm{L^{-1}\,M\,T^{-2}}$	$\mathrm{J\,m^{-3}}$
entropy	S	$\mathrm{L^2\,M\,T^{-2}\,\Theta^{-1}}$	$\mathrm{J\,K^{-1}}$
Faraday constant	F	$\mathrm{T\,I\,N^{-1}}$	$\mathrm{C\,mol^{-1}}$
force	$\boldsymbol{F}$	$\mathrm{L\,M\,T^{-2}}$	N
frequency	ν, f	$\mathrm{T^{-1}}$	Hz
gravitational constant	G	$\mathrm{L^3\,M^{-1}\,T^{-2}}$	$\mathrm{m^3\,kg^{-1}\,s^{-2}}$
Hall coefficient	R_{H}	$\mathrm{L^3\,T^{-1}\,I^{-1}}$	$\mathrm{m^3\,C^{-1}}$
Hamiltonian	H	$\mathrm{L^2\,M\,T^{-2}}$	J
heat capacity	C	$\mathrm{L^2\,M\,T^{-2}\,\Theta^{-1}}$	$\mathrm{J\,K^{-1}}$
Hubble constant[1]	H	$\mathrm{T^{-1}}$	$\mathrm{s^{-1}}$
illuminance	E_{v}	$\mathrm{L^{-2}\,J}$	lx
impedance	Z	$\mathrm{L^2\,M\,T^{-3}\,I^{-2}}$	Ω

continued on next page ...

[1]The Hubble constant is almost universally quoted in units of $\mathrm{km\,s^{-1}\,Mpc^{-1}}$. There are about 3.1×10^{19} kilometres in a megaparsec.

physical quantity	*symbol*	*dimensions*	*SI units*
impulse	I	$\mathrm{L\,M\,T^{-1}}$	N s
inductance	L	$\mathrm{L^2\,M\,T^{-2}\,I^{-2}}$	H
irradiance	E_e	$\mathrm{M\,T^{-3}}$	$\mathrm{W\,m^{-2}}$
Lagrangian	L	$\mathrm{L^2\,M\,T^{-2}}$	J
length	L, l	L	m
luminous intensity	I_v	J	cd
magnetic dipole moment	$\boldsymbol{m}$, $\boldsymbol{\mu}$	$\mathrm{L^2\,I}$	$\mathrm{A\,m^2}$
magnetic field strength	$\boldsymbol{H}$	$\mathrm{L^{-1}\,I}$	$\mathrm{A\,m^{-1}}$
magnetic flux	Φ	$\mathrm{L^2\,M\,T^{-2}\,I^{-1}}$	Wb
magnetic flux density	$\boldsymbol{B}$	$\mathrm{M\,T^{-2}\,I^{-1}}$	T
magnetic vector potential	$\boldsymbol{A}$	$\mathrm{L\,M\,T^{-2}\,I^{-1}}$	$\mathrm{Wb\,m^{-1}}$
magnetisation	$\boldsymbol{M}$	$\mathrm{L^{-1}\,I}$	$\mathrm{A\,m^{-1}}$
mass	m, M	M	kg
mobility	μ	$\mathrm{M^{-1}\,T^2\,I}$	$\mathrm{m^2\,V^{-1}\,s^{-1}}$
molar gas constant	R	$\mathrm{L^2\,M\,T^{-2}\,\Theta^{-1}\,N^{-1}}$	$\mathrm{J\,mol^{-1}\,K^{-1}}$
moment of inertia	I	$\mathrm{L^2\,M}$	$\mathrm{kg\,m^2}$
momentum	$\boldsymbol{p}$	$\mathrm{L\,M\,T^{-1}}$	$\mathrm{kg\,m\,s^{-1}}$
number density	n	$\mathrm{L^{-3}}$	$\mathrm{m^{-3}}$
permeability	μ	$\mathrm{L\,M\,T^{-2}\,I^{-2}}$	$\mathrm{H\,m^{-1}}$
permittivity	ϵ	$\mathrm{L^{-3}\,M^{-1}\,T^4\,I^2}$	$\mathrm{F\,m^{-1}}$
Planck constant	h	$\mathrm{L^2\,M\,T^{-1}}$	J s
power	P	$\mathrm{L^2\,M\,T^{-3}}$	W
Poynting vector	$\boldsymbol{S}$	$\mathrm{M\,T^{-3}}$	$\mathrm{W\,m^{-2}}$
pressure	p, P	$\mathrm{L^{-1}\,M\,T^{-2}}$	Pa
radiant intensity	I_e	$\mathrm{L^2\,M\,T^{-3}}$	$\mathrm{W\,sr^{-1}}$
resistance	R	$\mathrm{L^2\,M\,T^{-3}\,I^{-2}}$	Ω
Rydberg constant	R_∞	$\mathrm{L^{-1}}$	$\mathrm{m^{-1}}$
shear modulus	μ, G	$\mathrm{L^{-1}\,M\,T^{-2}}$	Pa
specific heat capacity	c	$\mathrm{L^2\,T^{-2}\,\Theta^{-1}}$	$\mathrm{J\,kg^{-1}\,K^{-1}}$
speed	u, v, c	$\mathrm{L\,T^{-1}}$	$\mathrm{m\,s^{-1}}$
Stefan–Boltzmann constant	σ	$\mathrm{M\,T^{-3}\,\Theta^{-4}}$	$\mathrm{W\,m^{-2}\,K^{-4}}$
stress	σ, τ	$\mathrm{L^{-1}\,M\,T^{-2}}$	Pa
surface tension	σ, γ	$\mathrm{M\,T^{-2}}$	$\mathrm{N\,m^{-1}}$
temperature	T	Θ	K
thermal conductivity	λ	$\mathrm{L\,M\,T^{-3}\,\Theta^{-1}}$	$\mathrm{W\,m^{-1}\,K^{-1}}$
time	t	T	s
velocity	$\boldsymbol{v}$, $\boldsymbol{u}$	$\mathrm{L\,T^{-1}}$	$\mathrm{m\,s^{-1}}$
viscosity (dynamic)	η, μ	$\mathrm{L^{-1}\,M\,T^{-1}}$	Pa s
viscosity (kinematic)	ν	$\mathrm{L^2\,T^{-1}}$	$\mathrm{m^2\,s^{-1}}$
volume	V, v	$\mathrm{L^3}$	$\mathrm{m^3}$
wavevector	$\boldsymbol{k}$	$\mathrm{L^{-1}}$	$\mathrm{m^{-1}}$
weight	W	$\mathrm{L\,M\,T^{-2}}$	N
work	W	$\mathrm{L^2\,M\,T^{-2}}$	J
Young modulus	E	$\mathrm{L^{-1}\,M\,T^{-2}}$	Pa

1.6 Miscellaneous

Greek alphabet

A	α		alpha	N	ν		nu
B	β		beta	Ξ	ξ		xi
Γ	γ		gamma	O	o		omicron
Δ	δ		delta	Π	π	ϖ	pi
E	ϵ	ε	epsilon	P	ρ	ϱ	rho
Z	ζ		zeta	Σ	σ	ς	sigma
H	η		eta	T	τ		tau
Θ	θ	ϑ	theta	Υ	υ		upsilon
I	ι		iota	Φ	ϕ	φ	phi
K	κ		kappa	X	χ		chi
Λ	λ		lambda	Ψ	ψ		psi
M	μ		mu	Ω	ω		omega

Pi (π) to 1 000 decimal places

3.1415926535	8979323846	2643383279	5028841971	6939937510	5820974944	5923078164	0628620899	8628034825	3421170679
8214808651	3282306647	0938446095	5058223172	5359408128	4811174502	8410270193	8521105559	6446229489	5493038196
4428810975	6659334461	2847564823	3786783165	2712019091	4564856692	3460348610	4543266482	1339360726	0249141273
7245870066	0631558817	4881520920	9628292540	9171536436	7892590360	0113305305	4882046652	1384146951	9415116094
3305727036	5759591953	0921861173	8193261179	3105118548	0744623799	6274956735	1885752724	8912279381	8301194912
9833673362	4406566430	8602139494	6395224737	1907021798	6094370277	0539217176	2931767523	8467481846	7669405132
0005681271	4526356082	7785771342	7577896091	7363717872	1468440901	2249534301	4654958537	1050792279	6892589235
4201995611	2129021960	8640344181	5981362977	4771309960	5187072113	4999999837	2978049951	0597317328	1609631859
5024459455	3469083026	4252230825	3344685035	2619311881	7101000313	7838752886	5875332083	8142061717	7669147303
5982534904	2875546873	1159562863	8823537875	9375195778	1857780532	1712268066	1300192787	6611195909	2164201989

e to 1 000 decimal places

2.7182818284	5904523536	0287471352	6624977572	4709369995	9574966967	6277240766	3035354759	4571382178	5251664274
2746639193	2003059921	8174135966	2904357290	0334295260	5956307381	3232862794	3490763233	8298807531	9525101901
1573834187	9307021540	8914993488	4167509244	7614606680	8226480016	8477411853	7423454424	3710753907	7744992069
5517027618	3860626133	1384583000	7520449338	2656029760	6737113200	7093287091	2744374704	7230696977	2093101416
9283681902	5515108657	4637721112	5238978442	5056953696	7707854499	6996794686	4454905987	9316368892	3009879312
7736178215	4249992295	7635148220	8269895193	6680331825	2886939849	6465105820	9392398294	8879332036	2509443117
3012381970	6841614039	7019837679	3206832823	7646480429	5311802328	7825098194	5581530175	6717361332	0698112509
9618188159	3041690351	5988885193	4580727386	6738589422	8792284998	9208680582	5749279610	4841984443	6346324496
8487560233	6248270419	7862320900	2160990235	3043699418	4914631409	3431738143	6405462531	5209618369	0888707016
7683964243	7814059271	4563549061	3031072085	1038375051	0115747704	1718986106	8739696552	1267154688	9570350354

Chapter 2 Mathematics

2.1 Notation

Mathematics is, of course, a vast subject, and so here we concentrate on those mathematical methods and relationships that are most often applied in the physical sciences and engineering.

Although there is a high degree of consistency in accepted mathematical notation, there is some variation. For example the spherical harmonics, Y_l^m, can be written Y_{lm}, and there is some freedom with their signs. In general, the conventions chosen here follow common practice as closely as possible, whilst maintaining consistency with the rest of the handbook.

In particular:

scalars	a	general vectors	$\boldsymbol{a}$
unit vectors	$\hat{\boldsymbol{a}}$	scalar product	$\boldsymbol{a}\cdot\boldsymbol{b}$
vector cross-product	$\boldsymbol{a}\times\boldsymbol{b}$	gradient operator	∇
Laplacian operator	∇^2	derivative	$\frac{\mathrm{d}f}{\mathrm{d}x}$ etc.
partial derivatives	$\frac{\partial f}{\partial x}$ etc.	derivative of r with respect to t	$\dot{r}$
nth derivative	$\frac{\mathrm{d}^n f}{\mathrm{d}x^n}$	closed loop integral	$\oint_L \mathrm{d}\boldsymbol{l}$
closed surface integral	$\oint_S \mathrm{d}\boldsymbol{s}$	matrix	A or a_{ij}
mean value (of x)	$\langle x\rangle$	binomial coefficient	$\binom{n}{r}$
factorial	!	unit imaginary ($\mathrm{i}^2=-1$)	i
exponential constant	e	modulus (of x)	$\lvert x\rvert$
natural logarithm	ln	log to base 10	$\log_{10}$

2.2 Vectors and matrices

Vector algebra

Scalar product[a]	$\boldsymbol{a}\cdot\boldsymbol{b}=\lvert\boldsymbol{a}\rvert\lvert\boldsymbol{b}\rvert\cos\theta$	(2.1)
Vector product[b]	$\boldsymbol{a}\times\boldsymbol{b}=\lvert\boldsymbol{a}\rvert\lvert\boldsymbol{b}\rvert\sin\theta\,\hat{\boldsymbol{n}}=\begin{vmatrix}\hat{\boldsymbol{x}} & \hat{\boldsymbol{y}} & \hat{\boldsymbol{z}}\\ a_x & a_y & a_z\\ b_x & b_y & b_z\end{vmatrix}$	(2.2)
Product rules	$\boldsymbol{a}\cdot\boldsymbol{b}=\boldsymbol{b}\cdot\boldsymbol{a}$	(2.3)
	$\boldsymbol{a}\times\boldsymbol{b}=-\boldsymbol{b}\times\boldsymbol{a}$	(2.4)
	$\boldsymbol{a}\cdot(\boldsymbol{b}+\boldsymbol{c})=(\boldsymbol{a}\cdot\boldsymbol{b})+(\boldsymbol{a}\cdot\boldsymbol{c})$	(2.5)
	$\boldsymbol{a}\times(\boldsymbol{b}+\boldsymbol{c})=(\boldsymbol{a}\times\boldsymbol{b})+(\boldsymbol{a}\times\boldsymbol{c})$	(2.6)
Lagrange's identity	$(\boldsymbol{a}\times\boldsymbol{b})\cdot(\boldsymbol{c}\times\boldsymbol{d})=(\boldsymbol{a}\cdot\boldsymbol{c})(\boldsymbol{b}\cdot\boldsymbol{d})-(\boldsymbol{a}\cdot\boldsymbol{d})(\boldsymbol{b}\cdot\boldsymbol{c})$	(2.7)
Scalar triple product	$(\boldsymbol{a}\times\boldsymbol{b})\cdot\boldsymbol{c}=\begin{vmatrix}a_x & a_y & a_z\\ b_x & b_y & b_z\\ c_x & c_y & c_z\end{vmatrix}$	(2.8)
	$=(\boldsymbol{b}\times\boldsymbol{c})\cdot\boldsymbol{a}=(\boldsymbol{c}\times\boldsymbol{a})\cdot\boldsymbol{b}$	(2.9)
	$=$ volume of parallelepiped	(2.10)
Vector triple product	$(\boldsymbol{a}\times\boldsymbol{b})\times\boldsymbol{c}=(\boldsymbol{a}\cdot\boldsymbol{c})\boldsymbol{b}-(\boldsymbol{b}\cdot\boldsymbol{c})\boldsymbol{a}$	(2.11)
	$\boldsymbol{a}\times(\boldsymbol{b}\times\boldsymbol{c})=(\boldsymbol{a}\cdot\boldsymbol{c})\boldsymbol{b}-(\boldsymbol{a}\cdot\boldsymbol{b})\boldsymbol{c}$	(2.12)
Reciprocal vectors	$\boldsymbol{a}'=(\boldsymbol{b}\times\boldsymbol{c})/[(\boldsymbol{a}\times\boldsymbol{b})\cdot\boldsymbol{c}]$	(2.13)
	$\boldsymbol{b}'=(\boldsymbol{c}\times\boldsymbol{a})/[(\boldsymbol{a}\times\boldsymbol{b})\cdot\boldsymbol{c}]$	(2.14)
	$\boldsymbol{c}'=(\boldsymbol{a}\times\boldsymbol{b})/[(\boldsymbol{a}\times\boldsymbol{b})\cdot\boldsymbol{c}]$	(2.15)
	$(\boldsymbol{a}'\cdot\boldsymbol{a})=(\boldsymbol{b}'\cdot\boldsymbol{b})=(\boldsymbol{c}'\cdot\boldsymbol{c})=1$	(2.16)
Vector $\boldsymbol{a}$ with respect to a nonorthogonal basis $\{\boldsymbol{e}_1,\boldsymbol{e}_2,\boldsymbol{e}_3\}$[c]	$\boldsymbol{a}=(\boldsymbol{e}_1'\cdot\boldsymbol{a})\boldsymbol{e}_1+(\boldsymbol{e}_2'\cdot\boldsymbol{a})\boldsymbol{e}_2+(\boldsymbol{e}_3'\cdot\boldsymbol{a})\boldsymbol{e}_3$	(2.17)

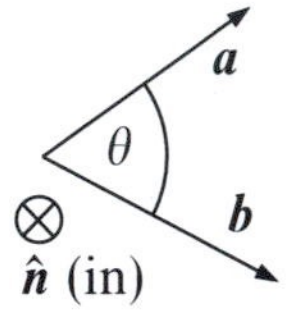

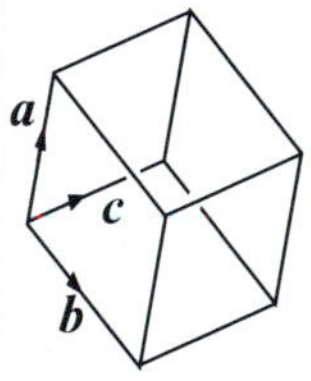

[a] Also known as the "dot product" or the "inner product."
[b] Also known as the "cross-product." $\hat{\boldsymbol{n}}$ is a unit vector making a right-handed set with $\boldsymbol{a}$ and $\boldsymbol{b}$.
[c] The prime (′) denotes a reciprocal vector.

Common three-dimensional coordinate systems

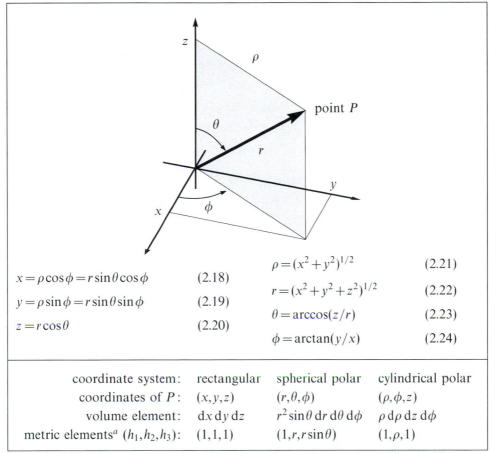

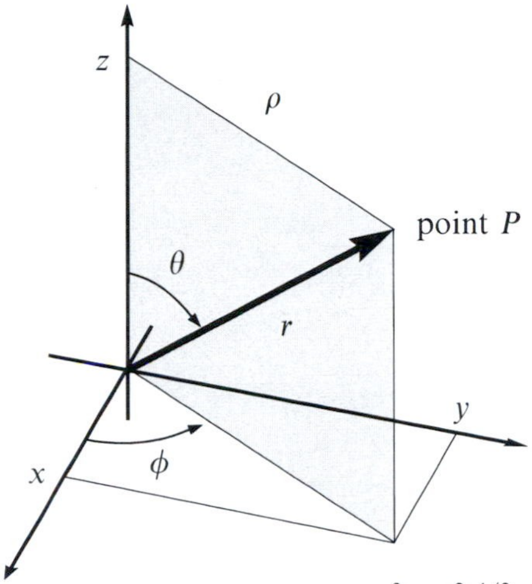

$$x = \rho\cos\phi = r\sin\theta\cos\phi \tag{2.18}$$

$$y = \rho\sin\phi = r\sin\theta\sin\phi \tag{2.19}$$

$$z = r\cos\theta \tag{2.20}$$

$$\rho = (x^2+y^2)^{1/2} \tag{2.21}$$

$$r = (x^2+y^2+z^2)^{1/2} \tag{2.22}$$

$$\theta = \arccos(z/r) \tag{2.23}$$

$$\phi = \arctan(y/x) \tag{2.24}$$

coordinate system:	rectangular	spherical polar	cylindrical polar
coordinates of P:	(x,y,z)	(r,θ,ϕ)	(ρ,ϕ,z)
volume element:	$\mathrm{d}x\,\mathrm{d}y\,\mathrm{d}z$	$r^2\sin\theta\,\mathrm{d}r\,\mathrm{d}\theta\,\mathrm{d}\phi$	$\rho\,\mathrm{d}\rho\,\mathrm{d}z\,\mathrm{d}\phi$
metric elements[a] (h_1,h_2,h_3):	$(1,1,1)$	$(1,r,r\sin\theta)$	$(1,\rho,1)$

[a] In an orthogonal coordinate system (parameterised by coordinates q_1,q_2,q_3), the differential line element $\mathrm{d}l$ is obtained from $(\mathrm{d}l)^2=(h_1\,\mathrm{d}q_1)^2+(h_2\,\mathrm{d}q_2)^2+(h_3\,\mathrm{d}q_3)^2$.

Gradient

Rectangular coordinates	$\nabla f = \frac{\partial f}{\partial x}\hat{\boldsymbol{x}} + \frac{\partial f}{\partial y}\hat{\boldsymbol{y}} + \frac{\partial f}{\partial z}\hat{\boldsymbol{z}}$	(2.25)	f $\hat{}$	scalar field unit vector
Cylindrical coordinates	$\nabla f = \frac{\partial f}{\partial \rho}\hat{\boldsymbol{\rho}} + \frac{1}{r}\frac{\partial f}{\partial \phi}\hat{\boldsymbol{\phi}} + \frac{\partial f}{\partial z}\hat{\boldsymbol{z}}$	(2.26)	ρ	distance from the z axis
Spherical polar coordinates	$\nabla f = \frac{\partial f}{\partial r}\hat{\boldsymbol{r}} + \frac{1}{r}\frac{\partial f}{\partial \theta}\hat{\boldsymbol{\theta}} + \frac{1}{r\sin\theta}\frac{\partial f}{\partial \phi}\hat{\boldsymbol{\phi}}$	(2.27)		
General orthogonal coordinates	$\nabla f = \frac{\hat{\boldsymbol{q}}_1}{h_1}\frac{\partial f}{\partial q_1} + \frac{\hat{\boldsymbol{q}}_2}{h_2}\frac{\partial f}{\partial q_2} + \frac{\hat{\boldsymbol{q}}_3}{h_3}\frac{\partial f}{\partial q_3}$	(2.28)	q_i h_i	basis metric elements

Divergence

Rectangular coordinates	$\nabla\cdot\boldsymbol{A}=\frac{\partial A_x}{\partial x}+\frac{\partial A_y}{\partial y}+\frac{\partial A_z}{\partial z}$ (2.29)	$\boldsymbol{A}$ vector field; A_i ith component of $\boldsymbol{A}$
Cylindrical coordinates	$\nabla\cdot\boldsymbol{A}=\frac{1}{\rho}\frac{\partial(\rho A_\rho)}{\partial\rho}+\frac{1}{\rho}\frac{\partial A_\phi}{\partial\phi}+\frac{\partial A_z}{\partial z}$ (2.30)	ρ distance from the z axis
Spherical polar coordinates	$\nabla\cdot\boldsymbol{A}=\frac{1}{r^2}\frac{\partial(r^2A_r)}{\partial r}+\frac{1}{r\sin\theta}\frac{\partial(A_\theta\sin\theta)}{\partial\theta}+\frac{1}{r\sin\theta}\frac{\partial A_\phi}{\partial\phi}$ (2.31)	
General orthogonal coordinates	$\nabla\cdot\boldsymbol{A}=\frac{1}{h_1h_2h_3}\left[\frac{\partial}{\partial q_1}(A_1h_2h_3)+\frac{\partial}{\partial q_2}(A_2h_3h_1)+\frac{\partial}{\partial q_3}(A_3h_1h_2)\right]$ (2.32)	q_i basis; h_i metric elements

Curl

Rectangular coordinates	$\nabla\times\boldsymbol{A}=\begin{vmatrix}\hat{\boldsymbol{x}} & \hat{\boldsymbol{y}} & \hat{\boldsymbol{z}}\\ \partial/\partial x & \partial/\partial y & \partial/\partial z\\ A_x & A_y & A_z\end{vmatrix}$ (2.33)	$\hat{}$ unit vector; $\boldsymbol{A}$ vector field; A_i ith component of $\boldsymbol{A}$
Cylindrical coordinates	$\nabla\times\boldsymbol{A}=\begin{vmatrix}\hat{\boldsymbol{\rho}}/\rho & \hat{\boldsymbol{\phi}} & \hat{\boldsymbol{z}}/\rho\\ \partial/\partial\rho & \partial/\partial\phi & \partial/\partial z\\ A_\rho & \rho A_\phi & A_z\end{vmatrix}$ (2.34)	ρ distance from the z axis
Spherical polar coordinates	$\nabla\times\boldsymbol{A}=\begin{vmatrix}\hat{\boldsymbol{r}}/(r^2\sin\theta) & \hat{\boldsymbol{\theta}}/(r\sin\theta) & \hat{\boldsymbol{\phi}}/r\\ \partial/\partial r & \partial/\partial\theta & \partial/\partial\phi\\ A_r & rA_\theta & rA_\phi\sin\theta\end{vmatrix}$ (2.35)	
General orthogonal coordinates	$\nabla\times\boldsymbol{A}=\frac{1}{h_1h_2h_3}\begin{vmatrix}\hat{\boldsymbol{q}}_1h_1 & \hat{\boldsymbol{q}}_2h_2 & \hat{\boldsymbol{q}}_3h_3\\ \partial/\partial q_1 & \partial/\partial q_2 & \partial/\partial q_3\\ h_1A_1 & h_2A_2 & h_3A_3\end{vmatrix}$ (2.36)	q_i basis; h_i metric elements

Radial forms[a]

$\nabla r=\frac{\boldsymbol{r}}{r}$	(2.37)	$\nabla(1/r)=\frac{-\boldsymbol{r}}{r^3}$	(2.41)
$\nabla\cdot\boldsymbol{r}=3$	(2.38)	$\nabla\cdot(\boldsymbol{r}/r^2)=\frac{1}{r^2}$	(2.42)
$\nabla r^2=2\boldsymbol{r}$	(2.39)	$\nabla(1/r^2)=\frac{-2\boldsymbol{r}}{r^4}$	(2.43)
$\nabla\cdot(r\boldsymbol{r})=4r$	(2.40)	$\nabla\cdot(\boldsymbol{r}/r^3)=4\pi\delta(\boldsymbol{r})$	(2.44)

[a]Note that the curl of any purely radial function is zero. $\delta(\boldsymbol{r})$ is the Dirac delta function.

Laplacian (scalar)

Rectangular coordinates	$$\nabla^2 f = \frac{\partial^2 f}{\partial x^2} + \frac{\partial^2 f}{\partial y^2} + \frac{\partial^2 f}{\partial z^2} \quad (2.45)$$	f scalar field
Cylindrical coordinates	$$\nabla^2 f = \frac{1}{\rho}\frac{\partial}{\partial \rho}\left(\rho \frac{\partial f}{\partial \rho}\right) + \frac{1}{\rho^2}\frac{\partial^2 f}{\partial \phi^2} + \frac{\partial^2 f}{\partial z^2} \quad (2.46)$$	ρ distance from the z axis
Spherical polar coordinates	$$\nabla^2 f = \frac{1}{r^2}\frac{\partial}{\partial r}\left(r^2 \frac{\partial f}{\partial r}\right) + \frac{1}{r^2 \sin\theta}\frac{\partial}{\partial \theta}\left(\sin\theta \frac{\partial f}{\partial \theta}\right) + \frac{1}{r^2 \sin^2\theta}\frac{\partial^2 f}{\partial \phi^2} \quad (2.47)$$	
General orthogonal coordinates	$$\nabla^2 f = \frac{1}{h_1 h_2 h_3}\left[\frac{\partial}{\partial q_1}\left(\frac{h_2 h_3}{h_1}\frac{\partial f}{\partial q_1}\right) + \frac{\partial}{\partial q_2}\left(\frac{h_3 h_1}{h_2}\frac{\partial f}{\partial q_2}\right) + \frac{\partial}{\partial q_3}\left(\frac{h_1 h_2}{h_3}\frac{\partial f}{\partial q_3}\right)\right] \quad (2.48)$$	q_i basis h_i metric elements

Differential operator identities

$$\nabla(fg) \equiv f\nabla g + g\nabla f \quad (2.49)$$	
$$\nabla\cdot(f\boldsymbol{A}) \equiv f\nabla\cdot\boldsymbol{A} + \boldsymbol{A}\cdot\nabla f \quad (2.50)$$	
$$\nabla\times(f\boldsymbol{A}) \equiv f\nabla\times\boldsymbol{A} + (\nabla f)\times\boldsymbol{A} \quad (2.51)$$	
$$\nabla(\boldsymbol{A}\cdot\boldsymbol{B}) \equiv \boldsymbol{A}\times(\nabla\times\boldsymbol{B}) + (\boldsymbol{A}\cdot\nabla)\boldsymbol{B} + \boldsymbol{B}\times(\nabla\times\boldsymbol{A}) + (\boldsymbol{B}\cdot\nabla)\boldsymbol{A} \quad (2.52)$$	
$$\nabla\cdot(\boldsymbol{A}\times\boldsymbol{B}) \equiv \boldsymbol{B}\cdot(\nabla\times\boldsymbol{A}) - \boldsymbol{A}\cdot(\nabla\times\boldsymbol{B}) \quad (2.53)$$	f,g scalar fields $\boldsymbol{A},\boldsymbol{B}$ vector fields
$$\nabla\times(\boldsymbol{A}\times\boldsymbol{B}) \equiv \boldsymbol{A}(\nabla\cdot\boldsymbol{B}) - \boldsymbol{B}(\nabla\cdot\boldsymbol{A}) + (\boldsymbol{B}\cdot\nabla)\boldsymbol{A} - (\boldsymbol{A}\cdot\nabla)\boldsymbol{B} \quad (2.54)$$	
$$\nabla\cdot(\nabla f) \equiv \nabla^2 f \equiv \triangle f \quad (2.55)$$	
$$\nabla\times(\nabla f) \equiv \boldsymbol{0} \quad (2.56)$$	
$$\nabla\cdot(\nabla\times\boldsymbol{A}) \equiv 0 \quad (2.57)$$	
$$\nabla\times(\nabla\times\boldsymbol{A}) \equiv \nabla(\nabla\cdot\boldsymbol{A}) - \nabla^2\boldsymbol{A} \quad (2.58)$$	

Vector integral transformations

Gauss's (Divergence) theorem	$$\int_V (\nabla\cdot\boldsymbol{A})\,\mathrm{d}V = \oint_{S_c} \boldsymbol{A}\cdot\mathrm{d}\boldsymbol{s} \quad (2.59)$$	$\boldsymbol{A}$ vector field $\mathrm{d}V$ volume element S_c closed surface V volume enclosed
Stokes's theorem	$$\int_S (\nabla\times\boldsymbol{A})\cdot\mathrm{d}\boldsymbol{s} = \oint_L \boldsymbol{A}\cdot\mathrm{d}\boldsymbol{l} \quad (2.60)$$	S surface $\mathrm{d}\boldsymbol{s}$ surface element L loop bounding S $\mathrm{d}\boldsymbol{l}$ line element
Green's first theorem	$$\oint_S (f\nabla g)\cdot\mathrm{d}\boldsymbol{s} = \int_V \nabla\cdot(f\nabla g)\,\mathrm{d}V \quad (2.61)$$ $$= \int_V [f\nabla^2 g + (\nabla f)\cdot(\nabla g)]\,\mathrm{d}V \quad (2.62)$$	f,g scalar fields
Green's second theorem	$$\oint_S [f(\nabla g) - g(\nabla f)]\cdot\mathrm{d}\boldsymbol{s} = \int_V (f\nabla^2 g - g\nabla^2 f)\,\mathrm{d}V \quad (2.63)$$	

Matrix algebra[a]

Matrix definition	$\mathbf{A}=\begin{pmatrix} a_{11} & a_{12} & \cdots & a_{1n} \\ a_{21} & a_{22} & \cdots & a_{2n} \\ \vdots & \vdots & \cdots & \vdots \\ a_{m1} & a_{m2} & \cdots & a_{mn} \end{pmatrix}$ (2.64)		**A** m by n matrix a_{ij} matrix elements
Matrix addition	$\mathbf{C}=\mathbf{A}+\mathbf{B}$ if $c_{ij}=a_{ij}+b_{ij}$	(2.65)	
Matrix multiplication	$\mathbf{C}=\mathbf{AB}$ if $c_{ij}=a_{ik}b_{kj}$ $(\mathbf{AB})\mathbf{C}=\mathbf{A}(\mathbf{BC})$ $\mathbf{A}(\mathbf{B}+\mathbf{C})=\mathbf{AB}+\mathbf{AC}$	(2.66) (2.67) (2.68)	
Transpose matrix[b]	$\tilde{a}_{ij}=a_{ji}$ $(\widetilde{\mathbf{AB}\ldots\mathbf{N}})=\tilde{\mathbf{N}}\ldots\tilde{\mathbf{B}}\tilde{\mathbf{A}}$	(2.69) (2.70)	$\tilde{a}_{ij}$ transpose matrix (sometimes a^{T}_{ij}, or a'_{ij})
Adjoint matrix (definition 1)[c]	$\mathbf{A}^{\dagger}=\tilde{\mathbf{A}}^{*}$ $(\mathbf{AB}\ldots\mathbf{N})^{\dagger}=\mathbf{N}^{\dagger}\ldots\mathbf{B}^{\dagger}\mathbf{A}^{\dagger}$	(2.71) (2.72)	* complex conjugate (of each component) † adjoint (or Hermitian conjugate)
Hermitian matrix[d]	$\mathbf{H}^{\dagger}=\mathbf{H}$	(2.73)	**H** Hermitian (or self-adjoint) matrix

examples:

$$\mathbf{A}=\begin{pmatrix} a_{11} & a_{12} & a_{13} \\ a_{21} & a_{22} & a_{23} \\ a_{31} & a_{32} & a_{33} \end{pmatrix} \qquad \mathbf{B}=\begin{pmatrix} b_{11} & b_{12} & b_{13} \\ b_{21} & b_{22} & b_{23} \\ b_{31} & b_{32} & b_{33} \end{pmatrix}$$

$$\tilde{\mathbf{A}}=\begin{pmatrix} a_{11} & a_{21} & a_{31} \\ a_{12} & a_{22} & a_{32} \\ a_{13} & a_{23} & a_{33} \end{pmatrix} \qquad \mathbf{A}+\mathbf{B}=\begin{pmatrix} a_{11}+b_{11} & a_{12}+b_{12} & a_{13}+b_{13} \\ a_{21}+b_{21} & a_{22}+b_{22} & a_{23}+b_{23} \\ a_{31}+b_{31} & a_{32}+b_{32} & a_{33}+b_{33} \end{pmatrix}$$

$$\mathbf{AB}=\begin{pmatrix} a_{11}b_{11}+a_{12}b_{21}+a_{13}b_{31} & a_{11}b_{12}+a_{12}b_{22}+a_{13}b_{32} & a_{11}b_{13}+a_{12}b_{23}+a_{13}b_{33} \\ a_{21}b_{11}+a_{22}b_{21}+a_{23}b_{31} & a_{21}b_{12}+a_{22}b_{22}+a_{23}b_{32} & a_{21}b_{13}+a_{22}b_{23}+a_{23}b_{33} \\ a_{31}b_{11}+a_{32}b_{21}+a_{33}b_{31} & a_{31}b_{12}+a_{32}b_{22}+a_{33}b_{32} & a_{31}b_{13}+a_{32}b_{23}+a_{33}b_{33} \end{pmatrix}$$

[a]Terms are implicitly summed over repeated suffices; hence $a_{ik}b_{kj}$ equals $\sum_k a_{ik}b_{kj}$.
[b]See also Equation (2.85).
[c]Or "Hermitian conjugate matrix." The term "adjoint" is used in quantum physics for the transpose conjugate of a matrix and in linear algebra for the transpose matrix of its cofactors. These definitions are not compatible, but both are widely used [cf. Equation (2.80)].
[d]Hermitian matrices must also be square (see next table).

Square matrices[a]

Trace	$\operatorname{tr}\mathbf{A} = a_{ii}$ (2.74) $\operatorname{tr}(\mathbf{AB}) = \operatorname{tr}(\mathbf{BA})$ (2.75)	$\mathbf{A}$ a_{ij} a_{ii}	square matrix matrix elements implicitly $=\sum_i a_{ii}$
Determinant[b]	$\det\mathbf{A} = \epsilon_{ijk\ldots} a_{1i} a_{2j} a_{3k} \ldots$ (2.76) $= (-1)^{i+1} a_{i1} M_{i1}$ (2.77) $= a_{i1} C_{i1}$ (2.78) $\det(\mathbf{AB}\ldots\mathbf{N}) = \det\mathbf{A}\det\mathbf{B}\ldots\det\mathbf{N}$ (2.79)	tr det M_{ij} C_{ij}	trace determinant (or $\lvert\mathbf{A}\rvert$) minor of element a_{ij} cofactor of the element a_{ij}
Adjoint matrix (definition 2)[c]	$\operatorname{adj}\mathbf{A} = \tilde{C}_{ij} = C_{ji}$ (2.80)	adj $\sim$	adjoint (sometimes written $\hat{\mathbf{A}}$) transpose
Inverse matrix ($\det\mathbf{A} \neq 0$)	$a_{ij}^{-1} = \dfrac{C_{ji}}{\det\mathbf{A}} = \dfrac{\operatorname{adj}\mathbf{A}}{\det\mathbf{A}}$ (2.81) $\mathbf{A}\mathbf{A}^{-1} = \mathbf{1}$ (2.82) $(\mathbf{AB}\ldots\mathbf{N})^{-1} = \mathbf{N}^{-1}\ldots\mathbf{B}^{-1}\mathbf{A}^{-1}$ (2.83)	$\mathbf{1}$	unit matrix
Orthogonality condition	$a_{ij} a_{ik} = \delta_{jk}$ (2.84) i.e., $\tilde{\mathbf{A}} = \mathbf{A}^{-1}$ (2.85)	δ_{jk}	Kronecker delta ($=1$ if $i=j$, $=0$ otherwise)
Symmetry	If $\mathbf{A} = \tilde{\mathbf{A}}$, $\mathbf{A}$ is symmetric (2.86) If $\mathbf{A} = -\tilde{\mathbf{A}}$, $\mathbf{A}$ is antisymmetric (2.87)		
Unitary matrix	$\mathbf{U}^\dagger = \mathbf{U}^{-1}$ (2.88)	$\mathbf{U}$ $\dagger$	unitary matrix Hermitian conjugate

examples:

$$\mathbf{A} = \begin{pmatrix} a_{11} & a_{12} & a_{13} \\ a_{21} & a_{22} & a_{23} \\ a_{31} & a_{32} & a_{33} \end{pmatrix} \qquad \mathbf{B} = \begin{pmatrix} b_{11} & b_{12} \\ b_{21} & b_{22} \end{pmatrix}$$

$$\operatorname{tr}\mathbf{A} = a_{11} + a_{22} + a_{33} \qquad \operatorname{tr}\mathbf{B} = b_{11} + b_{22}$$

$$\det\mathbf{A} = a_{11}a_{22}a_{33} - a_{11}a_{23}a_{32} - a_{21}a_{12}a_{33} + a_{21}a_{13}a_{32} + a_{31}a_{12}a_{23} - a_{31}a_{13}a_{22}$$

$$\det\mathbf{B} = b_{11}b_{22} - b_{12}b_{21}$$

$$\mathbf{A}^{-1} = \frac{1}{\det\mathbf{A}} \begin{pmatrix} a_{22}a_{33} - a_{23}a_{32} & -a_{12}a_{33} + a_{13}a_{32} & a_{12}a_{23} - a_{13}a_{22} \\ -a_{21}a_{33} + a_{23}a_{31} & a_{11}a_{33} - a_{13}a_{31} & -a_{11}a_{23} + a_{13}a_{21} \\ a_{21}a_{32} - a_{22}a_{31} & -a_{11}a_{32} + a_{12}a_{31} & a_{11}a_{22} - a_{12}a_{21} \end{pmatrix}$$

$$\mathbf{B}^{-1} = \frac{1}{\det\mathbf{B}} \begin{pmatrix} b_{22} & -b_{12} \\ -b_{21} & b_{11} \end{pmatrix}$$

[a]Terms are implicitly summed over repeated suffices; hence $a_{ik}b_{kj}$ equals $\sum_k a_{ik}b_{kj}$.
[b]$\epsilon_{ijk\ldots}$ is defined as the natural extension of Equation (2.443) to n-dimensions (see page 50). M_{ij} is the determinant of the matrix $\mathbf{A}$ with the ith row and the jth column deleted. The cofactor $C_{ij} = (-1)^{i+j} M_{ij}$.
[c]Or "adjugate matrix." See the footnote to Equation (2.71) for a discussion of the term "adjoint."

Commutators

Commutator definition	$[\mathbf{A},\mathbf{B}]=\mathbf{AB}-\mathbf{BA}=-[\mathbf{B},\mathbf{A}]$	(2.89)	$[\cdot,\cdot]$	commutator
Adjoint	$[\mathbf{A},\mathbf{B}]^\dagger=[\mathbf{B}^\dagger,\mathbf{A}^\dagger]$	(2.90)	$\dagger$	adjoint
Distribution	$[\mathbf{A}+\mathbf{B},\mathbf{C}]=[\mathbf{A},\mathbf{C}]+[\mathbf{B},\mathbf{C}]$	(2.91)		
Association	$[\mathbf{AB},\mathbf{C}]=\mathbf{A}[\mathbf{B},\mathbf{C}]+[\mathbf{A},\mathbf{C}]\mathbf{B}$	(2.92)		
Jacobi identity	$[\mathbf{A},[\mathbf{B},\mathbf{C}]]=[\mathbf{B},[\mathbf{A},\mathbf{C}]]-[\mathbf{C},[\mathbf{A},\mathbf{B}]]$	(2.93)		

Pauli matrices

Pauli matrices	$\boldsymbol{\sigma}_1=\begin{pmatrix}0&1\\1&0\end{pmatrix}$ $\boldsymbol{\sigma}_2=\begin{pmatrix}0&-\mathbf{i}\\\mathbf{i}&0\end{pmatrix}$ $\boldsymbol{\sigma}_3=\begin{pmatrix}1&0\\0&-1\end{pmatrix}$ $\mathbf{1}=\begin{pmatrix}1&0\\0&1\end{pmatrix}$	(2.94)	$\boldsymbol{\sigma}_i$ $\mathbf{1}$ $\mathbf{i}$	Pauli spin matrices 2×2 unit matrix $\mathbf{i}^2=-1$
Anticommutation	$\boldsymbol{\sigma}_i\boldsymbol{\sigma}_j+\boldsymbol{\sigma}_j\boldsymbol{\sigma}_i=2\delta_{ij}\mathbf{1}$	(2.95)	δ_{ij}	Kronecker delta
Cyclic permutation	$\boldsymbol{\sigma}_i\boldsymbol{\sigma}_j=\mathbf{i}\boldsymbol{\sigma}_k$ $(\boldsymbol{\sigma}_i)^2=\mathbf{1}$	(2.96) (2.97)		

Rotation matrices[a]

Rotation about x_1	$\mathbf{R}_1(\theta)=\begin{pmatrix}1&0&0\\0&\cos\theta&\sin\theta\\0&-\sin\theta&\cos\theta\end{pmatrix}$	(2.98)	$\mathbf{R}_i(\theta)$ θ	matrix for rotation about the ith axis rotation angle
Rotation about x_2	$\mathbf{R}_2(\theta)=\begin{pmatrix}\cos\theta&0&-\sin\theta\\0&1&0\\\sin\theta&0&\cos\theta\end{pmatrix}$	(2.99)		
Rotation about x_3	$\mathbf{R}_3(\theta)=\begin{pmatrix}\cos\theta&\sin\theta&0\\-\sin\theta&\cos\theta&0\\0&0&1\end{pmatrix}$	(2.100)	α β γ	rotation about x_3 rotation about x_2' rotation about x_3''
Euler angles	$\mathbf{R}(\alpha,\beta,\gamma)=\begin{pmatrix}\cos\gamma\cos\beta\cos\alpha-\sin\gamma\sin\alpha & \cos\gamma\cos\beta\sin\alpha+\sin\gamma\cos\alpha & -\cos\gamma\sin\beta\\-\sin\gamma\cos\beta\cos\alpha-\cos\gamma\sin\alpha & -\sin\gamma\cos\beta\sin\alpha+\cos\gamma\cos\alpha & \sin\gamma\sin\beta\\\sin\beta\cos\alpha & \sin\beta\sin\alpha & \cos\beta\end{pmatrix}$	(2.101)	$\mathbf{R}$	rotation matrix

[a]Angles are in the right-handed sense for rotation of axes, or the left-handed sense for rotation of vectors. i.e., a vector $\boldsymbol{v}$ is given a right-handed rotation of θ about the x_3-axis using $\mathbf{R}_3(-\theta)\boldsymbol{v}\mapsto\boldsymbol{v}'$. Conventionally, $x_1\equiv x$, $x_2\equiv y$, and $x_3\equiv z$.

2.3 Series, summations, and progressions

Progressions and summations

Arithmetic progression	$S_n = a+(a+d)+(a+2d)+\cdots$ $+[a+(n-1)d]$	(2.102)	n number of terms; S_n sum of n successive terms; a first term; d common difference; l last term
	$=\frac{n}{2}[2a+(n-1)d]$	(2.103)	
	$=\frac{n}{2}(a+l)$	(2.104)	
Geometric progression	$S_n = a+ar+ar^2+\cdots+ar^{n-1}$	(2.105)	
	$=a\frac{1-r^n}{1-r}$	(2.106)	r common ratio
	$S_\infty = \frac{a}{1-r} \quad (\lvert r\rvert<1)$	(2.107)	
Arithmetic mean	$\langle x\rangle_{\rm a} = \frac{1}{n}(x_1+x_2+\cdots+x_n)$	(2.108)	$\langle\cdot\rangle_{\rm a}$ arithmetic mean
Geometric mean	$\langle x\rangle_{\rm g} = (x_1x_2x_3\ldots x_n)^{1/n}$	(2.109)	$\langle\cdot\rangle_{\rm g}$ geometric mean
Harmonic mean	$\langle x\rangle_{\rm h} = n\left(\frac{1}{x_1}+\frac{1}{x_2}+\cdots+\frac{1}{x_n}\right)^{-1}$	(2.110)	$\langle\cdot\rangle_{\rm h}$ harmonic mean
Relative mean magnitudes	$\langle x\rangle_{\rm a} \ge \langle x\rangle_{\rm g} \ge \langle x\rangle_{\rm h}$ if $x_i>0$ for all i	(2.111)	
Summation formulas	$\sum_{i=1}^{n} i = \frac{n}{2}(n+1)$	(2.112)	
	$\sum_{i=1}^{n} i^2 = \frac{n}{6}(n+1)(2n+1)$	(2.113)	
	$\sum_{i=1}^{n} i^3 = \frac{n^2}{4}(n+1)^2$	(2.114)	
	$\sum_{i=1}^{n} i^4 = \frac{n}{30}(n+1)(2n+1)(3n^2+3n-1)$	(2.115)	i dummy integer
	$\sum_{i=1}^{\infty} \frac{(-1)^{i+1}}{i} = 1-\frac{1}{2}+\frac{1}{3}-\frac{1}{4}+\ldots = \ln 2$	(2.116)	
	$\sum_{i=1}^{\infty} \frac{(-1)^{i+1}}{2i-1} = 1-\frac{1}{3}+\frac{1}{5}-\frac{1}{7}+\ldots = \frac{\pi}{4}$	(2.117)	
	$\sum_{i=1}^{\infty} \frac{1}{i^2} = 1+\frac{1}{4}+\frac{1}{9}+\frac{1}{16}+\ldots = \frac{\pi^2}{6}$	(2.118)	
Euler's constant[a]	$\gamma = \lim_{n\to\infty}\left(1+\frac{1}{2}+\frac{1}{3}+\cdots+\frac{1}{n}-\ln n\right)$	(2.119)	γ Euler's constant

[a] $\gamma \simeq 0.577215664\ldots$

Power series

Binomial series[a]	$(1+x)^n = 1+nx+\frac{n(n-1)}{2!}x^2+\frac{n(n-1)(n-2)}{3!}x^3+\cdots$	(2.120)
Binomial coefficient[b]	${}^nC_r \equiv \binom{n}{r} \equiv \frac{n!}{r!(n-r)!}$	(2.121)
Binomial theorem	$(a+b)^n = \sum_{k=0}^{n} \binom{n}{k} a^{n-k}b^k$	(2.122)
Taylor series (about a)[c]	$f(a+x)=f(a)+xf^{(1)}(a)+\frac{x^2}{2!}f^{(2)}(a)+\cdots+\frac{x^{n-1}}{(n-1)!}f^{(n-1)}(a)+\cdots$	(2.123)
Taylor series (3-D)	$f(\boldsymbol{a}+\boldsymbol{x})=f(\boldsymbol{a})+(\boldsymbol{x}\cdot\nabla)f\|_{\boldsymbol{a}}+\frac{(\boldsymbol{x}\cdot\nabla)^2}{2!}f\|_{\boldsymbol{a}}+\frac{(\boldsymbol{x}\cdot\nabla)^3}{3!}f\|_{\boldsymbol{a}}+\cdots$	(2.124)
Maclaurin series	$f(x)=f(0)+xf^{(1)}(0)+\frac{x^2}{2!}f^{(2)}(0)+\cdots+\frac{x^{n-1}}{(n-1)!}f^{(n-1)}(0)+\cdots$	(2.125)

[a]If n is a positive integer the series terminates and is valid for all x. Otherwise the (infinite) series is convergent for $|x|<1$.
[b]The coefficient of x^r in the binomial series.
[c]$xf^{(n)}(a)$ is x times the nth derivative of the function $f(x)$ with respect to x evaluated at a, taken as well behaved around a. $(\boldsymbol{x}\cdot\nabla)^n f|_{\boldsymbol{a}}$ is its extension to three dimensions.

Limits

$n^c x^n \to 0$ as $n\to\infty$ if $\|x\|<1$ (for any fixed c)	(2.126)
$x^n/n! \to 0$ as $n\to\infty$ (for any fixed x)	(2.127)
$(1+x/n)^n \to \mathrm{e}^x$ as $n\to\infty$	(2.128)
$x\ln x \to 0$ as $x\to 0$	(2.129)
$\frac{\sin x}{x} \to 1$ as $x\to 0$	(2.130)
If $f(a)=g(a)=0$ or ∞ then $\lim_{x\to a}\frac{f(x)}{g(x)}=\frac{f^{(1)}(a)}{g^{(1)}(a)}$ (l'Hôpital's rule)	(2.131)

2

Series expansions

$\exp(x)$	$1+x+\frac{x^2}{2!}+\frac{x^3}{3!}+\cdots$	(2.132)	(for all x)
$\ln(1+x)$	$x-\frac{x^2}{2}+\frac{x^3}{3}-\frac{x^4}{4}+\cdots$	(2.133)	$(-1<x\leq 1)$
$\ln\left(\frac{1+x}{1-x}\right)$	$2\left(x+\frac{x^3}{3}+\frac{x^5}{5}+\frac{x^7}{7}+\cdots\right)$	(2.134)	$(\lvert x\rvert<1)$
$\cos(x)$	$1-\frac{x^2}{2!}+\frac{x^4}{4!}-\frac{x^6}{6!}+\cdots$	(2.135)	(for all x)
$\sin(x)$	$x-\frac{x^3}{3!}+\frac{x^5}{5!}-\frac{x^7}{7!}+\cdots$	(2.136)	(for all x)
$\tan(x)$	$x+\frac{x^3}{3}+\frac{2x^5}{15}+\frac{17x^7}{315}\cdots$	(2.137)	$(\lvert x\rvert<\pi/2)$
$\sec(x)$	$1+\frac{x^2}{2}+\frac{5x^4}{24}+\frac{61x^6}{720}+\cdots$	(2.138)	$(\lvert x\rvert<\pi/2)$
$\csc(x)$	$\frac{1}{x}+\frac{x}{6}+\frac{7x^3}{360}+\frac{31x^5}{15120}+\cdots$	(2.139)	$(\lvert x\rvert<\pi)$
$\cot(x)$	$\frac{1}{x}-\frac{x}{3}-\frac{x^3}{45}-\frac{2x^5}{945}-\cdots$	(2.140)	$(\lvert x\rvert<\pi)$
$\arcsin(x)^a$	$x+\frac{1}{2}\frac{x^3}{3}+\frac{1\cdot 3}{2\cdot 4}\frac{x^5}{5}+\frac{1\cdot 3\cdot 5}{2\cdot 4\cdot 6}\frac{x^7}{7}\cdots$	(2.141)	$(\lvert x\rvert<1)$
$\arctan(x)^b$	$x-\frac{x^3}{3}+\frac{x^5}{5}-\frac{x^7}{7}+\cdots$	(2.142)	$(\lvert x\rvert\leq 1)$
	$\frac{\pi}{2}-\frac{1}{x}+\frac{1}{3x^3}-\frac{1}{5x^5}+\cdots$		$(x>1)$
	$-\frac{\pi}{2}-\frac{1}{x}+\frac{1}{3x^3}-\frac{1}{5x^5}+\cdots$		$(x<-1)$
$\cosh(x)$	$1+\frac{x^2}{2!}+\frac{x^4}{4!}+\frac{x^6}{6!}+\cdots$	(2.143)	(for all x)
$\sinh(x)$	$x+\frac{x^3}{3!}+\frac{x^5}{5!}+\frac{x^7}{7!}+\cdots$	(2.144)	(for all x)
$\tanh(x)$	$x-\frac{x^3}{3}+\frac{2x^5}{15}-\frac{17x^7}{315}+\cdots$	(2.145)	$(\lvert x\rvert<\pi/2)$

[a] $\arccos(x)=\pi/2-\arcsin(x)$. Note that $\arcsin(x)\equiv\sin^{-1}(x)$ etc.

[b] $\operatorname{arccot}(x)=\pi/2-\arctan(x)$.

Inequalities

Triangle inequality	$\lvert a_1\rvert - \lvert a_2\rvert \le \lvert a_1 + a_2\rvert \le \lvert a_1\rvert + \lvert a_2\rvert$;	(2.146)
	$\left\lvert \sum_{i=1}^{n} a_i \right\rvert \le \sum_{i=1}^{n} \lvert a_i\rvert$	(2.147)
Chebyshev inequality	if $a_1 \ge a_2 \ge a_3 \ge \ldots \ge a_n$	(2.148)
	and $b_1 \ge b_2 \ge b_3 \ge \ldots \ge b_n$	(2.149)
	then $n\sum_{i=1}^{n} a_i b_i \ge \left(\sum_{i=1}^{n} a_i\right)\left(\sum_{i=1}^{n} b_i\right)$	(2.150)
Cauchy inequality	$\left(\sum_{i=1}^{n} a_i b_i\right)^2 \le \sum_{i=1}^{n} a_i^2 \sum_{i=1}^{n} b_i^2$	(2.151)
Schwarz inequality	$\left[\int_a^b f(x)g(x)\,\mathrm{d}x\right]^2 \le \int_a^b [f(x)]^2\,\mathrm{d}x \int_a^b [g(x)]^2\,\mathrm{d}x$	(2.152)

2.4 Complex variables

Complex numbers

Cartesian form	$z = x + \mathbf{i}y$	(2.153)	z	complex variable
			$\mathbf{i}$	$\mathbf{i}^2 = -1$
			x, y	real variables
Polar form	$z = r\mathrm{e}^{\mathrm{i}\theta} = r(\cos\theta + \mathbf{i}\sin\theta)$	(2.154)	r	amplitude (real)
			θ	phase (real)
Modulus[a]	$\lvert z\rvert = r = (x^2 + y^2)^{1/2}$	(2.155)	$\lvert z\rvert$	modulus of z
	$\lvert z_1 \cdot z_2\rvert = \lvert z_1\rvert \cdot \lvert z_2\rvert$	(2.156)		
Argument	$\theta = \arg z = \arctan\frac{y}{x}$	(2.157)	$\arg z$	argument of z
	$\arg(z_1 z_2) = \arg z_1 + \arg z_2$	(2.158)		
Complex conjugate	$z^* = x - \mathbf{i}y = r\mathrm{e}^{-\mathrm{i}\theta}$	(2.159)	z^*	complex conjugate of $z = r\mathrm{e}^{\mathrm{i}\theta}$
	$\arg(z^*) = -\arg z$	(2.160)		
	$z \cdot z^* = \lvert z\rvert^2$	(2.161)		
Logarithm[b]	$\ln z = \ln r + \mathbf{i}(\theta + 2\pi n)$	(2.162)	n	integer

[a] Or "magnitude."

[b] The principal value of $\ln z$ is given by $n = 0$ and $-\pi < \theta \le \pi$.

Complex analysis[a]

Cauchy–Riemann equations[b]	if $f(z)=u(x,y)+\mathrm{i}v(x,y)$ then $\dfrac{\partial u}{\partial x}=\dfrac{\partial v}{\partial y}$	(2.163)
	$\dfrac{\partial u}{\partial y}=-\dfrac{\partial v}{\partial x}$	(2.164)
Cauchy–Goursat theorem[c]	$\oint_c f(z)\,\mathrm{d}z=0$	(2.165)
Cauchy integral formula[d]	$f(z_0)=\dfrac{1}{2\pi\mathbf{i}}\oint_c \dfrac{f(z)}{z-z_0}\,\mathrm{d}z$	(2.166)
	$f^{(n)}(z_0)=\dfrac{n!}{2\pi\mathbf{i}}\oint_c \dfrac{f(z)}{(z-z_0)^{n+1}}\,\mathrm{d}z$	(2.167)
Laurent expansion[e]	$f(z)=\sum_{n=-\infty}^{\infty} a_n(z-z_0)^n$	(2.168)
	where $a_n=\dfrac{1}{2\pi\mathbf{i}}\oint_c \dfrac{f(z')}{(z'-z_0)^{n+1}}\,\mathrm{d}z'$	(2.169)
Residue theorem	$\oint_c f(z)\,\mathrm{d}z=2\pi\mathbf{i}\sum$ enclosed residues	(2.170)

z	complex variable
$\mathbf{i}$	$\mathrm{i}^2=-1$
x,y	real variables
$f(z)$	function of z
u,v	real functions
$^{(n)}$	nth derivative
a_n	Laurent coefficients
a_{-1}	residue of $f(z)$ at z_0
z'	dummy variable

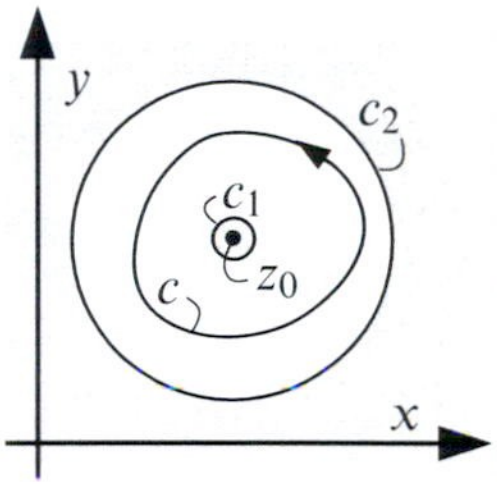

[a]Closed contour integrals are taken in the counterclockwise sense, once.
[b]Necessary condition for $f(z)$ to be analytic at a given point.
[c]If $f(z)$ is analytic within and on a simple closed curve c. Sometimes called "Cauchy's theorem."
[d]If $f(z)$ is analytic within and on a simple closed curve c, encircling z_0.
[e]Of $f(z)$, (analytic) in the annular region between concentric circles, c_1 and c_2, centred on z_0. c is any closed curve in this region encircling z_0.

2.5 Trigonometric and hyperbolic formulas

Trigonometric relationships

$$\sin(A \pm B) = \sin A \cos B \pm \cos A \sin B \qquad (2.171)$$

$$\cos(A \pm B) = \cos A \cos B \mp \sin A \sin B \qquad (2.172)$$

$$\tan(A \pm B) = \frac{\tan A \pm \tan B}{1 \mp \tan A \tan B} \qquad (2.173)$$

$$\cos A \cos B = \frac{1}{2}[\cos(A+B) + \cos(A-B)] \qquad (2.174)$$

$$\sin A \cos B = \frac{1}{2}[\sin(A+B) + \sin(A-B)] \qquad (2.175)$$

$$\sin A \sin B = \frac{1}{2}[\cos(A-B) - \cos(A+B)] \qquad (2.176)$$

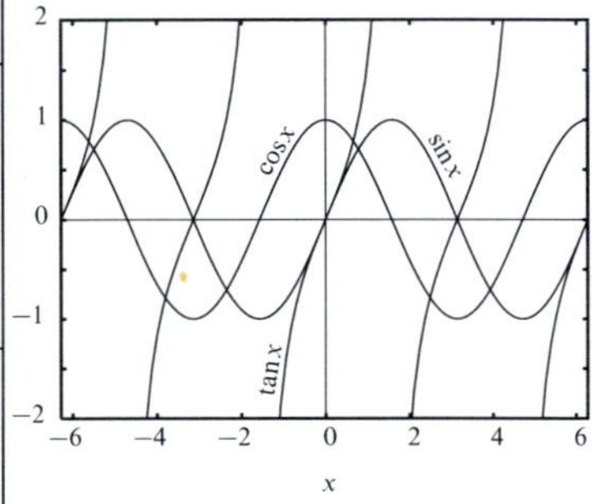

$$\cos^2 A + \sin^2 A = 1 \qquad (2.177)$$

$$\sec^2 A - \tan^2 A = 1 \qquad (2.178)$$

$$\csc^2 A - \cot^2 A = 1 \qquad (2.179)$$

$$\sin 2A = 2 \sin A \cos A \qquad (2.180)$$

$$\cos 2A = \cos^2 A - \sin^2 A \qquad (2.181)$$

$$\tan 2A = \frac{2 \tan A}{1 - \tan^2 A} \qquad (2.182)$$

$$\sin 3A = 3 \sin A - 4 \sin^3 A \qquad (2.183)$$

$$\cos 3A = 4 \cos^3 A - 3 \cos A \qquad (2.184)$$

$$\sin A + \sin B = 2 \sin \frac{A+B}{2} \cos \frac{A-B}{2} \qquad (2.185)$$

$$\sin A - \sin B = 2 \cos \frac{A+B}{2} \sin \frac{A-B}{2} \qquad (2.186)$$

$$\cos A + \cos B = 2 \cos \frac{A+B}{2} \cos \frac{A-B}{2} \qquad (2.187)$$

$$\cos A - \cos B = -2 \sin \frac{A+B}{2} \sin \frac{A-B}{2} \qquad (2.188)$$

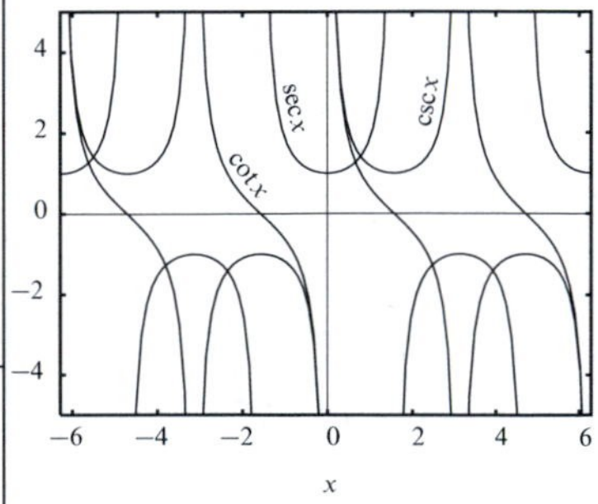

$$\cos^2 A = \frac{1}{2}(1 + \cos 2A) \qquad (2.189)$$

$$\sin^2 A = \frac{1}{2}(1 - \cos 2A) \qquad (2.190)$$

$$\cos^3 A = \frac{1}{4}(3 \cos A + \cos 3A) \qquad (2.191)$$

$$\sin^3 A = \frac{1}{4}(3 \sin A - \sin 3A) \qquad (2.192)$$

Hyperbolic relationships[a]

$$\sinh(x \pm y) = \sinh x \cosh y \pm \cosh x \sinh y \quad (2.193)$$

$$\cosh(x \pm y) = \cosh x \cosh y \pm \sinh x \sinh y \quad (2.194)$$

$$\tanh(x \pm y) = \frac{\tanh x \pm \tanh y}{1 \pm \tanh x \tanh y} \quad (2.195)$$

$$\cosh x \cosh y = \frac{1}{2}[\cosh(x+y) + \cosh(x-y)] \quad (2.196)$$

$$\sinh x \cosh y = \frac{1}{2}[\sinh(x+y) + \sinh(x-y)] \quad (2.197)$$

$$\sinh x \sinh y = \frac{1}{2}[\cosh(x+y) - \cosh(x-y)] \quad (2.198)$$

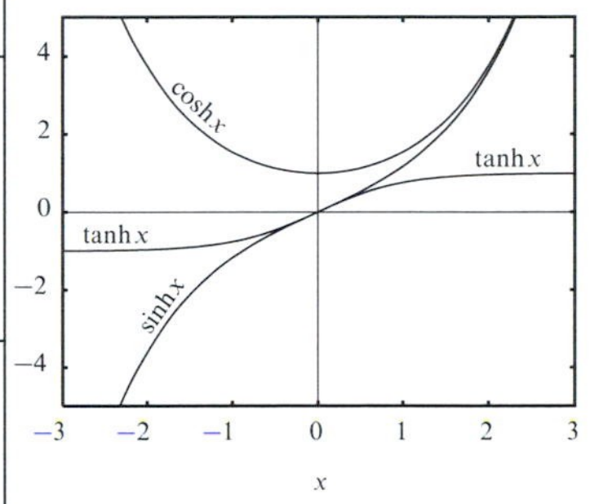

$$\cosh^2 x - \sinh^2 x = 1 \quad (2.199)$$

$$\mathrm{sech}^2 x + \tanh^2 x = 1 \quad (2.200)$$

$$\coth^2 x - \mathrm{csch}^2 x = 1 \quad (2.201)$$

$$\sinh 2x = 2 \sinh x \cosh x \quad (2.202)$$

$$\cosh 2x = \cosh^2 x + \sinh^2 x \quad (2.203)$$

$$\tanh 2x = \frac{2 \tanh x}{1 + \tanh^2 x} \quad (2.204)$$

$$\sinh 3x = 3 \sinh x + 4 \sinh^3 x \quad (2.205)$$

$$\cosh 3x = 4 \cosh^3 x - 3 \cosh x \quad (2.206)$$

$$\sinh x + \sinh y = 2 \sinh \frac{x+y}{2} \cosh \frac{x-y}{2} \quad (2.207)$$

$$\sinh x - \sinh y = 2 \cosh \frac{x+y}{2} \sinh \frac{x-y}{2} \quad (2.208)$$

$$\cosh x + \cosh y = 2 \cosh \frac{x+y}{2} \cosh \frac{x-y}{2} \quad (2.209)$$

$$\cosh x - \cosh y = 2 \sinh \frac{x+y}{2} \sinh \frac{x-y}{2} \quad (2.210)$$

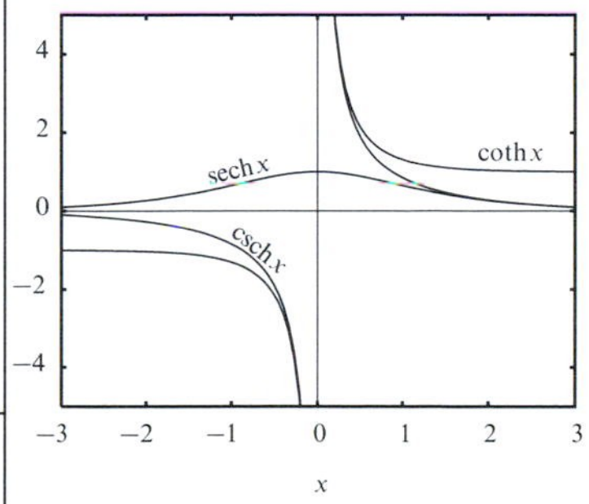

$$\cosh^2 x = \frac{1}{2}(\cosh 2x + 1) \quad (2.211)$$

$$\sinh^2 x = \frac{1}{2}(\cosh 2x - 1) \quad (2.212)$$

$$\cosh^3 x = \frac{1}{4}(3 \cosh x + \cosh 3x) \quad (2.213)$$

$$\sinh^3 x = \frac{1}{4}(\sinh 3x - 3 \sinh x) \quad (2.214)$$

[a]These can be derived from trigonometric relationships by using the substitutions $\cos x \mapsto \cosh x$ and $\sin x \mapsto \mathbf{i} \sinh x$.

Trigonometric and hyperbolic definitions

de Moivre's theorem $(\cos x+\mathrm{i}\sin x)^n=\mathrm{e}^{\mathrm{i}nx}=\cos nx+\mathrm{i}\sin nx$	(2.215)		
$\cos x=\frac{1}{2}(\mathrm{e}^{\mathrm{i}x}+\mathrm{e}^{-\mathrm{i}x})$	(2.216)	$\cosh x=\frac{1}{2}(\mathrm{e}^{x}+\mathrm{e}^{-x})$	(2.217)
$\sin x=\frac{1}{2\mathrm{i}}(\mathrm{e}^{\mathrm{i}x}-\mathrm{e}^{-\mathrm{i}x})$	(2.218)	$\sinh x=\frac{1}{2}(\mathrm{e}^{x}-\mathrm{e}^{-x})$	(2.219)
$\tan x=\frac{\sin x}{\cos x}$	(2.220)	$\tanh x=\frac{\sinh x}{\cosh x}$	(2.221)
$\cos \mathrm{i}x=\cosh x$	(2.222)	$\cosh \mathrm{i}x=\cos x$	(2.223)
$\sin \mathrm{i}x=\mathrm{i}\sinh x$	(2.224)	$\sinh \mathrm{i}x=\mathrm{i}\sin x$	(2.225)
$\cot x=(\tan x)^{-1}$	(2.226)	$\coth x=(\tanh x)^{-1}$	(2.227)
$\sec x=(\cos x)^{-1}$	(2.228)	$\mathrm{sech}\, x=(\cosh x)^{-1}$	(2.229)
$\csc x=(\sin x)^{-1}$	(2.230)	$\mathrm{csch}\, x=(\sinh x)^{-1}$	(2.231)

Inverse trigonometric functions[a]

$\arcsin x=\arctan\left[\frac{x}{(1-x^2)^{1/2}}\right]$	(2.232)
$\arccos x=\arctan\left[\frac{(1-x^2)^{1/2}}{x}\right]$	(2.233)
$\mathrm{arccsc}\, x=\arctan\left[\frac{1}{(x^2-1)^{1/2}}\right]$	(2.234)
$\mathrm{arcsec}\, x=\arctan\left[(x^2-1)^{1/2}\right]$	(2.235)
$\mathrm{arccot}\, x=\arctan\left(\frac{1}{x}\right)$	(2.236)
$\arccos x=\frac{\pi}{2}-\arcsin x$	(2.237)

[a]Valid in the angle range $0\le\theta\le\pi/2$. Note that $\arcsin x\equiv\sin^{-1}x$ etc.

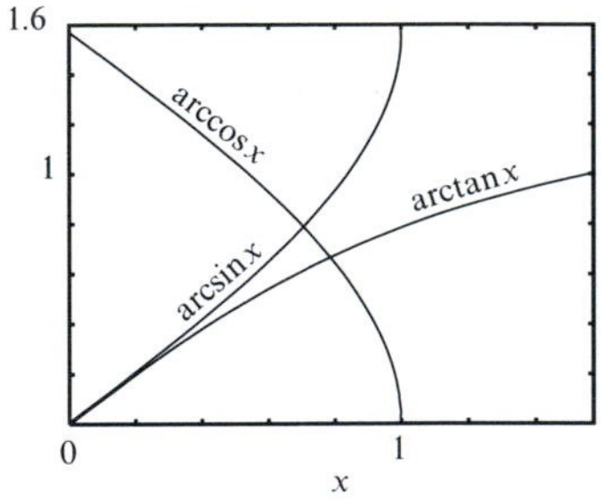

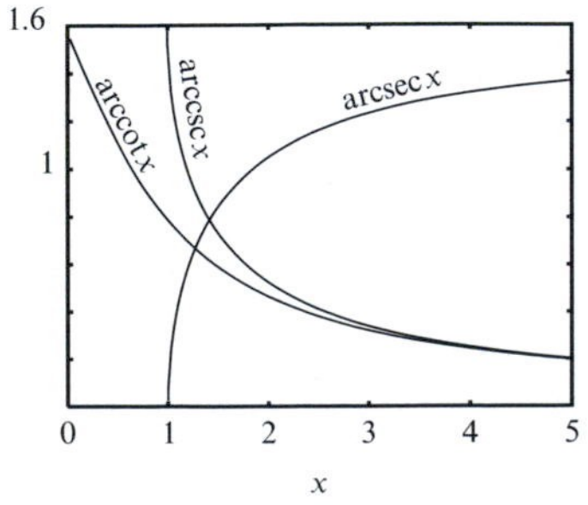

Inverse hyperbolic functions

$\operatorname{arsinh} x \equiv \sinh^{-1} x = \ln\left[x + (x^2+1)^{1/2}\right]$	(2.238)	for all x
$\operatorname{arcosh} x \equiv \cosh^{-1} x = \ln\left[x + (x^2-1)^{1/2}\right]$	(2.239)	$x \geq 1$
$\operatorname{artanh} x \equiv \tanh^{-1} x = \frac{1}{2}\ln\left(\frac{1+x}{1-x}\right)$	(2.240)	$\|x\| < 1$
$\operatorname{arcoth} x \equiv \coth^{-1} x = \frac{1}{2}\ln\left(\frac{x+1}{x-1}\right)$	(2.241)	$\|x\| > 1$
$\operatorname{arsech} x \equiv \operatorname{sech}^{-1} x = \ln\left[\frac{1}{x} + \frac{(1-x^2)^{1/2}}{x}\right]$	(2.242)	$0 < x \leq 1$
$\operatorname{arcsch} x \equiv \operatorname{csch}^{-1} x = \ln\left[\frac{1}{x} + \frac{(1+x^2)^{1/2}}{x}\right]$	(2.243)	$x \neq 0$

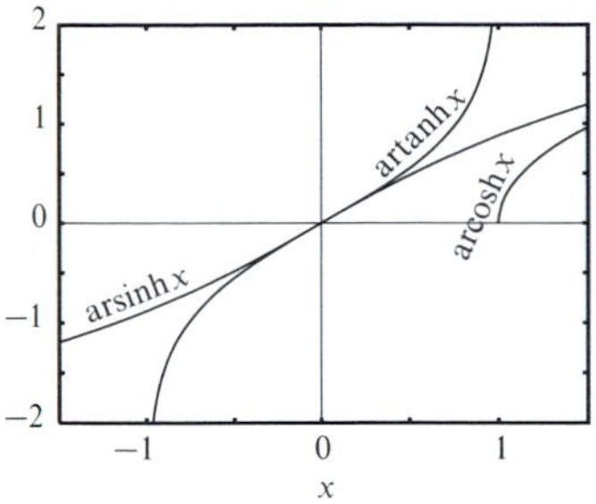

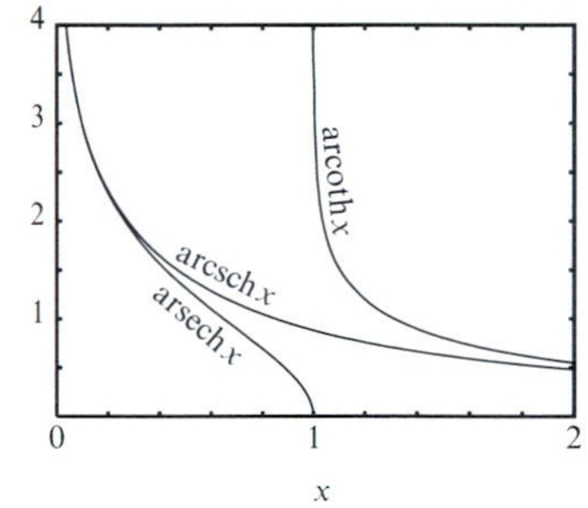

2.6 Mensuration

Moiré fringes[a]

Parallel pattern	$d_{\mathrm{M}} = \left\| \frac{1}{d_1} - \frac{1}{d_2} \right\|^{-1}$	(2.244)	d_{M} Moiré fringe spacing $d_{1,2}$ grating spacings
Rotational pattern[b]	$d_{\mathrm{M}} = \frac{d}{2\|\sin(\theta/2)\|}$	(2.245)	d common grating spacing θ relative rotation angle ($\|\theta\| \leq \pi/2$)

[a]From overlapping linear gratings.
[b]From identical gratings, spacing d, with a relative rotation θ.

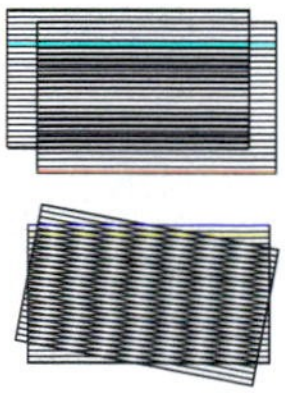

Plane triangles

Sine formula[a]	$\frac{a}{\sin A}=\frac{b}{\sin B}=\frac{c}{\sin C}$	(2.246)
Cosine formulas	$a^2=b^2+c^2-2bc\cos A$	(2.247)
	$\cos A=\frac{b^2+c^2-a^2}{2bc}$	(2.248)
	$a=b\cos C+c\cos B$	(2.249)
Tangent formula	$\tan\frac{A-B}{2}=\frac{a-b}{a+b}\cot\frac{C}{2}$	(2.250)
Area	area $=\frac{1}{2}ab\sin C$	(2.251)
	$=\frac{a^2}{2}\frac{\sin B\sin C}{\sin A}$	(2.252)
	$=[s(s-a)(s-b)(s-c)]^{1/2}$	(2.253)
	where $s=\frac{1}{2}(a+b+c)$	(2.254)

[a]The diameter of the circumscribed circle equals $a/\sin A$.

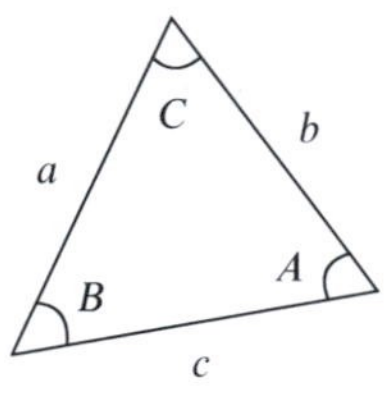

Spherical triangles[a]

Sine formula	$\frac{\sin a}{\sin A}=\frac{\sin b}{\sin B}=\frac{\sin c}{\sin C}$	(2.255)
Cosine formulas	$\cos a=\cos b\cos c+\sin b\sin c\cos A$	(2.256)
	$\cos A=-\cos B\cos C+\sin B\sin C\cos a$	(2.257)
Analogue formula	$\sin a\cos B=\cos b\sin c-\sin b\cos c\cos A$	(2.258)
Four-parts formula	$\cos a\cos C=\sin a\cot b-\sin C\cot B$	(2.259)
Area[b]	$E=A+B+C-\pi$	(2.260)

[a]On a unit sphere.
[b]Also called the "spherical excess."

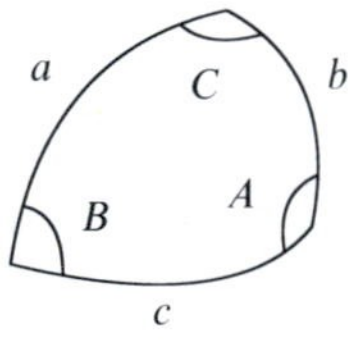

Perimeter, area, and volume

Perimeter of circle	$P=2\pi r$	(2.261)	P r	perimeter radius
Area of circle	$A=\pi r^2$	(2.262)	A	area
Surface area of sphere[a]	$A=4\pi R^2$	(2.263)	R	sphere radius
Volume of sphere	$V=\frac{4}{3}\pi R^3$	(2.264)	V	volume
Perimeter of ellipse[b]	$P=4a\mathrm{E}(\pi/2,e)$ $\simeq 2\pi\left(\frac{a^2+b^2}{2}\right)^{1/2}$	(2.265) (2.266)	a b E e	semi-major axis semi-minor axis elliptic integral of the second kind (p. 45) eccentricity $(=1-b^2/a^2)$
Area of ellipse	$A=\pi ab$	(2.267)		
Volume of ellipsoid[c]	$V=4\pi\frac{abc}{3}$	(2.268)	c	third semi-axis
Surface area of cylinder	$A=2\pi r(h+r)$	(2.269)	h	height
Volume of cylinder	$V=\pi r^2 h$	(2.270)		
Area of circular cone[d]	$A=\pi rl$	(2.271)	l	slant height
Volume of cone or pyramid	$V=A_\mathrm{b}h/3$	(2.272)	A_b	base area
Surface area of torus	$A=\pi^2(r_1+r_2)(r_2-r_1)$	(2.273)	r_1 r_2	inner radius outer radius
Volume of torus	$V=\frac{\pi^2}{4}(r_2^2-r_1^2)(r_2-r_1)$	(2.274)		
Area[d] of spherical cap, depth d	$A=2\pi Rd$	(2.275)	d	cap depth
Volume of spherical cap, depth d	$V=\pi d^2\left(R-\frac{d}{3}\right)$	(2.276)	Ω z α	solid angle distance from centre half-angle subtended
Solid angle of a circle from a point on its axis, z from centre	$\Omega=2\pi\left[1-\frac{z}{(z^2+r^2)^{1/2}}\right]$ $=2\pi(1-\cos\alpha)$	(2.277) (2.278)	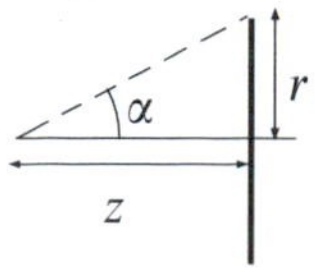	

[a]Sphere defined by $x^2+y^2+z^2=R^2$.
[b]The approximation is exact when $e=0$ and $e\simeq 0.91$, giving a maximum error of 11% at $e=1$.
[c]Ellipsoid defined by $x^2/a^2+y^2/b^2+z^2/c^2=1$.
[d]Curved surface only.

Conic sections

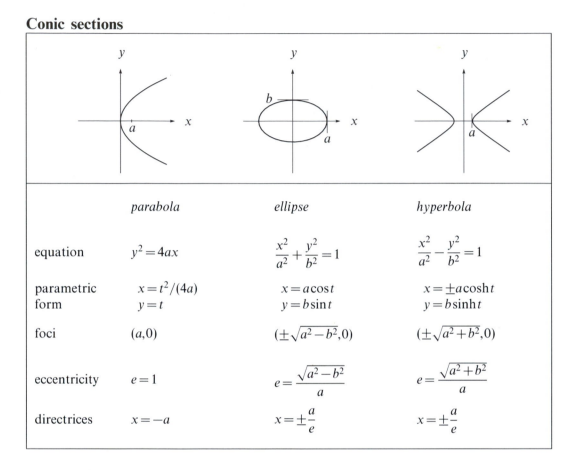

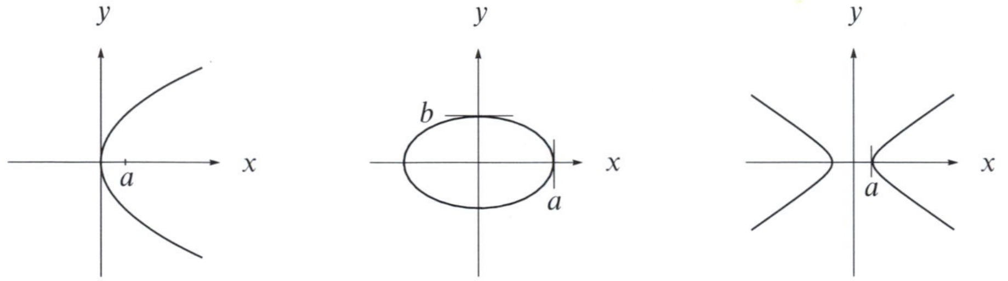

	parabola	ellipse	hyperbola
equation	$y^2=4ax$	$\frac{x^2}{a^2}+\frac{y^2}{b^2}=1$	$\frac{x^2}{a^2}-\frac{y^2}{b^2}=1$
parametric form	$x=t^2/(4a)$ $y=t$	$x=a\cos t$ $y=b\sin t$	$x=\pm a\cosh t$ $y=b\sinh t$
foci	$(a,0)$	$(\pm\sqrt{a^2-b^2},0)$	$(\pm\sqrt{a^2+b^2},0)$
eccentricity	$e=1$	$e=\frac{\sqrt{a^2-b^2}}{a}$	$e=\frac{\sqrt{a^2+b^2}}{a}$
directrices	$x=-a$	$x=\pm\frac{a}{e}$	$x=\pm\frac{a}{e}$

Platonic solids[a]

solid *(faces,edges,vertices)*	*volume*	*surface area*	*circumradius*	*inradius*
tetrahedron (4,6,4)	$\frac{a^3\sqrt{2}}{12}$	$a^2\sqrt{3}$	$\frac{a\sqrt{6}}{4}$	$\frac{a\sqrt{6}}{12}$
cube (6,12,8)	a^3	$6a^2$	$\frac{a\sqrt{3}}{2}$	$\frac{a}{2}$
octahedron (8,12,6)	$\frac{a^3\sqrt{2}}{3}$	$2a^2\sqrt{3}$	$\frac{a}{\sqrt{2}}$	$\frac{a}{\sqrt{6}}$
dodecahedron (12,30,20)	$\frac{a^3(15+7\sqrt{5})}{4}$	$3a^2\sqrt{5(5+2\sqrt{5})}$	$\frac{a}{4}\sqrt{3}(1+\sqrt{5})$	$\frac{a}{4}\sqrt{\frac{50+22\sqrt{5}}{5}}$
icosahedron (20,30,12)	$\frac{5a^3(3+\sqrt{5})}{12}$	$5a^2\sqrt{3}$	$\frac{a}{4}\sqrt{2(5+\sqrt{5})}$	$\frac{a}{4}\left(\sqrt{3}+\sqrt{\frac{5}{3}}\right)$

[a]Of side a. Both regular and irregular polyhedra follow the Euler relation, faces − edges + vertices = 2.

Curve measure

Length of plane curve	$l=\int_a^b \left[1+\left(\frac{\mathrm{d}y}{\mathrm{d}x}\right)^2\right]^{1/2} \mathrm{d}x$	(2.279)	a b $y(x)$ l	start point end point plane curve length
Surface of revolution	$A=2\pi \int_a^b y \left[1+\left(\frac{\mathrm{d}y}{\mathrm{d}x}\right)^2\right]^{1/2} \mathrm{d}x$	(2.280)	A	surface area
Volume of revolution	$V=\pi \int_a^b y^2 \, \mathrm{d}x$	(2.281)	V	volume
Radius of curvature	$\rho=\left[1+\left(\frac{\mathrm{d}y}{\mathrm{d}x}\right)^2\right]^{3/2} \left(\frac{\mathrm{d}^2y}{\mathrm{d}x^2}\right)^{-1}$	(2.282)	ρ	radius of curvature

Differential geometry[a]

Unit tangent	$\hat{\boldsymbol{\tau}}=\frac{\dot{\boldsymbol{r}}}{\lvert\dot{\boldsymbol{r}}\rvert}=\frac{\dot{\boldsymbol{r}}}{v}$	(2.283)	τ $\boldsymbol{r}$ v	tangent curve parameterised by $\boldsymbol{r}(t)$ $\lvert\dot{\boldsymbol{r}}(t)\rvert$
Unit principal normal	$\hat{\boldsymbol{n}}=\frac{\ddot{\boldsymbol{r}}-\dot{v}\hat{\boldsymbol{\tau}}}{\lvert\ddot{\boldsymbol{r}}-\dot{v}\hat{\boldsymbol{\tau}}\rvert}$	(2.284)	$\boldsymbol{n}$	principal normal
Unit binormal	$\hat{\boldsymbol{b}}=\hat{\boldsymbol{\tau}}\times\hat{\boldsymbol{n}}$	(2.285)	$\boldsymbol{b}$	binormal
Curvature	$\kappa=\frac{\lvert\dot{\boldsymbol{r}}\times\ddot{\boldsymbol{r}}\rvert}{\lvert\dot{\boldsymbol{r}}\rvert^3}$	(2.286)	κ	curvature
Radius of curvature	$\rho=\frac{1}{\kappa}$	(2.287)	ρ	radius of curvature
Torsion	$\lambda=\frac{\dot{\boldsymbol{r}}\cdot(\ddot{\boldsymbol{r}}\times\dddot{\boldsymbol{r}})}{\lvert\dot{\boldsymbol{r}}\times\ddot{\boldsymbol{r}}\rvert^2}$	(2.288)	λ	torsion
Frenet's formulas	$\dot{\hat{\boldsymbol{\tau}}}=\kappa v\hat{\boldsymbol{n}}$ (2.289) $\dot{\hat{\boldsymbol{n}}}=-\kappa v\hat{\boldsymbol{\tau}}+\lambda v\hat{\boldsymbol{b}}$ (2.290) $\dot{\hat{\boldsymbol{b}}}=-\lambda v\hat{\boldsymbol{n}}$ (2.291)		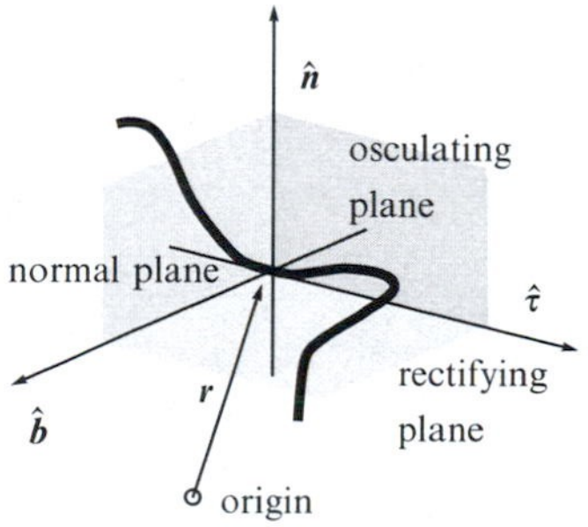	

[a]For a continuous curve in three dimensions, traced by the position vector $\boldsymbol{r}(t)$.

2.7 Differentiation

Derivatives (general)

Power	$\frac{d}{dx}(u^n)=nu^{n-1}\frac{du}{dx}$	(2.292)	n	power index
Product	$\frac{d}{dx}(uv)=u\frac{dv}{dx}+v\frac{du}{dx}$	(2.293)	u,v	functions of x
Quotient	$\frac{d}{dx}\left(\frac{u}{v}\right)=\frac{1}{v}\frac{du}{dx}-\frac{u}{v^2}\frac{dv}{dx}$	(2.294)		
Function of a function[a]	$\frac{d}{dx}[f(u)]=\frac{d}{du}[f(u)]\cdot\frac{du}{dx}$	(2.295)	$f(u)$	function of $u(x)$
Leibniz theorem	$\frac{d^n}{dx^n}[uv]=\binom{n}{0}v\frac{d^nu}{dx^n}+\binom{n}{1}\frac{dv}{dx}\frac{d^{n-1}u}{dx^{n-1}}+\cdots+\binom{n}{k}\frac{d^kv}{dx^k}\frac{d^{n-k}u}{dx^{n-k}}+\cdots+\binom{n}{n}u\frac{d^nv}{dx^n}$	(2.296)	$\binom{n}{k}$	binomial coefficient
Differentiation under the integral sign	$\frac{d}{dq}\left[\int_p^q f(x)\,dx\right]=f(q)$ (p constant)	(2.297)		
	$\frac{d}{dp}\left[\int_p^q f(x)\,dx\right]=-f(p)$ (q constant)	(2.298)		
General integral	$\frac{d}{dx}\left[\int_{u(x)}^{v(x)} f(t)\,dt\right]=f(v)\frac{dv}{dx}-f(u)\frac{du}{dx}$	(2.299)		
Logarithm	$\frac{d}{dx}(\log_b\lvert ax\rvert)=(x\ln b)^{-1}$	(2.300)	b a	log base constant
Exponential	$\frac{d}{dx}(e^{ax})=ae^{ax}$	(2.301)		
Inverse functions	$\frac{dx}{dy}=\left(\frac{dy}{dx}\right)^{-1}$	(2.302)		
	$\frac{d^2x}{dy^2}=-\frac{d^2y}{dx^2}\left(\frac{dy}{dx}\right)^{-3}$	(2.303)		
	$\frac{d^3x}{dy^3}=\left[3\left(\frac{d^2y}{dx^2}\right)^2-\frac{dy}{dx}\frac{d^3y}{dx^3}\right]\left(\frac{dy}{dx}\right)^{-5}$	(2.304)		

[a]The "chain rule."

Trigonometric derivatives[a]

$\frac{\mathrm{d}}{\mathrm{d}x}(\sin ax) = a\cos ax$	(2.305)	$\frac{\mathrm{d}}{\mathrm{d}x}(\cos ax) = -a\sin ax$	(2.306)
$\frac{\mathrm{d}}{\mathrm{d}x}(\tan ax) = a\sec^2 ax$	(2.307)	$\frac{\mathrm{d}}{\mathrm{d}x}(\csc ax) = -a\csc ax \cdot \cot ax$	(2.308)
$\frac{\mathrm{d}}{\mathrm{d}x}(\sec ax) = a\sec ax \cdot \tan ax$	(2.309)	$\frac{\mathrm{d}}{\mathrm{d}x}(\cot ax) = -a\csc^2 ax$	(2.310)
$\frac{\mathrm{d}}{\mathrm{d}x}(\arcsin ax) = a(1-a^2x^2)^{-1/2}$	(2.311)	$\frac{\mathrm{d}}{\mathrm{d}x}(\arccos ax) = -a(1-a^2x^2)^{-1/2}$	(2.312)
$\frac{\mathrm{d}}{\mathrm{d}x}(\arctan ax) = a(1+a^2x^2)^{-1}$	(2.313)	$\frac{\mathrm{d}}{\mathrm{d}x}(\mathrm{arccsc}\, ax) = -\frac{a}{\lvert ax \rvert}(a^2x^2-1)^{-1/2}$	(2.314)
$\frac{\mathrm{d}}{\mathrm{d}x}(\mathrm{arcsec}\, ax) = \frac{a}{\lvert ax \rvert}(a^2x^2-1)^{-1/2}$	(2.315)	$\frac{\mathrm{d}}{\mathrm{d}x}(\mathrm{arccot}\, ax) = -a(a^2x^2+1)^{-1}$	(2.316)

[a] a is a constant.

Hyperbolic derivatives[a]

$\frac{\mathrm{d}}{\mathrm{d}x}(\sinh ax) = a\cosh ax$	(2.317)	$\frac{\mathrm{d}}{\mathrm{d}x}(\cosh ax) = a\sinh ax$	(2.318)
$\frac{\mathrm{d}}{\mathrm{d}x}(\tanh ax) = a\,\mathrm{sech}^2 ax$	(2.319)	$\frac{\mathrm{d}}{\mathrm{d}x}(\mathrm{csch}\, ax) = -a\,\mathrm{csch}\, ax \cdot \coth ax$	(2.320)
$\frac{\mathrm{d}}{\mathrm{d}x}(\mathrm{sech}\, ax) = -a\,\mathrm{sech}\, ax \cdot \tanh ax$	(2.321)	$\frac{\mathrm{d}}{\mathrm{d}x}(\coth ax) = -a\,\mathrm{csch}^2 ax$	(2.322)
$\frac{\mathrm{d}}{\mathrm{d}x}(\mathrm{arsinh}\, ax) = a(a^2x^2+1)^{-1/2}$	(2.323)	$\frac{\mathrm{d}}{\mathrm{d}x}(\mathrm{arcosh}\, ax) = a(a^2x^2-1)^{-1/2}$	(2.324)
$\frac{\mathrm{d}}{\mathrm{d}x}(\mathrm{artanh}\, ax) = a(1-a^2x^2)^{-1}$	(2.325)	$\frac{\mathrm{d}}{\mathrm{d}x}(\mathrm{arcsch}\, ax) = -\frac{a}{\lvert ax \rvert}(1+a^2x^2)^{-1/2}$	(2.326)
$\frac{\mathrm{d}}{\mathrm{d}x}(\mathrm{arsech}\, ax) = -\frac{a}{\lvert ax \rvert}(1-a^2x^2)^{-1/2}$	(2.327)	$\frac{\mathrm{d}}{\mathrm{d}x}(\mathrm{arcoth}\, ax) = a(1-a^2x^2)^{-1}$	(2.328)

[a] a is a constant.

Partial derivatives

Total differential	$\mathrm{d}f=\frac{\partial f}{\partial x}\mathrm{d}x+\frac{\partial f}{\partial y}\mathrm{d}y+\frac{\partial f}{\partial z}\mathrm{d}z$	(2.329)	f $f(x,y,z)$
Reciprocity	$\left.\frac{\partial g}{\partial x}\right\|_y\left.\frac{\partial x}{\partial y}\right\|_g\left.\frac{\partial y}{\partial g}\right\|_x=-1$	(2.330)	g $g(x,y)$
Chain rule	$\frac{\partial f}{\partial u}=\frac{\partial f}{\partial x}\frac{\partial x}{\partial u}+\frac{\partial f}{\partial y}\frac{\partial y}{\partial u}+\frac{\partial f}{\partial z}\frac{\partial z}{\partial u}$	(2.331)	
Jacobian	$J=\frac{\partial(x,y,z)}{\partial(u,v,w)}=\begin{vmatrix}\frac{\partial x}{\partial u}&\frac{\partial x}{\partial v}&\frac{\partial x}{\partial w}\\\frac{\partial y}{\partial u}&\frac{\partial y}{\partial v}&\frac{\partial y}{\partial w}\\\frac{\partial z}{\partial u}&\frac{\partial z}{\partial v}&\frac{\partial z}{\partial w}\end{vmatrix}$	(2.332)	J Jacobian u $u(x,y,z)$ v $v(x,y,z)$ w $w(x,y,z)$
Change of variable	$\int_V f(x,y,z)\,\mathrm{d}x\,\mathrm{d}y\,\mathrm{d}z=\int_{V'} f(u,v,w)J\,\mathrm{d}u\,\mathrm{d}v\,\mathrm{d}w$	(2.333)	V volume in (x,y,z) V' volume in (u,v,w) mapped to by V
Euler–Lagrange equation	if $I=\int_a^b F(x,y,y')\,\mathrm{d}x$ then $\delta I=0$ when $\frac{\partial F}{\partial y}=\frac{\mathrm{d}}{\mathrm{d}x}\left(\frac{\partial F}{\partial y'}\right)$	(2.334)	y' $\mathrm{d}y/\mathrm{d}x$ a,b fixed end points

Stationary points[a]

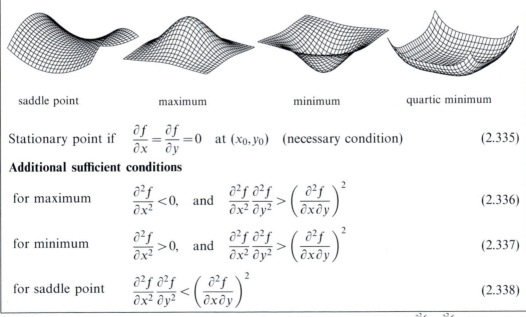

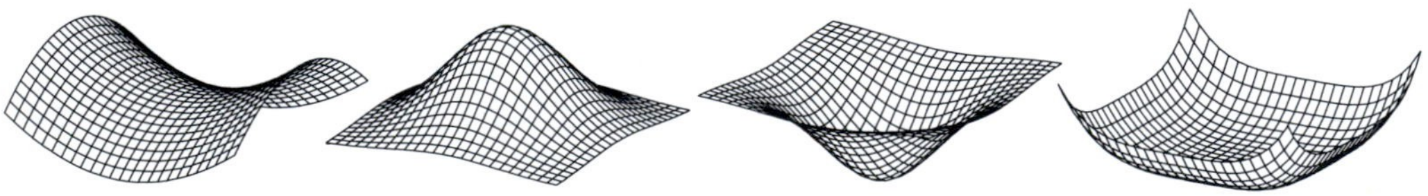

saddle point | maximum | minimum | quartic minimum

Stationary point if $\frac{\partial f}{\partial x}=\frac{\partial f}{\partial y}=0$ at (x_0,y_0) (necessary condition) (2.335)

Additional sufficient conditions

for maximum $\frac{\partial^2 f}{\partial x^2}<0$, and $\frac{\partial^2 f}{\partial x^2}\frac{\partial^2 f}{\partial y^2}>\left(\frac{\partial^2 f}{\partial x\partial y}\right)^2$ (2.336)

for minimum $\frac{\partial^2 f}{\partial x^2}>0$, and $\frac{\partial^2 f}{\partial x^2}\frac{\partial^2 f}{\partial y^2}>\left(\frac{\partial^2 f}{\partial x\partial y}\right)^2$ (2.337)

for saddle point $\frac{\partial^2 f}{\partial x^2}\frac{\partial^2 f}{\partial y^2}<\left(\frac{\partial^2 f}{\partial x\partial y}\right)^2$ (2.338)

[a]Of a function $f(x,y)$ at the point (x_0,y_0). Note that at, for example, a *quartic* minimum $\frac{\partial^2 f}{\partial x^2}=\frac{\partial^2 f}{\partial y^2}=0$.

Differential equations

Laplace	$\nabla^2 f = 0$	(2.339)	f	$f(x,y,z)$
Diffusion[a]	$\frac{\partial f}{\partial t} = D\nabla^2 f$	(2.340)	D	diffusion coefficient
Helmholtz	$\nabla^2 f + \alpha^2 f = 0$	(2.341)	α	constant
Wave	$\nabla^2 f = \frac{1}{c^2}\frac{\partial^2 f}{\partial t^2}$	(2.342)	c	wave speed
Legendre	$\frac{\mathrm{d}}{\mathrm{d}x}\left[(1-x^2)\frac{\mathrm{d}y}{\mathrm{d}x}\right] + l(l+1)y = 0$	(2.343)	l	integer
Associated Legendre	$\frac{\mathrm{d}}{\mathrm{d}x}\left[(1-x^2)\frac{\mathrm{d}y}{\mathrm{d}x}\right] + \left[l(l+1) - \frac{m^2}{1-x^2}\right]y = 0$	(2.344)	m	integer
Bessel	$x^2\frac{\mathrm{d}^2y}{\mathrm{d}x^2} + x\frac{\mathrm{d}y}{\mathrm{d}x} + (x^2 - m^2)y = 0$	(2.345)		
Hermite	$\frac{\mathrm{d}^2y}{\mathrm{d}x^2} - 2x\frac{\mathrm{d}y}{\mathrm{d}x} + 2\alpha y = 0$	(2.346)		
Laguerre	$x\frac{\mathrm{d}^2y}{\mathrm{d}x^2} + (1-x)\frac{\mathrm{d}y}{\mathrm{d}x} + \alpha y = 0$	(2.347)		
Associated Laguerre	$x\frac{\mathrm{d}^2y}{\mathrm{d}x^2} + (1+k-x)\frac{\mathrm{d}y}{\mathrm{d}x} + \alpha y = 0$	(2.348)	k	integer
Chebyshev	$(1-x^2)\frac{\mathrm{d}^2y}{\mathrm{d}x^2} - x\frac{\mathrm{d}y}{\mathrm{d}x} + n^2 y = 0$	(2.349)	n	integer
Euler (or Cauchy)	$x^2\frac{\mathrm{d}^2y}{\mathrm{d}x^2} + ax\frac{\mathrm{d}y}{\mathrm{d}x} + by = f(x)$	(2.350)	a,b	constants
Bernoulli	$\frac{\mathrm{d}y}{\mathrm{d}x} + p(x)y = q(x)y^a$	(2.351)	p,q	functions of x
Airy	$\frac{\mathrm{d}^2y}{\mathrm{d}x^2} = xy$	(2.352)		

[a]Also known as the "conduction equation." For thermal conduction, $f \equiv T$ and D, the thermal diffusivity, $\equiv \kappa \equiv \lambda/(\rho c_p)$, where T is the temperature distribution, λ the thermal conductivity, ρ the density, and c_p the specific heat capacity of the material.

2

2.8 Integration

Standard forms[a]

$\int u\,dv = [uv] - \int v\,du$	(2.353)	$\int uv\,dx = v\int u\,dx - \int\left(\int u\,dx\right)\frac{dv}{dx}\,dx$	(2.354)
$\int x^n\,dx = \frac{x^{n+1}}{n+1} \quad (n \neq -1)$	(2.355)	$\int \frac{1}{x}\,dx = \ln\lvert x\rvert$	(2.356)
$\int e^{ax}\,dx = \frac{1}{a}e^{ax}$	(2.357)	$\int xe^{ax}\,dx = e^{ax}\left(\frac{x}{a} - \frac{1}{a^2}\right)$	(2.358)
$\int \ln ax\,dx = x(\ln ax - 1)$	(2.359)	$\int \frac{f'(x)}{f(x)}\,dx = \ln f(x)$	(2.360)
$\int x\ln ax\,dx = \frac{x^2}{2}\left(\ln ax - \frac{1}{2}\right)$	(2.361)	$\int b^{ax}\,dx = \frac{b^{ax}}{a\ln b} \quad (b>0)$	(2.362)
$\int \frac{1}{a+bx}\,dx = \frac{1}{b}\ln(a+bx)$	(2.363)	$\int \frac{1}{x(a+bx)}\,dx = -\frac{1}{a}\ln\frac{a+bx}{x}$	(2.364)
$\int \frac{1}{(a+bx)^2}\,dx = \frac{-1}{b(a+bx)}$	(2.365)	$\int \frac{1}{a^2+b^2x^2}\,dx = \frac{1}{ab}\arctan\left(\frac{bx}{a}\right)$	(2.366)
$\int \frac{1}{x(x^n+a)}\,dx = \frac{1}{an}\ln\left\lvert\frac{x^n}{x^n+a}\right\rvert$	(2.367)	$\int \frac{1}{x^2-a^2}\,dx = \frac{1}{2a}\ln\left\lvert\frac{x-a}{x+a}\right\rvert$	(2.368)
$\int \frac{x}{x^2 \pm a^2}\,dx = \frac{1}{2}\ln\lvert x^2 \pm a^2\rvert$	(2.369)	$\int \frac{x}{(x^2 \pm a^2)^n}\,dx = \frac{-1}{2(n-1)(x^2 \pm a^2)^{n-1}}$	(2.370)
$\int \frac{1}{(a^2-x^2)^{1/2}}\,dx = \arcsin\left(\frac{x}{a}\right)$	(2.371)	$\int \frac{1}{(x^2 \pm a^2)^{1/2}}\,dx = \ln\lvert x + (x^2 \pm a^2)^{1/2}\rvert$	(2.372)
$\int \frac{x}{(x^2 \pm a^2)^{1/2}}\,dx = (x^2 \pm a^2)^{1/2}$	(2.373)	$\int \frac{1}{x(x^2-a^2)^{1/2}}\,dx = \frac{1}{a}\operatorname{arcsec}\left(\frac{x}{a}\right)$	(2.374)

[a] a and b are non-zero constants.

Trigonometric and hyperbolic integrals

$\int \sin x \, dx = -\cos x$	(2.375)	$\int \sinh x \, dx = \cosh x$	(2.376)
$\int \cos x \, dx = \sin x$	(2.377)	$\int \cosh x \, dx = \sinh x$	(2.378)
$\int \tan x \, dx = -\ln\|\cos x\|$	(2.379)	$\int \tanh x \, dx = \ln(\cosh x)$	(2.380)
$\int \csc x \, dx = \ln\left\|\tan \frac{x}{2}\right\|$	(2.381)	$\int \operatorname{csch} x \, dx = \ln\left\|\tanh \frac{x}{2}\right\|$	(2.382)
$\int \sec x \, dx = \ln\|\sec x + \tan x\|$	(2.383)	$\int \operatorname{sech} x \, dx = 2\arctan(e^x)$	(2.384)
$\int \cot x \, dx = \ln\|\sin x\|$	(2.385)	$\int \coth x \, dx = \ln\|\sinh x\|$	(2.386)

$$\int \sin mx \cdot \sin nx \, dx = \frac{\sin(m-n)x}{2(m-n)} - \frac{\sin(m+n)x}{2(m+n)} \quad (m^2 \neq n^2) \qquad (2.387)$$

$$\int \sin mx \cdot \cos nx \, dx = -\frac{\cos(m-n)x}{2(m-n)} - \frac{\cos(m+n)x}{2(m+n)} \quad (m^2 \neq n^2) \qquad (2.388)$$

$$\int \cos mx \cdot \cos nx \, dx = \frac{\sin(m-n)x}{2(m-n)} + \frac{\sin(m+n)x}{2(m+n)} \quad (m^2 \neq n^2) \qquad (2.389)$$

Named integrals

Error function	$\operatorname{erf}(x) = \frac{2}{\pi^{1/2}} \int_0^x \exp(-t^2) \, dt$	(2.390)
Complementary error function	$\operatorname{erfc}(x) = 1 - \operatorname{erf}(x) = \frac{2}{\pi^{1/2}} \int_x^\infty \exp(-t^2) \, dt$	(2.391)
Fresnel integrals[a]	$C(x) = \int_0^x \cos \frac{\pi t^2}{2} \, dt; \quad S(x) = \int_0^x \sin \frac{\pi t^2}{2} \, dt$	(2.392)
	$C(x) + \mathbf{i} S(x) = \frac{1+\mathbf{i}}{2} \operatorname{erf}\left[\frac{\pi^{1/2}}{2}(1-\mathbf{i})x\right]$	(2.393)
Exponential integral	$\operatorname{Ei}(x) = \int_{-\infty}^x \frac{e^t}{t} \, dt \quad (x > 0)$	(2.394)
Gamma function	$\Gamma(x) = \int_0^\infty t^{x-1} e^{-t} \, dt \quad (x > 0)$	(2.395)
Elliptic integrals (trigonometric form)	$F(\phi, k) = \int_0^\phi \frac{1}{(1-k^2 \sin^2\theta)^{1/2}} \, d\theta$ (first kind)	(2.396)
	$E(\phi, k) = \int_0^\phi (1-k^2 \sin^2\theta)^{1/2} \, d\theta$ (second kind)	(2.397)

[a]See also page 167.

Definite integrals

$$\int_0^\infty e^{-ax^2}\,dx=\frac{1}{2}\left(\frac{\pi}{a}\right)^{1/2} \quad (a>0)$$	(2.398)
$$\int_0^\infty xe^{-ax^2}\,dx=\frac{1}{2a} \quad (a>0)$$	(2.399)
$$\int_0^\infty x^n e^{-ax}\,dx=\frac{n!}{a^{n+1}} \quad (a>0;n=0,1,2,\ldots)$$	(2.400)
$$\int_{-\infty}^\infty \exp(2bx-ax^2)\,dx=\left(\frac{\pi}{a}\right)^{1/2}\exp\left(\frac{b^2}{a}\right) \quad (a>0)$$	(2.401)
$$\int_0^\infty x^n e^{-ax^2}\,dx=\begin{cases}1\cdot 3\cdot 5\cdot\ldots\cdot(n-1)(2a)^{-(n+1)/2}(\pi/2)^{1/2} & n>0 \text{ and even}\\ 2\cdot 4\cdot 6\cdot\ldots\cdot(n-1)(2a)^{-(n+1)/2} & n>1 \text{ and odd}\end{cases}$$	(2.402)
$$\int_0^1 x^p(1-x)^q\,dx=\frac{p!q!}{(p+q+1)!} \quad (p,q \text{ integers} >0)$$	(2.403)
$$\int_0^\infty \cos(ax^2)\,dx=\int_0^\infty \sin(ax^2)\,dx=\frac{1}{2}\left(\frac{\pi}{2a}\right)^{1/2} \quad (a>0)$$	(2.404)
$$\int_0^\infty \frac{\sin x}{x}\,dx=\int_0^\infty \frac{\sin^2 x}{x^2}\,dx=\frac{\pi}{2}$$	(2.405)
$$\int_0^\infty \frac{1}{(1+x)x^a}\,dx=\frac{\pi}{\sin a\pi} \quad (0<a<1)$$	(2.406)

2.9 Special functions and polynomials

Gamma function

Definition	$$\Gamma(z)=\int_0^\infty t^{z-1}e^{-t}\,dt \quad [\Re(z)>0]$$	(2.407)
Relations	$$n!=\Gamma(n+1)=n\Gamma(n) \quad (n=0,1,2,\ldots)$$	(2.408)
	$$\Gamma(1/2)=\pi^{1/2}$$	(2.409)
	$$\binom{z}{w}=\frac{z!}{w!(z-w)!}=\frac{\Gamma(z+1)}{\Gamma(w+1)\Gamma(z-w+1)}$$	(2.410)
Stirling's formulas (for $\lvert z\rvert,n\gg 1$)	$$\Gamma(z)\simeq e^{-z}z^{z-(1/2)}(2\pi)^{1/2}\left(1+\frac{1}{12z}+\frac{1}{288z^2}-\cdots\right)$$	(2.411)
	$$n!\simeq n^{n+(1/2)}e^{-n}(2\pi)^{1/2}$$	(2.412)
	$$\ln(n!)\simeq n\ln n-n$$	(2.413)

Bessel functions

Series expansion	$J_\nu(x)=\left(\frac{x}{2}\right)^\nu\sum_{k=0}^{\infty}\frac{(-x^2/4)^k}{k!\Gamma(\nu+k+1)}$	(2.414)	$J_\nu(x)$ Bessel function of the first kind
	$Y_\nu(x)=\frac{J_\nu(x)\cos(\pi\nu)-J_{-\nu}(x)}{\sin(\pi\nu)}$	(2.415)	$Y_\nu(x)$ Bessel function of the second kind; $\Gamma(\nu)$ Gamma function; ν order ($\nu\geq0$)
Approximations	$J_\nu(x)\simeq\begin{cases}\frac{1}{\Gamma(\nu+1)}\left(\frac{x}{2}\right)^\nu & (0\leq x\ll\nu)\\ \left(\frac{2}{\pi x}\right)^{1/2}\cos\left(x-\frac{1}{2}\nu\pi-\frac{\pi}{4}\right) & (x\gg\nu)\end{cases}$	(2.416)	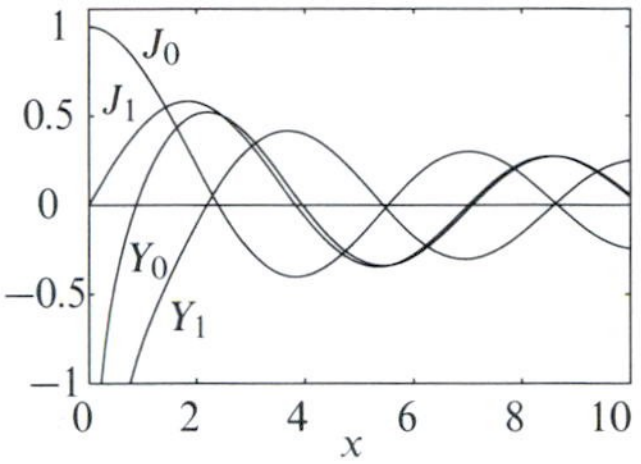
	$Y_\nu(x)\simeq\begin{cases}\frac{-\Gamma(\nu)}{\pi}\left(\frac{x}{2}\right)^{-\nu} & (0<x\ll\nu)\\ \left(\frac{2}{\pi x}\right)^{1/2}\sin\left(x-\frac{1}{2}\nu\pi-\frac{\pi}{4}\right) & (x\gg\nu)\end{cases}$	(2.417)	
Modified Bessel functions	$I_\nu(x)=(-\mathbf{i})^\nu J_\nu(\mathbf{i}x)$	(2.418)	$I_\nu(x)$ modified Bessel function of the first kind
	$K_\nu(x)=\frac{\pi}{2}\mathbf{i}^{\nu+1}[J_\nu(\mathbf{i}x)+\mathbf{i}Y_\nu(\mathbf{i}x)]$	(2.419)	$K_\nu(x)$ modified Bessel function of the second kind
Spherical Bessel function	$j_\nu(x)=\left(\frac{\pi}{2x}\right)^{1/2}J_{\nu+\frac{1}{2}}(x)$	(2.420)	$j_\nu(x)$ spherical Bessel function of the first kind [similarly for $y_\nu(x)$]

Legendre polynomials[a]

Legendre equation	$(1-x^2)\frac{\mathrm{d}^2P_l(x)}{\mathrm{d}x^2}-2x\frac{\mathrm{d}P_l(x)}{\mathrm{d}x}+l(l+1)P_l(x)=0$	(2.421)	P_l Legendre polynomials; l order ($l\geq0$)
Rodrigues' formula	$P_l(x)=\frac{1}{2^l l!}\frac{\mathrm{d}^l}{\mathrm{d}x^l}(x^2-1)^l$	(2.422)	
Recurrence relation	$(l+1)P_{l+1}(x)=(2l+1)xP_l(x)-lP_{l-1}(x)$	(2.423)	
Orthogonality	$\int_{-1}^{1}P_l(x)P_{l'}(x)\,\mathrm{d}x=\frac{2}{2l+1}\delta_{ll'}$	(2.424)	$\delta_{ll'}$ Kronecker delta
Explicit form	$P_l(x)=2^{-l}\sum_{m=0}^{l/2}(-1)^m\binom{l}{m}\binom{2l-2m}{l}x^{l-2m}$	(2.425)	$\binom{l}{m}$ binomial coefficients
Expansion of plane wave	$\exp(\mathbf{i}kz)=\exp(\mathbf{i}kr\cos\theta)$	(2.426)	k wavenumber; z propagation axis $z=r\cos\theta$
	$=\sum_{l=0}^{\infty}(2l+1)\mathbf{i}^l j_l(kr)P_l(\cos\theta)$	(2.427)	j_l spherical Bessel function of the first kind (order l)

$P_0(x)=1$	$P_2(x)=(3x^2-1)/2$	$P_4(x)=(35x^4-30x^2+3)/8$
$P_1(x)=x$	$P_3(x)=(5x^3-3x)/2$	$P_5(x)=(63x^5-70x^3+15x)/8$

[a]Of the first kind.

Associated Legendre functions[a]

Associated Legendre equation	$$\frac{\mathrm{d}}{\mathrm{d}x}\left[(1-x^2)\frac{\mathrm{d}P_l^m(x)}{\mathrm{d}x}\right]+\left[l(l+1)-\frac{m^2}{1-x^2}\right]P_l^m(x)=0 \quad (2.428)$$	P_l^m associated Legendre functions
From Legendre polynomials	$$P_l^m(x)=(1-x^2)^{m/2}\frac{\mathrm{d}^m}{\mathrm{d}x^m}P_l(x), \quad 0\le m\le l \quad (2.429)$$ $$P_l^{-m}(x)=(-1)^m\frac{(l-m)!}{(l+m)!}P_l^m(x) \quad (2.430)$$	P_l Legendre polynomials
Recurrence relations	$$P_{m+1}^m(x)=x(2m+1)P_m^m(x) \quad (2.431)$$ $$P_m^m(x)=(-1)^m(2m-1)!!(1-x^2)^{m/2} \quad (2.432)$$ $$(l-m+1)P_{l+1}^m(x)=(2l+1)xP_l^m(x)-(l+m)P_{l-1}^m(x) \quad (2.433)$$	!! $5!!=5\cdot3\cdot1$ etc.
Orthogonality	$$\int_{-1}^{1}P_l^m(x)P_{l'}^m(x)\,\mathrm{d}x=\frac{(l+m)!}{(l-m)!}\frac{2}{2l+1}\delta_{ll'} \quad (2.434)$$	$\delta_{ll'}$ Kronecker delta

$P_0^0(x)=1$	$P_1^0(x)=x$	$P_1^1(x)=-(1-x^2)^{1/2}$
$P_2^0(x)=(3x^2-1)/2$	$P_2^1(x)=-3x(1-x^2)^{1/2}$	$P_2^2(x)=3(1-x^2)$

[a]Of the first kind. $P_l^m(x)$ can be defined with a $(-1)^m$ factor in Equation (2.429) as well as Equation (2.430).

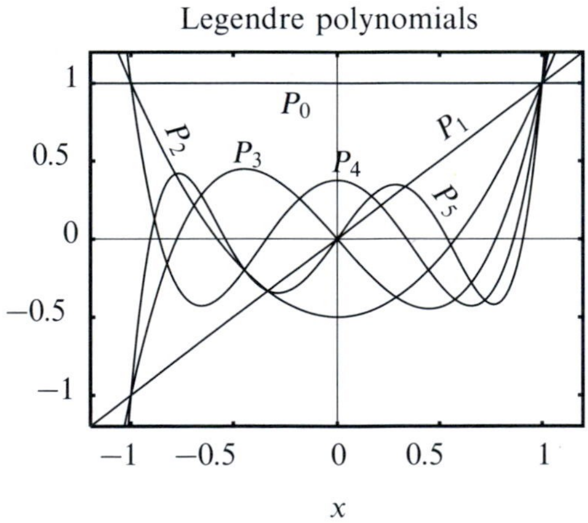

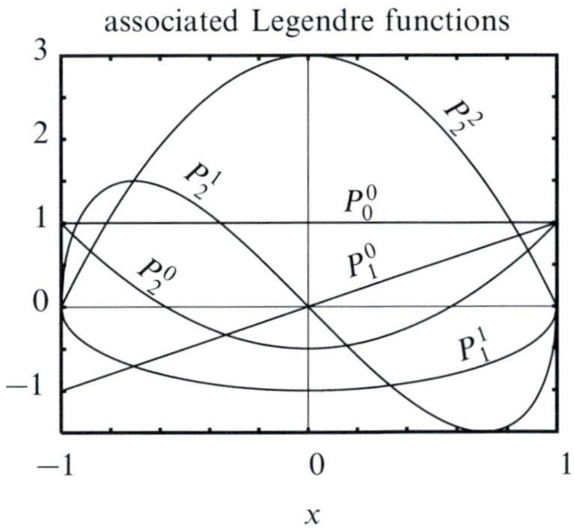

Spherical harmonics

Differential equation	$$\left[\frac{1}{\sin\theta}\frac{\partial}{\partial\theta}\left(\sin\theta\frac{\partial}{\partial\theta}\right)+\frac{1}{\sin^2\theta}\frac{\partial^2}{\partial\phi^2}\right]Y_l^m+l(l+1)Y_l^m=0 \quad (2.435)$$	Y_l^m spherical harmonics
Definition[a]	$$Y_l^m(\theta,\phi)=(-1)^m\left[\frac{2l+1}{4\pi}\frac{(l-m)!}{(l+m)!}\right]^{1/2}P_l^m(\cos\theta)\mathrm{e}^{\mathrm{i}m\phi} \quad (2.436)$$	P_l^m associated Legendre functions
Orthogonality	$$\int_{\phi=0}^{2\pi}\int_{\theta=0}^{\pi}Y_l^{m*}(\theta,\phi)Y_{l'}^{m'}(\theta,\phi)\sin\theta\,\mathrm{d}\theta\,\mathrm{d}\phi=\delta_{mm'}\delta_{ll'} \quad (2.437)$$	Y^* complex conjugate; $\delta_{ll'}$ Kronecker delta
Laplace series	$$f(\theta,\phi)=\sum_{l=0}^{\infty}\sum_{m=-l}^{l}a_{lm}Y_l^m(\theta,\phi) \quad (2.438)$$ where $$a_{lm}=\int_{\phi=0}^{2\pi}\int_{\theta=0}^{\pi}Y_l^{m*}(\theta,\phi)f(\theta,\phi)\sin\theta\,\mathrm{d}\theta\,\mathrm{d}\phi \quad (2.439)$$	f continuous function
Solution to Laplace equation	if $\nabla^2\psi(r,\theta,\phi)=0$, then $$\psi(r,\theta,\phi)=\sum_{l=0}^{\infty}\sum_{m=-l}^{l}Y_l^m(\theta,\phi)\cdot\left[a_{lm}r^l+b_{lm}r^{-(l+1)}\right] \quad (2.440)$$	ψ continuous function; a,b constants

$$Y_0^0(\theta,\phi)=\sqrt{\frac{1}{4\pi}} \qquad Y_1^0(\theta,\phi)=\sqrt{\frac{3}{4\pi}}\cos\theta$$

$$Y_1^{\pm1}(\theta,\phi)=\mp\sqrt{\frac{3}{8\pi}}\sin\theta\,\mathrm{e}^{\pm\mathrm{i}\phi} \qquad Y_2^0(\theta,\phi)=\sqrt{\frac{5}{4\pi}}\left(\frac{3}{2}\cos^2\theta-\frac{1}{2}\right)$$

$$Y_2^{\pm1}(\theta,\phi)=\mp\sqrt{\frac{15}{8\pi}}\sin\theta\cos\theta\,\mathrm{e}^{\pm\mathrm{i}\phi} \qquad Y_2^{\pm2}(\theta,\phi)=\sqrt{\frac{15}{32\pi}}\sin^2\theta\,\mathrm{e}^{\pm2\mathrm{i}\phi}$$

$$Y_3^0(\theta,\phi)=\frac{1}{2}\sqrt{\frac{7}{4\pi}}(5\cos^2\theta-3)\cos\theta \qquad Y_3^{\pm1}(\theta,\phi)=\mp\frac{1}{4}\sqrt{\frac{21}{4\pi}}\sin\theta(5\cos^2\theta-1)\mathrm{e}^{\pm\mathrm{i}\phi}$$

$$Y_3^{\pm2}(\theta,\phi)=\frac{1}{4}\sqrt{\frac{105}{2\pi}}\sin^2\theta\cos\theta\,\mathrm{e}^{\pm2\mathrm{i}\phi} \qquad Y_3^{\pm3}(\theta,\phi)=\mp\frac{1}{4}\sqrt{\frac{35}{4\pi}}\sin^3\theta\,\mathrm{e}^{\pm3\mathrm{i}\phi}$$

[a] Defined for $-l\le m\le l$, using the sign convention of the Condon–Shortley phase. Other sign conventions are possible.

Delta functions

Kronecker delta	$\delta_{ij}=\begin{cases}1 & \text{if } i=j\\ 0 & \text{if } i\neq j\end{cases}$	(2.441)	δ_{ij} Kronecker delta $i,j,k,\ldots$ indices (=1,2 or 3)
	$\delta_{ii}=3$	(2.442)	
Three-dimensional Levi–Civita symbol (permutation tensor)[a]	$\epsilon_{123}=\epsilon_{231}=\epsilon_{312}=1$ $\epsilon_{132}=\epsilon_{213}=\epsilon_{321}=-1$ all other $\epsilon_{ijk}=0$	(2.443)	
	$\epsilon_{ijk}\epsilon_{klm}=\delta_{il}\delta_{jm}-\delta_{im}\delta_{jl}$	(2.444)	ϵ_{ijk} Levi–Civita symbol (see also page 25)
	$\delta_{ij}\epsilon_{ijk}=0$	(2.445)	
	$\epsilon_{ilm}\epsilon_{jlm}=2\delta_{ij}$	(2.446)	
	$\epsilon_{ijk}\epsilon_{ijk}=6$	(2.447)	
Dirac delta function	$\int_a^b \delta(x)\,\mathrm{d}x=\begin{cases}1 & \text{if } a<0<b\\ 0 & \text{otherwise}\end{cases}$	(2.448)	
	$\int_a^b f(x)\delta(x-x_0)\,\mathrm{d}x=f(x_0)$	(2.449)	$\delta(x)$ Dirac delta function $f(x)$ smooth function of x a,b constants
	$\delta(x-x_0)f(x)=\delta(x-x_0)f(x_0)$	(2.450)	
	$\delta(-x)=\delta(x)$	(2.451)	
	$\delta(ax)=\lvert a\rvert^{-1}\delta(x)\quad(a\neq 0)$	(2.452)	
	$\delta(x)\simeq n\pi^{-1/2}\mathrm{e}^{-n^2x^2}\quad(n\gg 1)$	(2.453)	

[a]The general symbol $\epsilon_{ijk\ldots}$ is defined to be +1 for even permutations of the suffices, −1 for odd permutations, and 0 if a suffix is repeated. The sequence $(1,2,3,\ldots,n)$ is taken to be even. Swapping adjacent suffices an odd (or even) number of times gives an odd (or even) permutation.

2.10 Roots of quadratic and cubic equations

Quadratic equations

Equation	$ax^2+bx+c=0\quad(a\neq 0)$	(2.454)	x variable a,b,c real constants
Solutions	$x_{1,2}=\dfrac{-b\pm\sqrt{b^2-4ac}}{2a}$	(2.455)	x_1,x_2 quadratic roots
	$=\dfrac{-2c}{b\pm\sqrt{b^2-4ac}}$	(2.456)	
Solution combinations	$x_1+x_2=-b/a$	(2.457)	
	$x_1x_2=c/a$	(2.458)	

Cubic equations

Equation	$ax^3+bx^2+cx+d=0 \quad (a\neq 0)$	(2.459)	x a,b,c,d	variable real constants
Intermediate definitions	$p=\frac{1}{3}\left(\frac{3c}{a}-\frac{b^2}{a^2}\right)$	(2.460)		
	$q=\frac{1}{27}\left(\frac{2b^3}{a^3}-\frac{9bc}{a^2}+\frac{27d}{a}\right)$	(2.461)	D	discriminant
	$D=\left(\frac{p}{3}\right)^3+\left(\frac{q}{2}\right)^2$	(2.462)		

If $D\geq 0$, also define:		If $D<0$, also define:	
$u=\left(\frac{-q}{2}+D^{1/2}\right)^{1/3}$	(2.463)	$\phi=\arccos\left[\frac{-q}{2}\left(\frac{\lvert p\rvert}{3}\right)^{-3/2}\right]$	(2.467)
$v=\left(\frac{-q}{2}-D^{1/2}\right)^{1/3}$	(2.464)	$y_1=2\left(\frac{\lvert p\rvert}{3}\right)^{1/2}\cos\frac{\phi}{3}$	(2.468)
$y_1=u+v$	(2.465)	$y_{2,3}=-2\left(\frac{\lvert p\rvert}{3}\right)^{1/2}\cos\frac{\phi\pm\pi}{3}$	(2.469)
$y_{2,3}=\frac{-(u+v)}{2}\pm \mathrm{i}\frac{u-v}{2}3^{1/2}$	(2.466)		
1 real, 2 complex roots (if $D=0$: 3 real roots, at least 2 equal)		3 distinct real roots	

Solutions[a]	$x_n=y_n-\frac{b}{3a}$	(2.470)	x_n	cubic roots ($n=1,2,3$)
Solution combinations	$x_1+x_2+x_3=-b/a$	(2.471)		
	$x_1x_2+x_1x_3+x_2x_3=c/a$	(2.472)		
	$x_1x_2x_3=-d/a$	(2.473)		

[a] y_n are solutions to the reduced equation $y^3+py+q=0$.

2.11 Fourier series and transforms

Fourier series

Real form	$$f(x)=\frac{a_0}{2}+\sum_{n=1}^{\infty}\left(a_n\cos\frac{n\pi x}{L}+b_n\sin\frac{n\pi x}{L}\right)$$	(2.474)	$f(x)$ periodic function, period $2L$
	$$a_n=\frac{1}{L}\int_{-L}^{L}f(x)\cos\frac{n\pi x}{L}\,\mathrm{d}x$$	(2.475)	a_n,b_n Fourier coefficients
	$$b_n=\frac{1}{L}\int_{-L}^{L}f(x)\sin\frac{n\pi x}{L}\,\mathrm{d}x$$	(2.476)	
Complex form	$$f(x)=\sum_{n=-\infty}^{\infty}c_n\exp\left(\frac{\mathrm{i}n\pi x}{L}\right)$$	(2.477)	c_n complex Fourier coefficient
	$$c_n=\frac{1}{2L}\int_{-L}^{L}f(x)\exp\left(\frac{-\mathrm{i}n\pi x}{L}\right)\mathrm{d}x$$	(2.478)	
Parseval's theorem	$$\frac{1}{2L}\int_{-L}^{L}\lvert f(x)\rvert^2\,\mathrm{d}x=\frac{a_0^2}{4}+\frac{1}{2}\sum_{n=1}^{\infty}(a_n^2+b_n^2)$$	(2.479)	$\lvert\,\rvert$ modulus
	$$=\sum_{n=-\infty}^{\infty}\lvert c_n\rvert^2$$	(2.480)	

Fourier transform[a]

Definition 1	$$F(s)=\int_{-\infty}^{\infty}f(x)\mathrm{e}^{-2\pi\mathrm{i}xs}\,\mathrm{d}x$$	(2.481)	$f(x)$ function of x
	$$f(x)=\int_{-\infty}^{\infty}F(s)\mathrm{e}^{2\pi\mathrm{i}xs}\,\mathrm{d}s$$	(2.482)	$F(s)$ Fourier transform of $f(x)$
Definition 2	$$F(s)=\int_{-\infty}^{\infty}f(x)\mathrm{e}^{-\mathrm{i}xs}\,\mathrm{d}x$$	(2.483)	
	$$f(x)=\frac{1}{2\pi}\int_{-\infty}^{\infty}F(s)\mathrm{e}^{\mathrm{i}xs}\,\mathrm{d}s$$	(2.484)	
Definition 3	$$F(s)=\frac{1}{\sqrt{2\pi}}\int_{-\infty}^{\infty}f(x)\mathrm{e}^{-\mathrm{i}xs}\,\mathrm{d}x$$	(2.485)	
	$$f(x)=\frac{1}{\sqrt{2\pi}}\int_{-\infty}^{\infty}F(s)\mathrm{e}^{\mathrm{i}xs}\,\mathrm{d}s$$	(2.486)	

[a]All three (and more) definitions are used, but definition 1 is probably the best.

Fourier transform theorems[a]

Convolution	$f(x)*g(x)=\int_{-\infty}^{\infty} f(u)g(x-u)\,du$	(2.487)	f,g general functions; $*$ convolution
Convolution rules	$f*g=g*f$	(2.488)	f $f(x)\rightleftharpoons F(s)$
	$f*(g*h)=(f*g)*h$	(2.489)	g $g(x)\rightleftharpoons G(s)$
Convolution theorem	$f(x)g(x)\rightleftharpoons F(s)*G(s)$	(2.490)	$\rightleftharpoons$ Fourier transform relation
Autocorrelation	$f^*(x)\star f(x)=\int_{-\infty}^{\infty} f^*(u-x)f(u)\,du$	(2.491)	$\star$ correlation; f^* complex conjugate of f
Wiener–Khintchine theorem	$f^*(x)\star f(x)\rightleftharpoons \|F(s)\|^2$	(2.492)	
Cross-correlation	$f^*(x)\star g(x)=\int_{-\infty}^{\infty} f^*(u-x)g(u)\,du$	(2.493)	
Correlation theorem	$h(x)\star j(x)\rightleftharpoons H(s)J^*(s)$	(2.494)	h,j real functions; H $H(s)\rightleftharpoons h(x)$; J $J(s)\rightleftharpoons j(x)$
Parseval's relation[b]	$\int_{-\infty}^{\infty} f(x)g^*(x)\,dx=\int_{-\infty}^{\infty} F(s)G^*(s)\,ds$	(2.495)	
Parseval's theorem[c]	$\int_{-\infty}^{\infty} \|f(x)\|^2\,dx=\int_{-\infty}^{\infty} \|F(s)\|^2\,ds$	(2.496)	
Derivatives	$\frac{df(x)}{dx}\rightleftharpoons 2\pi i sF(s)$	(2.497)	
	$\frac{d}{dx}[f(x)*g(x)]=\frac{df(x)}{dx}*g(x)=\frac{dg(x)}{dx}*f(x)$	(2.498)	

[a]Defining the Fourier transform as $F(s)=\int_{-\infty}^{\infty} f(x)e^{-2\pi ixs}\,dx$.
[b]Also called the "power theorem."
[c]Also called "Rayleigh's theorem."

Fourier symmetry relationships

$f(x)$	$\rightleftharpoons$	$F(s)$	definitions
even	$\rightleftharpoons$	even	real: $f(x)=f^*(x)$
odd	$\rightleftharpoons$	odd	imaginary: $f(x)=-f^*(x)$
real, even	$\rightleftharpoons$	real, even	even: $f(x)=f(-x)$
real, odd	$\rightleftharpoons$	imaginary, odd	odd: $f(x)=-f(-x)$
imaginary, even	$\rightleftharpoons$	imaginary, even	Hermitian: $f(x)=f^*(-x)$
complex, even	$\rightleftharpoons$	complex, even	anti-Hermitian: $f(x)=-f^*(-x)$
complex, odd	$\rightleftharpoons$	complex, odd	
real, asymmetric	$\rightleftharpoons$	complex, Hermitian	
imaginary, asymmetric	$\rightleftharpoons$	complex, anti-Hermitian	

Fourier transform pairs[a]

$f(x)$	$\rightleftharpoons$	$F(s)=\int_{-\infty}^{\infty} f(x)\mathrm{e}^{-2\pi \mathrm{i} s x}\,\mathrm{d}x$	(2.499)
$f(ax)$	$\rightleftharpoons$	$\frac{1}{\lvert a\rvert}F(s/a) \qquad (a\neq 0,\ \text{real})$	(2.500)
$f(x-a)$	$\rightleftharpoons$	$\mathrm{e}^{-2\pi \mathrm{i} a s}F(s) \qquad (a\ \text{real})$	(2.501)
$\frac{\mathrm{d}^n}{\mathrm{d}x^n}f(x)$	$\rightleftharpoons$	$(2\pi \mathbf{i} s)^n F(s)$	(2.502)
$\delta(x)$	$\rightleftharpoons$	1	(2.503)
$\delta(x-a)$	$\rightleftharpoons$	$\mathrm{e}^{-2\pi \mathrm{i} a s}$	(2.504)
$\mathrm{e}^{-a\lvert x\rvert}$	$\rightleftharpoons$	$\frac{2a}{a^2+4\pi^2 s^2} \qquad (a>0)$	(2.505)
$x\mathrm{e}^{-a\lvert x\rvert}$	$\rightleftharpoons$	$\frac{8\mathbf{i}\pi a s}{(a^2+4\pi^2 s^2)^2} \qquad (a>0)$	(2.506)
e^{-x^2/a^2}	$\rightleftharpoons$	$a\sqrt{\pi}\mathrm{e}^{-\pi^2 a^2 s^2}$	(2.507)
$\sin ax$	$\rightleftharpoons$	$\frac{1}{2\mathbf{i}}\left[\delta\left(s-\frac{a}{2\pi}\right)-\delta\left(s+\frac{a}{2\pi}\right)\right]$	(2.508)
$\cos ax$	$\rightleftharpoons$	$\frac{1}{2}\left[\delta\left(s-\frac{a}{2\pi}\right)+\delta\left(s+\frac{a}{2\pi}\right)\right]$	(2.509)
$\sum_{m=-\infty}^{\infty}\delta(x-ma)$	$\rightleftharpoons$	$\frac{1}{a}\sum_{n=-\infty}^{\infty}\delta\left(s-\frac{n}{a}\right)$	(2.510)
$f(x)=\begin{cases}0 & x<0\\ 1 & x>0\end{cases}$ ("step")	$\rightleftharpoons$	$\frac{1}{2}\delta(s)-\frac{\mathbf{i}}{2\pi s}$	(2.511)
$f(x)=\begin{cases}1 & \lvert x\rvert\le a\\ 0 & \lvert x\rvert>a\end{cases}$ ("top hat")	$\rightleftharpoons$	$\frac{\sin 2\pi a s}{\pi s}=2a\,\mathrm{sinc}\,2as$	(2.512)
$f(x)=\begin{cases}\left(1-\frac{\lvert x\rvert}{a}\right) & \lvert x\rvert\le a\\ 0 & \lvert x\rvert>a\end{cases}$ ("triangle")	$\rightleftharpoons$	$\frac{1}{2\pi^2 a s^2}(1-\cos 2\pi a s)=a\,\mathrm{sinc}^2\,as$	(2.513)

[a]Equation (2.499) defines the Fourier transform used for these pairs. Note that $\mathrm{sinc}\,x\equiv(\sin\pi x)/(\pi x)$.

2.12 Laplace transforms

Laplace transform theorems

Definition[a]	$F(s)=\mathscr{L}\{f(t)\}=\int_0^{\infty} f(t)\mathrm{e}^{-st}\,\mathrm{d}t$	(2.514)	$\mathscr{L}\{\}$	Laplace transform
Convolution[b]	$F(s)\cdot G(s)=\mathscr{L}\left\{\int_0^{\infty} f(t-z)g(z)\,\mathrm{d}z\right\}$	(2.515)	$F(s)$	$\mathscr{L}\{f(t)\}$
	$=\mathscr{L}\{f(t)*g(t)\}$	(2.516)	$G(s)$	$\mathscr{L}\{g(t)\}$
			$*$	convolution
Inverse[c]	$f(t)=\frac{1}{2\pi\mathrm{i}}\int_{\gamma-\mathrm{i}\infty}^{\gamma+\mathrm{i}\infty} \mathrm{e}^{st}F(s)\,\mathrm{d}s$	(2.517)	γ	constant
	$=\sum$ residues (for $t>0$)	(2.518)		
Transform of derivative	$\mathscr{L}\left\{\frac{\mathrm{d}^n f(t)}{\mathrm{d}t^n}\right\}=s^n\mathscr{L}\{f(t)\}-\sum_{r=0}^{n-1} s^{n-r-1}\left.\frac{\mathrm{d}^r f(t)}{\mathrm{d}t^r}\right\|_{t=0}$	(2.519)	n	integer >0
Derivative of transform	$\frac{\mathrm{d}^n F(s)}{\mathrm{d}s^n}=\mathscr{L}\{(-t)^n f(t)\}$	(2.520)		
Substitution	$F(s-a)=\mathscr{L}\{\mathrm{e}^{at}f(t)\}$	(2.521)	a	constant
Translation	$\mathrm{e}^{-as}F(s)=\mathscr{L}\{u(t-a)f(t-a)\}$	(2.522)	$u(t)$	unit step function
	where $u(t)=\begin{cases}0 & (t<0)\\ 1 & (t>0)\end{cases}$	(2.523)		

[a]If $\|\mathrm{e}^{-s_0t}f(t)\|$ is finite for sufficiently large t, the Laplace transform exists for $s>s_0$.
[b]Also known as the "faltung (or folding) theorem."
[c]Also known as the "Bromwich integral." γ is chosen so that the singularities in $F(s)$ are left of the integral line.

Laplace transform pairs

$$f(t) \Longrightarrow F(s) = \mathscr{L}\{f(t)\} = \int_0^\infty f(t)\mathrm{e}^{-st}\,\mathrm{d}t \tag{2.524}$$

$$\delta(t) \Longrightarrow 1 \tag{2.525}$$

$$1 \Longrightarrow 1/s \qquad (s>0) \tag{2.526}$$

$$t^n \Longrightarrow \frac{n!}{s^{n+1}} \qquad (s>0, n>-1) \tag{2.527}$$

$$t^{1/2} \Longrightarrow \sqrt{\frac{\pi}{4s^3}} \tag{2.528}$$

$$t^{-1/2} \Longrightarrow \sqrt{\frac{\pi}{s}} \tag{2.529}$$

$$\mathrm{e}^{at} \Longrightarrow \frac{1}{s-a} \qquad (s>a) \tag{2.530}$$

$$t\mathrm{e}^{at} \Longrightarrow \frac{1}{(s-a)^2} \qquad (s>a) \tag{2.531}$$

$$(1-at)\mathrm{e}^{-at} \Longrightarrow \frac{s}{(s+a)^2} \tag{2.532}$$

$$t^2\mathrm{e}^{-at} \Longrightarrow \frac{2}{(s+a)^3} \tag{2.533}$$

$$\sin at \Longrightarrow \frac{a}{s^2+a^2} \qquad (s>0) \tag{2.534}$$

$$\cos at \Longrightarrow \frac{s}{s^2+a^2} \qquad (s>0) \tag{2.535}$$

$$\sinh at \Longrightarrow \frac{a}{s^2-a^2} \qquad (s>a) \tag{2.536}$$

$$\cosh at \Longrightarrow \frac{s}{s^2-a^2} \qquad (s>a) \tag{2.537}$$

$$e^{-bt}\sin at \Longrightarrow \frac{a}{(s+b)^2+a^2} \tag{2.538}$$

$$e^{-bt}\cos at \Longrightarrow \frac{s+b}{(s+b)^2+a^2} \tag{2.539}$$

$$\mathrm{e}^{-at}f(t) \Longrightarrow F(s+a) \tag{2.540}$$

2.13 Probability and statistics

Discrete statistics

Mean	$\langle x\rangle=\frac{1}{N}\sum_{i=1}^{N}x_i$	(2.541)	x_i N $\langle\cdot\rangle$	data series series length mean value
Variance[a]	$\mathrm{var}[x]=\frac{1}{N-1}\sum_{i=1}^{N}(x_i-\langle x\rangle)^2$	(2.542)	$\mathrm{var}[\cdot]$	unbiased variance
Standard deviation	$\sigma[x]=(\mathrm{var}[x])^{1/2}$	(2.543)	σ	standard deviation
Skewness	$\mathrm{skew}[x]=\frac{N}{(N-1)(N-2)}\sum_{i=1}^{N}\left(\frac{x_i-\langle x\rangle}{\sigma}\right)^3$	(2.544)		
Kurtosis	$\mathrm{kurt}[x]\simeq\left[\frac{1}{N}\sum_{i=1}^{N}\left(\frac{x_i-\langle x\rangle}{\sigma}\right)^4\right]-3$	(2.545)		
Correlation coefficient[b]	$r=\frac{\sum_{i=1}^{N}(x_i-\langle x\rangle)(y_i-\langle y\rangle)}{\sqrt{\sum_{i=1}^{N}(x_i-\langle x\rangle)^2}\sqrt{\sum_{i=1}^{N}(y_i-\langle y\rangle)^2}}$	(2.546)	x,y r	data series to correlate correlation coefficient

[a]If $\langle x\rangle$ is derived from the data, $\{x_i\}$, the relation is as shown. If $\langle x\rangle$ is known independently, then an unbiased estimate is obtained by dividing the right-hand side by N rather than $N-1$.
[b]Also known as "Pearson's r."

Discrete probability distributions

distribution	$\mathrm{pr}(x)$	*mean*	*variance*	*domain*			
Binomial	$\binom{n}{x}p^x(1-p)^{n-x}$	np	$np(1-p)$	$(x=0,1,\ldots,n)$	(2.547)	$\binom{n}{x}$	binomial coefficient
Geometric	$(1-p)^{x-1}p$	$1/p$	$(1-p)/p^2$	$(x=1,2,3,\ldots)$	(2.548)		
Poisson	$\lambda^x\exp(-\lambda)/x!$	λ	λ	$(x=0,1,2,\ldots)$	(2.549)		

Continuous probability distributions

distribution	pr(x)	*mean*	*variance*	*domain*	
Uniform	$\frac{1}{b-a}$	$\frac{a+b}{2}$	$\frac{(b-a)^2}{12}$	$(a \le x \le b)$	(2.550)
Exponential	$\lambda \exp(-\lambda x)$	$1/\lambda$	$1/\lambda^2$	$(x \ge 0)$	(2.551)
Normal/ Gaussian	$\frac{1}{\sigma\sqrt{2\pi}} \exp\left[\frac{-(x-\mu)^2}{2\sigma^2}\right]$	μ	σ^2	$(-\infty < x < \infty)$	(2.552)
Chi-squared[a]	$\frac{e^{-x/2} x^{(r/2)-1}}{2^{r/2}\Gamma(r/2)}$	r	$2r$	$(x \ge 0)$	(2.553)
Rayleigh	$\frac{x}{\sigma^2} \exp\left(\frac{-x^2}{2\sigma^2}\right)$	$\sigma\sqrt{\pi/2}$	$2\sigma^2\left(1-\frac{\pi}{4}\right)$	$(x \ge 0)$	(2.554)
Cauchy/ Lorentzian	$\frac{a}{\pi(a^2+x^2)}$	(none)	(none)	$(-\infty < x < \infty)$	(2.555)

[a]With r degrees of freedom. Γ is the gamma function.

Multivariate normal distribution

Density function	$\text{pr}(\boldsymbol{x}) = \frac{\exp\left[-\frac{1}{2}(\boldsymbol{x}-\boldsymbol{\mu})\mathbf{C}^{-1}(\boldsymbol{x}-\boldsymbol{\mu})^T\right]}{(2\pi)^{k/2}[\det(\mathbf{C})]^{1/2}}$	(2.556)	pr k $\mathbf{C}$ $\boldsymbol{x}$ $\boldsymbol{\mu}$ ^{T} det μ_i	probability density number of dimensions covariance matrix variable (k dimensional) vector of means transpose determinant mean of ith variable
Mean	$\boldsymbol{\mu} = (\mu_1, \mu_2, \ldots, \mu_k)$	(2.557)		
Covariance	$\mathbf{C} = \sigma_{ij} = \langle x_i x_j \rangle - \langle x_i \rangle \langle x_j \rangle$	(2.558)	σ_{ij}	components of $\mathbf{C}$
Correlation coefficient	$r = \frac{\sigma_{ij}}{\sigma_i \sigma_j}$	(2.559)	r	correlation coefficient
Box–Muller transformation	$x_1 = (-2\ln y_1)^{1/2} \cos 2\pi y_2$ $x_2 = (-2\ln y_1)^{1/2} \sin 2\pi y_2$	(2.560) (2.561)	x_i y_i	normally distributed deviates deviates distributed uniformly between 0 and 1

Random walk

One-dimensional	$\mathrm{pr}(x)=\frac{1}{(2\pi Nl^2)^{1/2}}\exp\left(\frac{-x^2}{2Nl^2}\right)$	(2.562)	x $\mathrm{pr}(x)$ N l	displacement after N steps (can be positive or negative) probability density of x ($\int_{-\infty}^{\infty}\mathrm{pr}(x)\,\mathrm{d}x=1$) number of steps step length (all equal)
rms displacement	$x_{\mathrm{rms}}=N^{1/2}l$	(2.563)	x_{rms}	root-mean-squared displacement from start point
Three-dimensional	$\mathrm{pr}(r)=\left(\frac{a}{\pi^{1/2}}\right)^3\exp(-a^2r^2)$ where $a=\left(\frac{3}{2Nl^2}\right)^{1/2}$	(2.564)	r $\mathrm{pr}(r)$ a	radial distance from start point probability density of r ($\int_0^{\infty}4\pi r^2\,\mathrm{pr}(r)\,\mathrm{d}r=1$) (most probable distance)$^{-1}$
Mean distance	$\langle r\rangle=\left(\frac{8}{3\pi}\right)^{1/2}N^{1/2}l$	(2.565)	$\langle r\rangle$	mean distance from start point
rms distance	$r_{\mathrm{rms}}=N^{1/2}l$	(2.566)	r_{rms}	root-mean-squared distance from start point

Bayesian inference

Conditional probability	$\mathrm{pr}(x)=\int\mathrm{pr}(x\|y')\,\mathrm{pr}(y')\,\mathrm{d}y'$	(2.567)	$\mathrm{pr}(x)$ $\mathrm{pr}(x\|y')$	probability (density) of x conditional probability of x given y'
Joint probability	$\mathrm{pr}(x,y)=\mathrm{pr}(x)\,\mathrm{pr}(y\|x)$	(2.568)	$\mathrm{pr}(x,y)$	joint probability of x and y
Bayes' theorem[a]	$\mathrm{pr}(y\|x)=\frac{\mathrm{pr}(x\|y)\,\mathrm{pr}(y)}{\mathrm{pr}(x)}$	(2.569)		

[a]In this expression, $\mathrm{pr}(y|x)$ is known as the posterior probability, $\mathrm{pr}(x|y)$ the likelihood, and $\mathrm{pr}(y)$ the prior probability.

2.14 Numerical methods

Straight-line fitting[a]

Data	$(\{x_i\},\{y_i\})$ n points	(2.570)
Weights[b]	$\{w_i\}$	(2.571)
Model	$y=mx+c$	(2.572)
Residuals	$d_i=y_i-mx_i-c$	(2.573)
Weighted centre	$(\overline{x},\overline{y})=\frac{1}{\sum w_i}\left(\sum w_i x_i, \sum w_i y_i\right)$	(2.574)
Weighted moment	$D=\sum w_i(x_i-\overline{x})^2$	(2.575)
Gradient	$m=\frac{1}{D}\sum w_i(x_i-\overline{x})y_i$	(2.576)
	$\mathrm{var}[m]\simeq\frac{1}{D}\frac{\sum w_i d_i^2}{n-2}$	(2.577)
Intercept	$c=\overline{y}-m\overline{x}$	(2.578)
	$\mathrm{var}[c]\simeq\left(\frac{1}{\sum w_i}+\frac{\overline{x}^2}{D}\right)\frac{\sum w_i d_i^2}{n-2}$	(2.579)

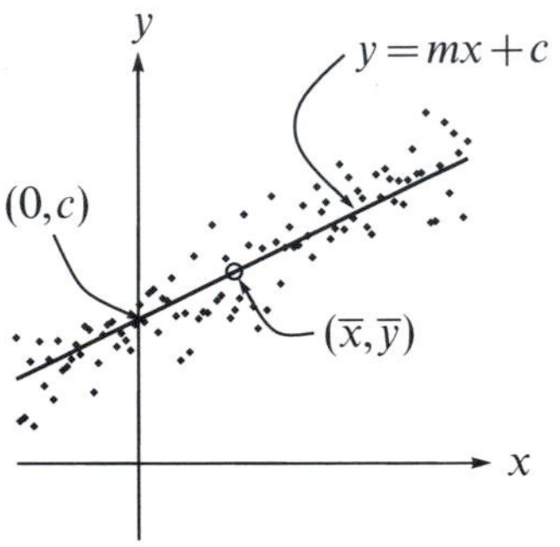

[a]Least-squares fit of data to $y=mx+c$. Errors on y-values only.
[b]If the errors on y_i are uncorrelated, then $w_i=1/\mathrm{var}[y_i]$.

Time series analysis[a]

Discrete convolution	$(r\star s)_j=\sum_{k=-(M/2)+1}^{M/2} s_{j-k}r_k$	(2.580)
Bartlett (triangular) window	$w_j=1-\left\lvert\frac{j-N/2}{N/2}\right\rvert$	(2.581)
Welch (quadratic) window	$w_j=1-\left[\frac{j-N/2}{N/2}\right]^2$	(2.582)
Hanning window	$w_j=\frac{1}{2}\left[1-\cos\left(\frac{2\pi j}{N}\right)\right]$	(2.583)
Hamming window	$w_j=0.54-0.46\cos\left(\frac{2\pi j}{N}\right)$	(2.584)

r_i response function
s_i time series
M response function duration

w_j windowing function
N length of time series

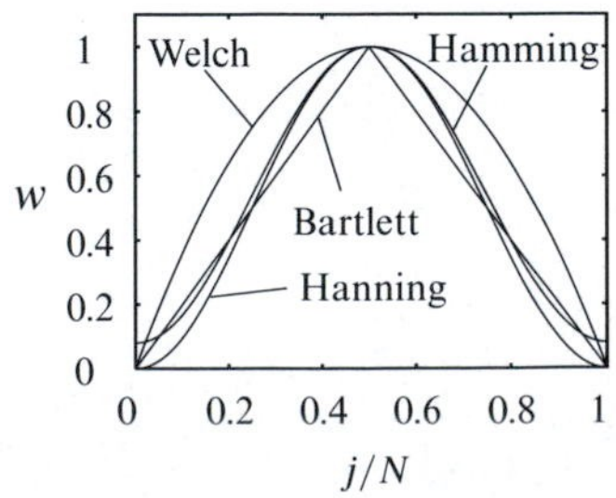

[a]The time series runs from $j=0\ldots(N-1)$, and the windowing functions peak at $j=N/2$.

Numerical integration

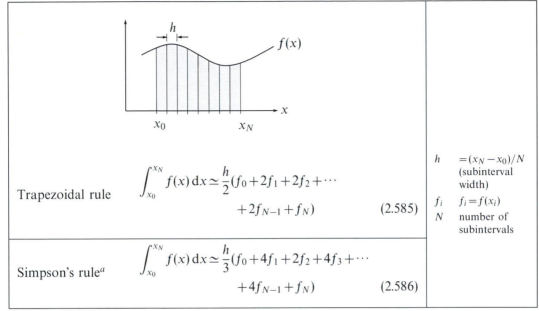

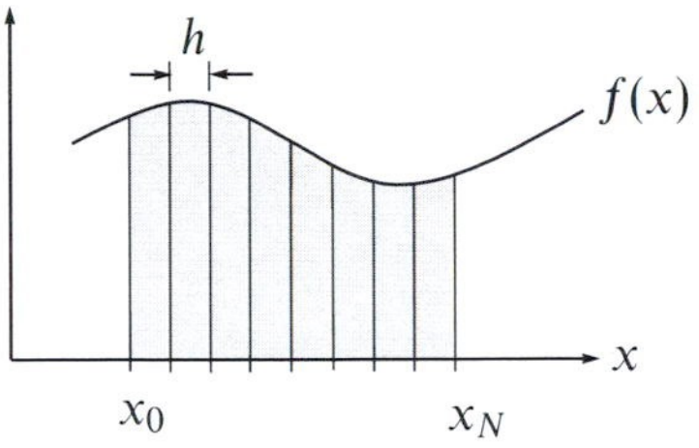

Trapezoidal rule	$\int_{x_0}^{x_N} f(x)\,\mathrm{d}x \simeq \frac{h}{2}(f_0+2f_1+2f_2+\cdots+2f_{N-1}+f_N)$ (2.585)	h $=(x_N-x_0)/N$ (subinterval width) f_i $f_i=f(x_i)$ N number of subintervals
Simpson's rule[a]	$\int_{x_0}^{x_N} f(x)\,\mathrm{d}x \simeq \frac{h}{3}(f_0+4f_1+2f_2+4f_3+\cdots+4f_{N-1}+f_N)$ (2.586)	

[a] N must be even. Simpson's rule is exact for quadratics and cubics.

Numerical differentiation[a]

$$\frac{\mathrm{d}f}{\mathrm{d}x} \simeq \frac{1}{12h}[-f(x+2h)+8f(x+h)-8f(x-h)+f(x-2h)] \quad (2.587)$$

$$\sim \frac{1}{2h}[f(x+h)-f(x-h)] \quad (2.588)$$

$$\frac{\mathrm{d}^2f}{\mathrm{d}x^2} \simeq \frac{1}{12h^2}[-f(x+2h)+16f(x+h)-30f(x)+16f(x-h)-f(x-2h)] \quad (2.589)$$

$$\sim \frac{1}{h^2}[f(x+h)-2f(x)+f(x-h)] \quad (2.590)$$

$$\frac{\mathrm{d}^3f}{\mathrm{d}x^3} \sim \frac{1}{2h^3}[f(x+2h)-2f(x+h)+2f(x-h)-f(x-2h)] \quad (2.591)$$

[a] Derivatives of $f(x)$ at x. h is a small interval in x.
Relations containing "$\simeq$" are $O(h^4)$; those containing "$\sim$" are $O(h^2)$.

Numerical solutions to $f(x)=0$

Secant method	$x_{n+1}=x_n-\frac{x_n-x_{n-1}}{f(x_n)-f(x_{n-1})}f(x_n)$	(2.592)	f function of x x_n $f(x_\infty)=0$
Newton–Raphson method	$x_{n+1}=x_n-\frac{f(x_n)}{f'(x_n)}$	(2.593)	f' $=\mathrm{d}f/\mathrm{d}x$

Numerical solutions to ordinary differential equations[a]

Euler's method	if	$\frac{dy}{dx}=f(x,y)$	(2.594)
	and	$h=x_{n+1}-x_n$	(2.595)
	then	$y_{n+1}=y_n+hf(x_n,y_n)+O(h^2)$	(2.596)
Runge–Kutta method (fourth-order)	if	$\frac{dy}{dx}=f(x,y)$	(2.597)
	and	$h=x_{n+1}-x_n$	(2.598)
		$k_1=hf(x_n,y_n)$	(2.599)
		$k_2=hf(x_n+h/2,y_n+k_1/2)$	(2.600)
		$k_3=hf(x_n+h/2,y_n+k_2/2)$	(2.601)
		$k_4=hf(x_n+h,y_n+k_3)$	(2.602)
	then	$y_{n+1}=y_n+\frac{k_1}{6}+\frac{k_2}{3}+\frac{k_3}{3}+\frac{k_4}{6}+O(h^5)$	(2.603)

[a]Ordinary differential equations (ODEs) of the form $\frac{dy}{dx}=f(x,y)$. Higher order equations should be reduced to a set of coupled first-order equations and solved in parallel.

Chapter 3 Dynamics and mechanics

3.1 Introduction

Unusually in physics, there is no pithy phrase that sums up the study of *dynamics* (the way in which forces produce motion), *kinematics* (the motion of matter), *mechanics* (the study of the forces and the motion they produce), and *statics* (the way forces combine to produce equilibrium). We will take the phrase *dynamics and mechanics* to encompass all the above, although it clearly does not!

To some extent this is because the equations governing the motion of matter include some of our oldest insights into the physical world and are consequentially steeped in tradition. One of the more delightful, or for some annoying, facets of this is the occasional use of arcane vocabulary in the description of motion. The epitome must be what Goldstein[1] calls "the jabberwockian sounding statement" *the polhode rolls without slipping on the herpolhode lying in the invariable plane*, describing "Poinsot's construction" – a method of visualising the free motion of a spinning rigid body. Despite this, dynamics and mechanics, including fluid mechanics, is arguably the most practically applicable of all the branches of physics.

Moreover, and in common with electromagnetism, the study of dynamics and mechanics has spawned a good deal of mathematical apparatus that has found uses in other fields. Most notably, the ideas behind the generalised dynamics of Lagrange and Hamilton lie behind much of quantum mechanics.

[1] H. Goldstein, *Classical Mechanics*, 2nd ed., 1980, Addison-Wesley.

3.2 Frames of reference

Galilean transformations

Time and position[a]	$\boldsymbol{r}=\boldsymbol{r}'+\boldsymbol{v}t$	(3.1)	$\boldsymbol{r},\boldsymbol{r}'$	position in frames S and S'
	$t=t'$	(3.2)	$\boldsymbol{v}$	velocity of S' in S
			t,t'	time in S and S'
Velocity	$\boldsymbol{u}=\boldsymbol{u}'+\boldsymbol{v}$	(3.3)	$\boldsymbol{u},\boldsymbol{u}'$	velocity in frames S and S'
Momentum	$\boldsymbol{p}=\boldsymbol{p}'+m\boldsymbol{v}$	(3.4)	$\boldsymbol{p},\boldsymbol{p}'$	particle momentum in frames S and S'
			m	particle mass
Angular momentum	$\boldsymbol{J}=\boldsymbol{J}'+m\boldsymbol{r}'\times\boldsymbol{v}+\boldsymbol{v}\times\boldsymbol{p}'t$	(3.5)	$\boldsymbol{J},\boldsymbol{J}'$	angular momentum in frames S and S'
Kinetic energy	$T=T'+m\boldsymbol{u}'\cdot\boldsymbol{v}+\frac{1}{2}mv^2$	(3.6)	T,T'	kinetic energy in frames S and S'

[a]Frames coincide at $t=0$.

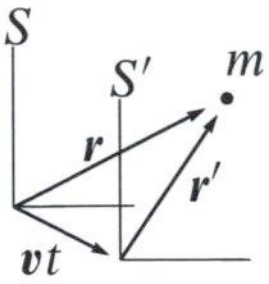

Lorentz (spacetime) transformations[a]

Lorentz factor	$\gamma=\left(1-\frac{v^2}{c^2}\right)^{-1/2}$	(3.7)	γ	Lorentz factor
			v	velocity of S' in S
			c	speed of light
Time and position	$x=\gamma(x'+vt')$; $\quad x'=\gamma(x-vt)$	(3.8)	x,x'	x-position in frames S and S' (similarly for y and z)
	$y=y'$; $\quad y'=y$	(3.9)		
	$z=z'$; $\quad z'=z$	(3.10)		
	$t=\gamma\left(t'+\frac{v}{c^2}x'\right)$; $\quad t'=\gamma\left(t-\frac{v}{c^2}x\right)$	(3.11)	t,t'	time in frames S and S'
Differential four-vector[b]	$\mathrm{d}\boldsymbol{X}=(c\,\mathrm{d}t,-\mathrm{d}x,-\mathrm{d}y,-\mathrm{d}z)$	(3.12)	$\boldsymbol{X}$	spacetime four-vector

[a]For frames S and S' coincident at $t=0$ in relative motion along x. See page 141 for the transformations of electromagnetic quantities.
[b]Covariant components, using the $(1,-1,-1,-1)$ signature.

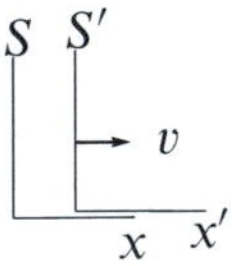

Velocity transformations[a]

Velocity			γ	Lorentz factor $=[1-(v/c)^2]^{-1/2}$
$u_x=\dfrac{u'_x+v}{1+u'_xv/c^2}$;	$u'_x=\dfrac{u_x-v}{1-u_xv/c^2}$	(3.13)	v	velocity of S' in S
$u_y=\dfrac{u'_y}{\gamma(1+u'_xv/c^2)}$;	$u'_y=\dfrac{u_y}{\gamma(1-u_xv/c^2)}$	(3.14)	c	speed of light
$u_z=\dfrac{u'_z}{\gamma(1+u'_xv/c^2)}$;	$u'_z=\dfrac{u_z}{\gamma(1-u_xv/c^2)}$	(3.15)	u_i,u'_i	particle velocity components in frames S and S'

[a]For frames S and S' coincident at $t=0$ in relative motion along x.

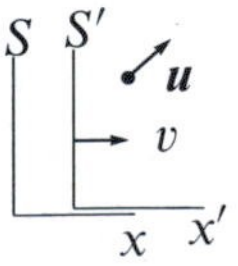

Momentum and energy transformations[a]

Momentum and energy			γ	Lorentz factor $=[1-(v/c)^2]^{-1/2}$
	$p_x=\gamma(p'_x+vE'/c^2)$; $\quad p'_x=\gamma(p_x-vE/c^2)$	(3.16)	v	velocity of S' in S
	$p_y=p'_y$; $\quad p'_y=p_y$	(3.17)	c	speed of light
	$p_z=p'_z$; $\quad p'_z=p_z$	(3.18)	p_x,p'_x	x components of momentum in S and S' (sim. for y and z)
	$E=\gamma(E'+vp'_x)$; $\quad E'=\gamma(E-vp_x)$	(3.19)	E,E'	energy in S and S'
	$E^2-p^2c^2=E'^2-p'^2c^2=m_0^2c^4$	(3.20)	m_0	(rest) mass
			p	total momentum in S
Four-vector[b]	$\boldsymbol{P}=(E/c,-p_x,-p_y,-p_z)$	(3.21)	$\boldsymbol{P}$	momentum four-vector

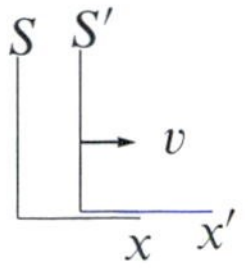

[a]For frames S and S' coincident at $t=0$ in relative motion along x.
[b]Covariant components, using the $(1,-1,-1,-1)$ signature.

Propagation of light[a]

Doppler effect	$\frac{\nu'}{\nu}=\gamma\left(1+\frac{v}{c}\cos\alpha\right)$	(3.22)	ν	frequency received in S
			ν'	frequency emitted in S'
			α	arrival angle in S
Aberration[b]	$\cos\theta=\frac{\cos\theta'+v/c}{1+(v/c)\cos\theta'}$	(3.23)	γ	Lorentz factor $=[1-(v/c)^2]^{-1/2}$
			v	velocity of S' in S
	$\cos\theta'=\frac{\cos\theta-v/c}{1-(v/c)\cos\theta}$	(3.24)	c	speed of light
			θ,θ'	emission angle of light in S and S'
Relativistic beaming[c]	$P(\theta)=\frac{\sin\theta}{2\gamma^2[1-(v/c)\cos\theta]^2}$	(3.25)	$P(\theta)$	angular distribution of photons in S

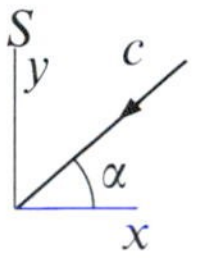

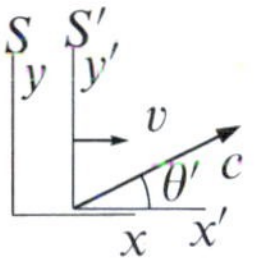

[a]For frames S and S' coincident at $t=0$ in relative motion along x.
[b]Light travelling in the opposite sense has a propagation angle of $\pi+\theta$ radians.
[c]Angular distribution of photons from a source, isotropic and stationary in S'. $\int_0^\pi P(\theta)\,d\theta=1$.

Four-vectors[a]

Covariant and contravariant components	$x_0=x^0 \quad x_1=-x^1$ $x_2=-x^2 \quad x_3=-x^3$	(3.26)	x_i	covariant vector components
			x^i	contravariant components
Scalar product	$x^iy_i=x^0y_0+x^1y_1+x^2y_2+x^3y_3$	(3.27)		
Lorentz transformations			x^i,x'^i	four-vector components in frames S and S'
	$x^0=\gamma[x'^0+(v/c)x'^1]$; $\quad x'^0=\gamma[x^0-(v/c)x^1]$	(3.28)	γ	Lorentz factor $=[1-(v/c)^2]^{-1/2}$
	$x^1=\gamma[x'^1+(v/c)x'^0]$; $\quad x'^1=\gamma[x^1-(v/c)x^0]$	(3.29)	v	velocity of S' in S
	$x^2=x'^2$; $\quad x'^3=x^3$	(3.30)	c	speed of light

[a]For frames S and S', coincident at $t=0$ in relative motion along the (1) direction. Note that the $(1,-1,-1,-1)$ signature used here is common in special relativity, whereas $(-1,1,1,1)$ is often used in connection with general relativity (page 67).

Rotating frames

Vector transformation	$\left[\frac{\mathrm{d}\boldsymbol{A}}{\mathrm{d}t}\right]_S=\left[\frac{\mathrm{d}\boldsymbol{A}}{\mathrm{d}t}\right]_{S'}+\boldsymbol{\omega}\times\boldsymbol{A}$	(3.31)	$\boldsymbol{A}$ any vector; S stationary frame; S' rotating frame; $\boldsymbol{\omega}$ angular velocity of S' in S
Acceleration	$\dot{\boldsymbol{v}}=\dot{\boldsymbol{v}}'+2\boldsymbol{\omega}\times\boldsymbol{v}'+\boldsymbol{\omega}\times(\boldsymbol{\omega}\times\boldsymbol{r}')$	(3.32)	$\dot{\boldsymbol{v}},\dot{\boldsymbol{v}}'$ accelerations in S and S'; $\boldsymbol{v}'$ velocity in S'; $\boldsymbol{r}'$ position in S'
Coriolis force	$\boldsymbol{F}'_{\mathrm{cor}}=-2m\boldsymbol{\omega}\times\boldsymbol{v}'$	(3.33)	$\boldsymbol{F}'_{\mathrm{cor}}$ coriolis force; m particle mass
Centrifugal force	$\boldsymbol{F}'_{\mathrm{cen}}=-m\boldsymbol{\omega}\times(\boldsymbol{\omega}\times\boldsymbol{r}')$	(3.34)	$\boldsymbol{F}'_{\mathrm{cen}}$ centrifugal force; $\boldsymbol{r}'_{\perp}$ perpendicular to particle from rotation axis
	$=+m\omega^2\boldsymbol{r}'_{\perp}$	(3.35)	
Motion relative to Earth	$m\ddot{x}=F_x+2m\omega_{\mathrm{e}}(\dot{y}\sin\lambda-\dot{z}\cos\lambda)$	(3.36)	F_i nongravitational force; λ latitude; z local vertical axis; y northerly axis; x easterly axis
	$m\ddot{y}=F_y-2m\omega_{\mathrm{e}}\dot{x}\sin\lambda$	(3.37)	
	$m\ddot{z}=F_z-mg+2m\omega_{\mathrm{e}}\dot{x}\cos\lambda$	(3.38)	
Foucault's pendulum[a]	$\Omega_{\mathrm{f}}=-\omega_{\mathrm{e}}\sin\lambda$	(3.39)	Ω_{f} pendulum's rate of turn; ω_{e} Earth's spin rate

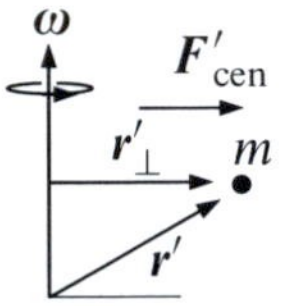

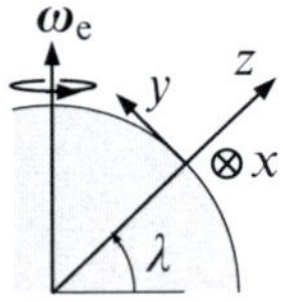

[a]The sign is such as to make the rotation clockwise in the northern hemisphere.

3.3 Gravitation

Newtonian gravitation

Newton's law of gravitation	$\boldsymbol{F}_1=\frac{Gm_1m_2}{r_{12}^2}\hat{\boldsymbol{r}}_{12}$	(3.40)	$m_{1,2}$ masses; $\boldsymbol{F}_1$ force on m_1 $(=-\boldsymbol{F}_2)$; $\boldsymbol{r}_{12}$ vector from m_1 to m_2; $\hat{}$ unit vector
Newtonian field equations[a]	$\boldsymbol{g}=-\nabla\phi$	(3.41)	G constant of gravitation; $\boldsymbol{g}$ gravitational field strength; ϕ gravitational potential; ρ mass density
	$\nabla^2\phi=-\nabla\cdot\boldsymbol{g}=4\pi G\rho$	(3.42)	
Fields from an isolated uniform sphere, mass M, $\boldsymbol{r}$ from the centre	$\boldsymbol{g}(\boldsymbol{r})=\begin{cases}-\frac{GM}{r^2}\hat{\boldsymbol{r}} & (r>a)\\ -\frac{GMr}{a^3}\hat{\boldsymbol{r}} & (r<a)\end{cases}$	(3.43)	$\boldsymbol{r}$ vector from sphere centre; M mass of sphere; a radius of sphere
	$\phi(\boldsymbol{r})=\begin{cases}-\frac{GM}{r} & (r>a)\\ \frac{GM}{2a^3}(r^2-3a^2) & (r<a)\end{cases}$	(3.44)	

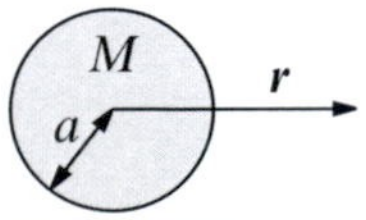

[a]The gravitational force on a mass m is $m\boldsymbol{g}$.

General relativity[a]

Line element	$ds^2 = g_{\mu\nu}\,dx^\mu\,dx^\nu = -d\tau^2$	(3.45)	ds	invariant interval
			$d\tau$	proper time interval
			$g_{\mu\nu}$	metric tensor
			dx^μ	differential of x^μ
Christoffel symbols and covariant differentiation	$\Gamma^\alpha{}_{\beta\gamma} = \frac{1}{2}g^{\alpha\delta}(g_{\delta\beta,\gamma} + g_{\delta\gamma,\beta} - g_{\beta\gamma,\delta})$	(3.46)	$\Gamma^\alpha{}_{\beta\gamma}$	Christoffel symbols
	$\phi_{;\gamma} = \phi_{,\gamma} \equiv \partial\phi/\partial x^\gamma$	(3.47)	$,\alpha$	partial diff. w.r.t. x^α
	$A^\alpha_{;\gamma} = A^\alpha_{,\gamma} + \Gamma^\alpha{}_{\beta\gamma}A^\beta$	(3.48)	$;\alpha$	covariant diff. w.r.t. x^α
	$B_{\alpha;\gamma} = B_{\alpha,\gamma} - \Gamma^\beta{}_{\alpha\gamma}B_\beta$	(3.49)	ϕ	scalar
			A^α	contravariant vector
			B_α	covariant vector
Riemann tensor	$R^\alpha{}_{\beta\gamma\delta} = \Gamma^\alpha{}_{\mu\gamma}\Gamma^\mu{}_{\beta\delta} - \Gamma^\alpha{}_{\mu\delta}\Gamma^\mu{}_{\beta\gamma} + \Gamma^\alpha{}_{\beta\delta,\gamma} - \Gamma^\alpha{}_{\beta\gamma,\delta}$	(3.50)	$R^\alpha{}_{\beta\gamma\delta}$	Riemann tensor
	$B_{\mu;\alpha;\beta} - B_{\mu;\beta;\alpha} = R^\gamma{}_{\mu\alpha\beta}B_\gamma$	(3.51)		
	$R_{\alpha\beta\gamma\delta} = -R_{\alpha\beta\delta\gamma}\,; \quad R_{\beta\alpha\gamma\delta} = -R_{\alpha\beta\gamma\delta}$	(3.52)		
	$R_{\alpha\beta\gamma\delta} + R_{\alpha\delta\beta\gamma} + R_{\alpha\gamma\delta\beta} = 0$	(3.53)		
Geodesic equation	$\frac{Dv^\mu}{D\lambda} = 0$	(3.54)	v^μ	tangent vector ($= dx^\mu/d\lambda$)
	where $\frac{DA^\mu}{D\lambda} \equiv \frac{dA^\mu}{d\lambda} + \Gamma^\mu{}_{\alpha\beta}A^\alpha v^\beta$	(3.55)	λ	affine parameter (e.g., τ for material particles)
Geodesic deviation	$\frac{D^2\xi^\mu}{D\lambda^2} = -R^\mu{}_{\alpha\beta\gamma}v^\alpha\xi^\beta v^\gamma$	(3.56)	ξ^μ	geodesic deviation
Ricci tensor	$R_{\alpha\beta} \equiv R^\sigma{}_{\alpha\sigma\beta} = g^{\sigma\delta}R_{\delta\alpha\sigma\beta} = R_{\beta\alpha}$	(3.57)	$R_{\alpha\beta}$	Ricci tensor
Einstein tensor	$G^{\mu\nu} = R^{\mu\nu} - \frac{1}{2}g^{\mu\nu}R$	(3.58)	$G^{\mu\nu}$	Einstein tensor
			R	Ricci scalar ($= g^{\mu\nu}R_{\mu\nu}$)
Einstein's field equations	$G^{\mu\nu} = 8\pi T^{\mu\nu}$	(3.59)	$T^{\mu\nu}$	stress-energy tensor
			p	pressure (in rest frame)
Perfect fluid	$T^{\mu\nu} = (p+\rho)u^\mu u^\nu + pg^{\mu\nu}$	(3.60)	ρ	density (in rest frame)
			u^ν	fluid four-velocity
Schwarzschild solution (exterior)	$ds^2 = -\left(1-\frac{2M}{r}\right)dt^2 + \left(1-\frac{2M}{r}\right)^{-1}dr^2 + r^2(d\theta^2 + \sin^2\theta\,d\phi^2)$	(3.61)	M	spherically symmetric mass (see page 183)
			(r,θ,ϕ)	spherical polar coords.
			t	time
Kerr solution (outside a spinning black hole)	$ds^2 = -\frac{\Delta - a^2\sin^2\theta}{\varrho^2}dt^2 - 2a\frac{2Mr\sin^2\theta}{\varrho^2}dt\,d\phi + \frac{(r^2+a^2)^2 - a^2\Delta\sin^2\theta}{\varrho^2}\sin^2\theta\,d\phi^2 + \frac{\varrho^2}{\Delta}dr^2 + \varrho^2\,d\theta^2$	(3.62)	J	angular momentum (along z)
			a	$\equiv J/M$
			Δ	$\equiv r^2 - 2Mr + a^2$
			ϱ^2	$\equiv r^2 + a^2\cos^2\theta$

[a]General relativity conventionally uses the $(-1,1,1,1)$ metric signature and "geometrized units" in which $G=1$ and $c=1$. Thus, $1\,\text{kg} = 7.425\times10^{-28}\,\text{m}$ etc. Contravariant indices are written as superscripts and covariant indices as subscripts. Note also that ds^2 means $(ds)^2$ etc.

3.4 Particle motion

Dynamics definitions[a]

Newtonian force	$\boldsymbol{F}=m\ddot{\boldsymbol{r}}=\dot{\boldsymbol{p}}$	(3.63)	$\boldsymbol{F}$ m $\boldsymbol{r}$	force mass of particle particle position vector
Momentum	$\boldsymbol{p}=m\dot{\boldsymbol{r}}$	(3.64)	$\boldsymbol{p}$	momentum
Kinetic energy	$T=\frac{1}{2}mv^2$	(3.65)	T v	kinetic energy particle velocity
Angular momentum	$\boldsymbol{J}=\boldsymbol{r}\times\boldsymbol{p}$	(3.66)	$\boldsymbol{J}$	angular momentum
Couple (or torque)	$\boldsymbol{G}=\boldsymbol{r}\times\boldsymbol{F}$	(3.67)	$\boldsymbol{G}$	couple
Centre of mass (ensemble of N particles)	$\boldsymbol{R}_0=\frac{\sum_{i=1}^{N}m_i\boldsymbol{r}_i}{\sum_{i=1}^{N}m_i}$	(3.68)	$\boldsymbol{R}_0$ m_i $\boldsymbol{r}_i$	position vector of centre of mass mass of ith particle position vector of ith particle

[a]In the Newtonian limit, $v \ll c$, assuming m is constant.

Relativistic dynamics[a]

Lorentz factor	$\gamma=\left(1-\frac{v^2}{c^2}\right)^{-1/2}$	(3.69)	γ $\boldsymbol{v}$ c	Lorentz factor particle velocity speed of light
Momentum	$\boldsymbol{p}=\gamma m_0\boldsymbol{v}$	(3.70)	$\boldsymbol{p}$ m_0	relativistic momentum particle (rest) mass
Force	$\boldsymbol{F}=\frac{\mathrm{d}\boldsymbol{p}}{\mathrm{d}t}$	(3.71)	$\boldsymbol{F}$ t	force on particle time
Rest energy	$E_\mathrm{r}=m_0c^2$	(3.72)	E_r	particle rest energy
Kinetic energy	$T=m_0c^2(\gamma-1)$	(3.73)	T	relativistic kinetic energy
Total energy	$E=\gamma m_0c^2$ $=(p^2c^2+m_0^2c^4)^{1/2}$	(3.74) (3.75)	E	total energy $(=E_\mathrm{r}+T)$

[a]It is now common to regard mass as a Lorentz invariant property and to drop the term "rest mass." The symbol m_0 is used here to avoid confusion with the idea of "relativistic mass" $(=\gamma m_0)$ used by some authors.

Constant acceleration

$v=u+at$	(3.76)	u	initial velocity
$v^2=u^2+2as$	(3.77)	v	final velocity
$s=ut+\frac{1}{2}at^2$	(3.78)	t	time
$s=\frac{u+v}{2}t$	(3.79)	s	distance travelled
		a	acceleration

Reduced mass (of two interacting bodies)

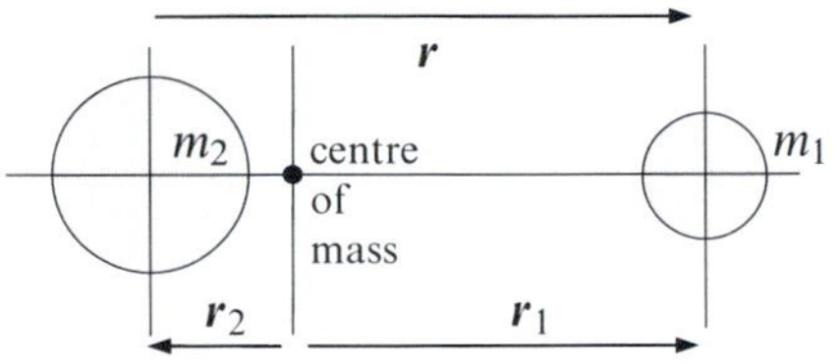

Reduced mass	$\mu = \frac{m_1 m_2}{m_1 + m_2}$	(3.80)	μ m_i	reduced mass interacting masses
Distances from centre of mass	$\boldsymbol{r}_1 = \frac{m_2}{m_1 + m_2}\boldsymbol{r}$	(3.81)	$\boldsymbol{r}_i$	position vectors from centre of mass
	$\boldsymbol{r}_2 = \frac{-m_1}{m_1 + m_2}\boldsymbol{r}$	(3.82)	$\boldsymbol{r}$ $\lvert\boldsymbol{r}\rvert$	$\boldsymbol{r} = \boldsymbol{r}_1 - \boldsymbol{r}_2$ distance between masses
Moment of inertia	$I = \mu\lvert\boldsymbol{r}\rvert^2$	(3.83)	I	moment of inertia
Total angular momentum	$\boldsymbol{J} = \mu\boldsymbol{r}\times\dot{\boldsymbol{r}}$	(3.84)	$\boldsymbol{J}$	angular momentum
Lagrangian	$L = \frac{1}{2}\mu\lvert\dot{\boldsymbol{r}}\rvert^2 - U(\lvert\boldsymbol{r}\rvert)$	(3.85)	L U	Lagrangian potential energy of interaction

Ballistics[a]

Velocity	$\boldsymbol{v} = v_0\cos\alpha\,\hat{\boldsymbol{x}} + (v_0\sin\alpha - gt)\hat{\boldsymbol{y}}$	(3.86)	v_0 $\boldsymbol{v}$ α	initial velocity velocity at t elevation angle
	$v^2 = v_0^2 - 2gy$	(3.87)	g	gravitational acceleration
Trajectory	$y = x\tan\alpha - \frac{gx^2}{2v_0^2\cos^2\alpha}$	(3.88)	$\hat{}$ t	unit vector time
Maximum height	$h = \frac{v_0^2}{2g}\sin^2\alpha$	(3.89)	h	maximum height
Horizontal range	$l = \frac{v_0^2}{g}\sin 2\alpha$	(3.90)	l	range

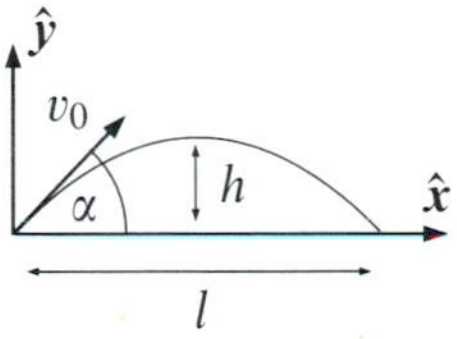

[a]Ignoring the curvature and rotation of the Earth and frictional losses. g is assumed constant.

Rocketry

Escape velocity[a]	$v_{\text{esc}} = \left(\frac{2GM}{r}\right)^{1/2}$	(3.91)
Specific impulse	$I_{\text{sp}} = \frac{u}{g}$	(3.92)
Exhaust velocity (into a vacuum)	$u = \left[\frac{2\gamma R T_{\text{c}}}{(\gamma-1)\mu}\right]^{1/2}$	(3.93)
Rocket equation ($g=0$)	$\Delta v = u \ln\left(\frac{M_{\text{i}}}{M_{\text{f}}}\right) \equiv u \ln \mathcal{M}$	(3.94)
Multistage rocket	$\Delta v = \sum_{i=1}^{N} u_i \ln \mathcal{M}_i$	(3.95)
In a constant gravitational field	$\Delta v = u \ln \mathcal{M} - gt\cos\theta$	(3.96)
Hohmann cotangential transfer[b]	$\Delta v_{ah} = \left(\frac{GM}{r_a}\right)^{1/2}\left[\left(\frac{2r_b}{r_a+r_b}\right)^{1/2} - 1\right]$	(3.97)
	$\Delta v_{hb} = \left(\frac{GM}{r_b}\right)^{1/2}\left[1 - \left(\frac{2r_a}{r_a+r_b}\right)^{1/2}\right]$	(3.98)

v_{esc}	escape velocity
G	constant of gravitation
M	mass of central body
r	central body radius
I_{sp}	specific impulse
u	effective exhaust velocity
g	acceleration due to gravity
R	molar gas constant
γ	ratio of heat capacities
T_{c}	combustion temperature
μ	effective molecular mass of exhaust gas
Δv	rocket velocity increment
M_{i}	pre-burn rocket mass
M_{f}	post-burn rocket mass
$\mathcal{M}$	mass ratio
N	number of stages
$\mathcal{M}_i$	mass ratio for ith burn
u_i	exhaust velocity of ith burn
t	burn time
θ	rocket zenith angle
Δv_{ah}	velocity increment, a to h
Δv_{hb}	velocity increment, h to b
r_a	radius of inner orbit
r_b	radius of outer orbit

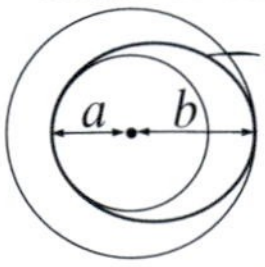

[a]From the surface of a spherically symmetric, nonrotating body, mass M.

[b]Transfer between coplanar, circular orbits a and b, via ellipse h with a minimal expenditure of energy.

Gravitationally bound orbital motion[a]

Potential energy of interaction	$U(r)=-\frac{GMm}{r}\equiv-\frac{\alpha}{r}$	(3.99)	$U(r)$ G M m α	potential energy constant of gravitation central mass orbiting mass ($\ll M$) GMm (for gravitation)
Total energy	$E=-\frac{\alpha}{r}+\frac{J^2}{2mr^2}=-\frac{\alpha}{2a}$	(3.100)	E J	total energy (constant) total angular momentum (constant)
Virial theorem ($1/r$ potential)	$E=\langle U\rangle/2=-\langle T\rangle$ $\langle U\rangle=-2\langle T\rangle$	(3.101) (3.102)	T $\langle\cdot\rangle$	kinetic energy mean value
Orbital equation (Kepler's 1st law)	$\frac{r_0}{r}=1+e\cos\phi$, or $r=\frac{a(1-e^2)}{1+e\cos\phi}$	(3.103) (3.104)	r_0 r e ϕ	semi-latus-rectum distance of m from M eccentricity phase (true anomaly)
Rate of sweeping area (Kepler's 2nd law)	$\frac{dA}{dt}=\frac{J}{2m}=\text{constant}$	(3.105)	A	area swept out by radius vector (total area $=\pi ab$)
Semi-major axis	$a=\frac{r_0}{1-e^2}=\frac{\alpha}{2\|E\|}$	(3.106)	a b	semi-major axis semi-minor axis
Semi-minor axis	$b=\frac{r_0}{(1-e^2)^{1/2}}=\frac{J}{(2m\|E\|)^{1/2}}$	(3.107)	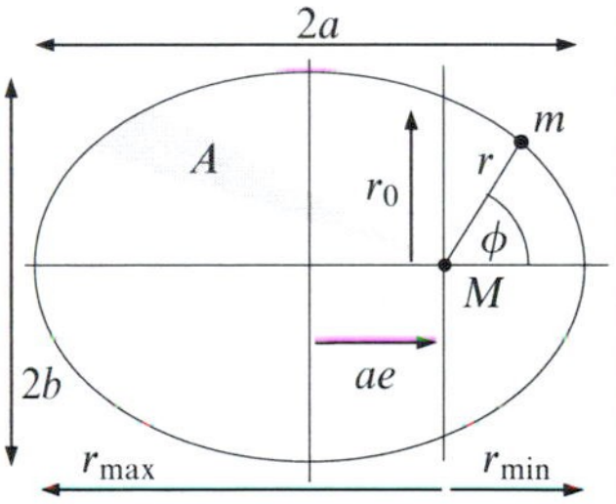	
Eccentricity[b]	$e=\left(1+\frac{2EJ^2}{m\alpha^2}\right)^{1/2}=\left(1-\frac{b^2}{a^2}\right)^{1/2}$	(3.108)		
Semi-latus-rectum	$r_0=\frac{J^2}{m\alpha}=\frac{b^2}{a}=a(1-e^2)$	(3.109)		
Pericentre	$r_{min}=\frac{r_0}{1+e}=a(1-e)$	(3.110)	r_{min}	pericentre distance
Apocentre	$r_{max}=\frac{r_0}{1-e}=a(1+e)$	(3.111)	r_{max}	apocentre distance
Speed	$v^2=GM\left(\frac{2}{r}-\frac{1}{a}\right)$	(3.112)	v	orbital speed
Period (Kepler's 3rd law)	$P=\pi\alpha\left(\frac{m}{2\|E\|^3}\right)^{1/2}=2\pi a^{3/2}\left(\frac{m}{\alpha}\right)^{1/2}$	(3.113)	P	orbital period

[a]For an inverse-square law of attraction between two isolated bodies in the nonrelativistic limit. If m is not $\ll M$, then the equations are valid with the substitutions $m\to\mu=Mm/(M+m)$ and $M\to(M+m)$ and with r taken as the body separation. The distance of mass m from the centre of mass is then $r\mu/m$ (see earlier table on *Reduced mass*). Other orbital dimensions scale similarly, and the two orbits have the same eccentricity.

[b]Note that if the total energy, E, is <0 then $e<1$ and the orbit is an ellipse (a circle if $e=0$). If $E=0$, then $e=1$ and the orbit is a parabola. If $E>0$ then $e>1$ and the orbit becomes a hyperbola (see *Rutherford scattering* on next page).

Rutherford scattering[a]

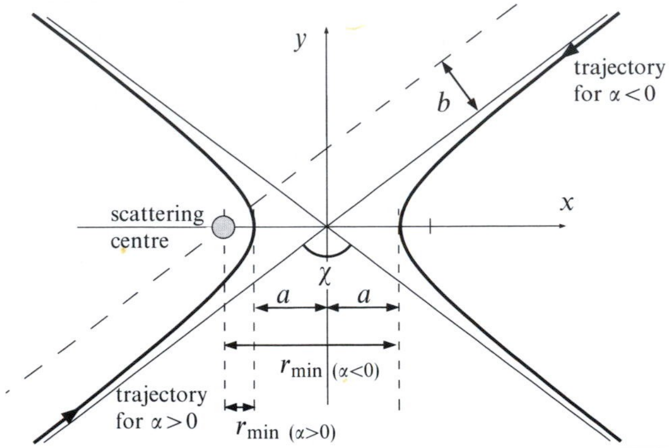

Scattering potential energy	$U(r) = -\frac{\alpha}{r}$ (3.114) $\alpha \begin{cases} <0 & \text{repulsive} \\ >0 & \text{attractive} \end{cases}$ (3.115)		$U(r)$ potential energy r particle separation α constant
Scattering angle	$\tan\frac{\chi}{2} = \frac{\lvert\alpha\rvert}{2Eb}$	(3.116)	χ scattering angle E total energy (>0) b impact parameter
Closest approach	$r_{\min} = \frac{\lvert\alpha\rvert}{2E}\left(\csc\frac{\chi}{2} - \frac{\alpha}{\lvert\alpha\rvert}\right)$ (3.117) $= a(e \pm 1)$ (3.118)		$r_{\min}$ closest approach a hyperbola semi-axis e eccentricity
Semi-axis	$a = \frac{\lvert\alpha\rvert}{2E}$	(3.119)	
Eccentricity	$e = \left(\frac{4E^2b^2}{\alpha^2} + 1\right)^{1/2} = \csc\frac{\chi}{2}$	(3.120)	
Motion trajectory[b]	$\frac{4E^2}{\alpha^2}x^2 - \frac{y^2}{b^2} = 1$	(3.121)	x, y position with respect to hyperbola centre
Scattering centre[c]	$x = \pm\left(\frac{\alpha^2}{4E^2} + b^2\right)^{1/2}$	(3.122)	
Rutherford scattering formula[d]	$\frac{d\sigma}{d\Omega} = \frac{1}{n}\frac{dN}{d\Omega}$ (3.123) $= \left(\frac{\alpha}{4E}\right)^2 \csc^4\frac{\chi}{2}$ (3.124)		$\frac{d\sigma}{d\Omega}$ differential scattering cross section n beam flux density dN number of particles scattered into $d\Omega$ Ω solid angle

[a]Nonrelativistic treatment for an inverse-square force law and a fixed scattering centre. Similar scattering results from either an attractive or repulsive force. See also *Conic sections* on page 38.
[b]The correct branch can be chosen by inspection.
[c]Also the focal points of the hyperbola.
[d]n is the number of particles per second passing through unit area perpendicular to the beam.

Inelastic collisions[a]

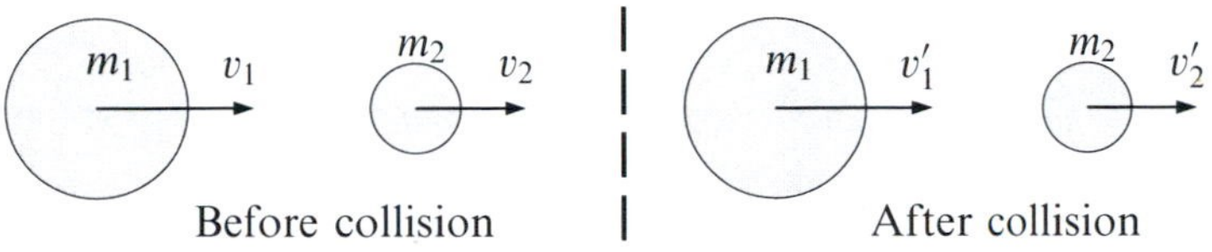

Coefficient of restitution	$v_2'-v_1'=\epsilon(v_1-v_2)$	(3.125)	ϵ	coefficient of restitution
	$\epsilon=1$ if perfectly elastic	(3.126)	v_i	pre-collision velocities
	$\epsilon=0$ if perfectly inelastic	(3.127)	v_i'	post-collision velocities
Loss of kinetic energy[b]	$\dfrac{T-T'}{T}=1-\epsilon^2$	(3.128)	T,T'	total KE in zero momentum frame before and after collision
Final velocities	$v_1'=\dfrac{m_1-\epsilon m_2}{m_1+m_2}v_1+\dfrac{(1+\epsilon)m_2}{m_1+m_2}v_2$	(3.129)	m_i	particle masses
	$v_2'=\dfrac{m_2-\epsilon m_1}{m_1+m_2}v_2+\dfrac{(1+\epsilon)m_1}{m_1+m_2}v_1$	(3.130)		

[a]Along the line of centres, $v_1,v_2 \ll c$.
[b]In zero momentum frame.

Oblique elastic collisions[a]

Before collision

Directions of motion	$\tan\theta_1'=\dfrac{m_2\sin 2\theta}{m_1-m_2\cos 2\theta}$	(3.131)	θ	angle between centre line and incident velocity
	$\theta_2'=\theta$	(3.132)	θ_i'	final trajectories
			m_i	sphere masses
Relative separation angle	$\theta_1'+\theta_2'\begin{cases}>\pi/2 & \text{if } m_1<m_2\\ =\pi/2 & \text{if } m_1=m_2\\ <\pi/2 & \text{if } m_1>m_2\end{cases}$	(3.133)		
Final velocities	$v_1'=\dfrac{(m_1^2+m_2^2-2m_1m_2\cos 2\theta)^{1/2}}{m_1+m_2}v$	(3.134)	v	incident velocity of m_1
	$v_2'=\dfrac{2m_1v}{m_1+m_2}\cos\theta$	(3.135)	v_i'	final velocities

[a]Collision between two perfectly elastic spheres: m_2 initially at rest, velocities $\ll c$.

3

3.5 Rigid body dynamics

Moment of inertia tensor

Moment of inertia tensor[a]	$I_{ij}=\int (r^2\delta_{ij}-x_ix_j)\,\mathrm{d}m$	(3.136)
	$\mathsf{I}=\begin{pmatrix} \int(y^2+z^2)\,\mathrm{d}m & -\int xy\,\mathrm{d}m & -\int xz\,\mathrm{d}m \\ -\int xy\,\mathrm{d}m & \int(x^2+z^2)\,\mathrm{d}m & -\int yz\,\mathrm{d}m \\ -\int xz\,\mathrm{d}m & -\int yz\,\mathrm{d}m & \int(x^2+y^2)\,\mathrm{d}m \end{pmatrix}$	(3.137)
Parallel axis theorem	$I_{12}=I_{12}^{\star}-ma_1a_2$	(3.138)
	$I_{11}=I_{11}^{\star}+m(a_2^2+a_3^2)$	(3.139)
	$I_{ij}=I_{ij}^{\star}+m(\|\boldsymbol{a}\|^2\delta_{ij}-a_ia_j)$	(3.140)
Angular momentum	$\boldsymbol{J}=\mathsf{I}\boldsymbol{\omega}$	(3.141)
Rotational kinetic energy	$T=\frac{1}{2}\boldsymbol{\omega}\cdot\boldsymbol{J}=\frac{1}{2}I_{ij}\omega_i\omega_j$	(3.142)

r	$r^2=x^2+y^2+z^2$
δ_{ij}	Kronecker delta
I	moment of inertia tensor
$\mathrm{d}m$	mass element
x_i	position vector of $\mathrm{d}m$
I_{ij}	components of I
$I_{ij}^{\star}$	tensor with respect to centre of mass
$a_i,\boldsymbol{a}$	position vector of centre of mass
m	mass of body
$\boldsymbol{J}$	angular momentum
$\boldsymbol{\omega}$	angular velocity
T	kinetic energy

[a] I_{ii} are the moments of inertia of the body. I_{ij} $(i\neq j)$ are its products of inertia. The integrals are over the body volume.

Principal axes

Principal moment of inertia tensor	$\mathsf{I}'=\begin{pmatrix} I_1 & 0 & 0 \\ 0 & I_2 & 0 \\ 0 & 0 & I_3 \end{pmatrix}$	(3.143)
Angular momentum	$\boldsymbol{J}=(I_1\omega_1,I_2\omega_2,I_3\omega_3)$	(3.144)
Rotational kinetic energy	$T=\frac{1}{2}(I_1\omega_1^2+I_2\omega_2^2+I_3\omega_3^2)$	(3.145)
Moment of inertia ellipsoid[a]	$T=T(\omega_1,\omega_2,\omega_3)$	(3.146)
	$J_i=\dfrac{\partial T}{\partial\omega_i}$ ($\boldsymbol{J}$ is $\perp$ ellipsoid surface)	(3.147)
Perpendicular axis theorem	$I_1+I_2\begin{cases}\geq I_3 & \text{generally}\\ =I_3 & \text{flat lamina}\perp\text{to 3-axis}\end{cases}$	(3.148)
Symmetries	$I_1\neq I_2\neq I_3$ asymmetric top $I_1=I_2\neq I_3$ symmetric top $I_1=I_2=I_3$ spherical top	(3.149)

I'	principal moment of inertia tensor
I_i	principal moments of inertia
$\boldsymbol{J}$	angular momentum
ω_i	components of $\boldsymbol{\omega}$ along principal axes
T	kinetic energy

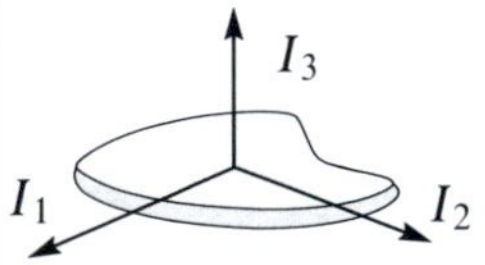

lamina

[a] The ellipsoid is defined by the surface of constant T.

Moments of inertia[a]

Thin rod, length l	$I_1 = I_2 = \dfrac{ml^2}{12}$ (3.150) $I_3 \simeq 0$ (3.151)		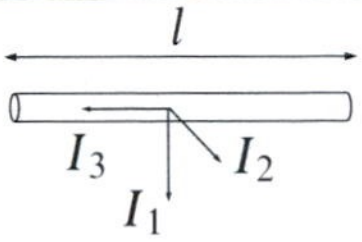
Solid sphere, radius r	$I_1 = I_2 = I_3 = \dfrac{2}{5}mr^2$	(3.152)	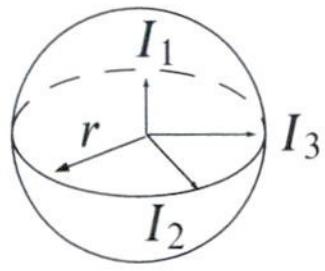
Spherical shell, radius r	$I_1 = I_2 = I_3 = \dfrac{2}{3}mr^2$	(3.153)	
Solid cylinder, radius r, length l	$I_1 = I_2 = \dfrac{m}{4}\left(r^2 + \dfrac{l^2}{3}\right)$ (3.154) $I_3 = \dfrac{1}{2}mr^2$ (3.155)		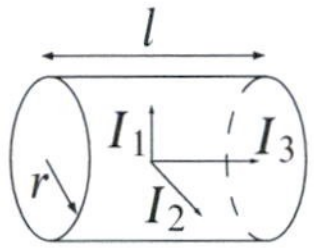
Solid cuboid, sides a,b,c	$I_1 = m(b^2 + c^2)/12$ (3.156) $I_2 = m(c^2 + a^2)/12$ (3.157) $I_3 = m(a^2 + b^2)/12$ (3.158)		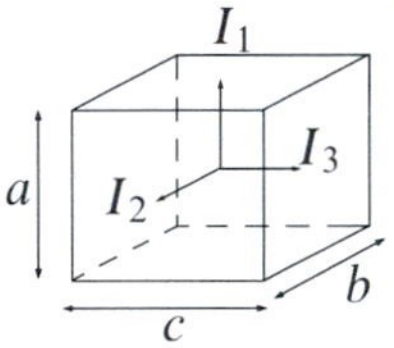
Solid circular cone, base radius r, height h[b]	$I_1 = I_2 = \dfrac{3}{20}m\left(r^2 + \dfrac{h^2}{4}\right)$ (3.159) $I_3 = \dfrac{3}{10}mr^2$ (3.160)		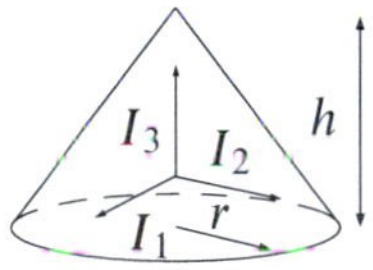
Solid ellipsoid, semi-axes a,b,c	$I_1 = m(b^2 + c^2)/5$ (3.161) $I_2 = m(c^2 + a^2)/5$ (3.162) $I_3 = m(a^2 + b^2)/5$ (3.163)		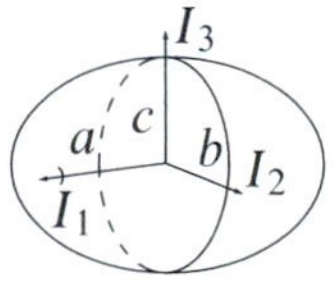
Elliptical lamina, semi-axes a,b	$I_1 = mb^2/4$ (3.164) $I_2 = ma^2/4$ (3.165) $I_3 = m(a^2 + b^2)/4$ (3.166)		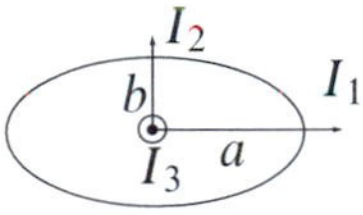
Disk, radius r	$I_1 = I_2 = mr^2/4$ (3.167) $I_3 = mr^2/2$ (3.168)		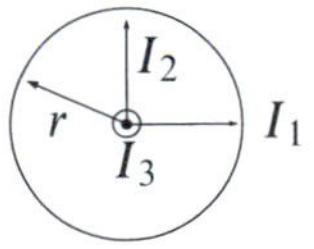
Triangular plate[c]	$I_3 = \dfrac{m}{36}(a^2 + b^2 + c^2)$	(3.169)	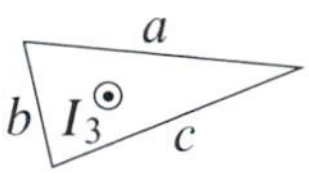

[a]With respect to principal axes for bodies of mass m and uniform density. The radius of gyration is defined as $k = (I/m)^{1/2}$.
[b]Origin of axes is at the centre of mass ($h/4$ above the base).
[c]Around an axis through the centre of mass and perpendicular to the plane of the plate.

Centres of mass

Solid hemisphere, radius r	$d=3r/8$ from sphere centre	(3.170)
Hemispherical shell, radius r	$d=r/2$ from sphere centre	(3.171)
Sector of disk, radius r, angle 2θ	$d=\frac{2}{3}r\frac{\sin\theta}{\theta}$ from disk centre	(3.172)
Arc of circle, radius r, angle 2θ	$d=r\frac{\sin\theta}{\theta}$ from circle centre	(3.173)
Arbitrary triangular lamina, height h[a]	$d=h/3$ perpendicular from base	(3.174)
Solid cone or pyramid, height h	$d=h/4$ perpendicular from base	(3.175)
Spherical cap, height h, sphere radius r	solid: $d=\frac{3}{4}\frac{(2r-h)^2}{3r-h}$ from sphere centre	(3.176)
	shell: $d=r-h/2$ from sphere centre	(3.177)
Semi-elliptical lamina, height h	$d=\frac{4h}{3\pi}$ from base	(3.178)

[a] h is the perpendicular distance between the base and apex of the triangle.

Pendulums

Simple pendulum	$P=2\pi\sqrt{\frac{l}{g}}\left(1+\frac{\theta_0^2}{16}+\cdots\right)$	(3.179)	P period g gravitational acceleration l length θ_0 maximum angular displacement	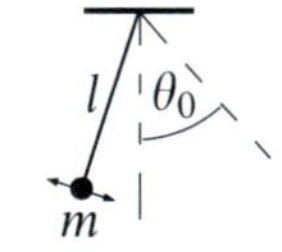
Conical pendulum	$P=2\pi\left(\frac{l\cos\alpha}{g}\right)^{1/2}$	(3.180)	α cone half-angle	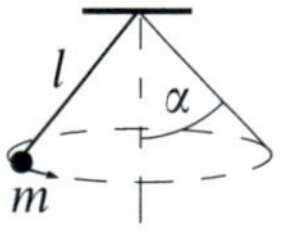
Torsional pendulum[a]	$P=2\pi\left(\frac{lI_0}{C}\right)^{1/2}$	(3.181)	I_0 moment of inertia of bob C torsional rigidity of wire (see page 81)	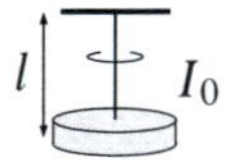
Compound pendulum[b]	$P\simeq 2\pi\left[\frac{1}{mga}(ma^2+I_1\cos^2\gamma_1+I_2\cos^2\gamma_2+I_3\cos^2\gamma_3)\right]^{1/2}$	(3.182)	a distance of rotation axis from centre of mass m mass of body I_i principal moments of inertia γ_i angles between rotation axis and principal axes	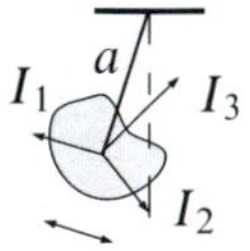
Equal double pendulum[c]	$P\simeq 2\pi\left[\frac{l}{(2\pm\sqrt{2})g}\right]^{1/2}$	(3.183)		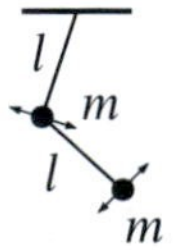

[a] Assuming the bob is supported parallel to a principal rotation axis.
[b] I.e., an arbitrary triaxial rigid body.
[c] For very small oscillations (two eigenmodes).

Tops and gyroscopes

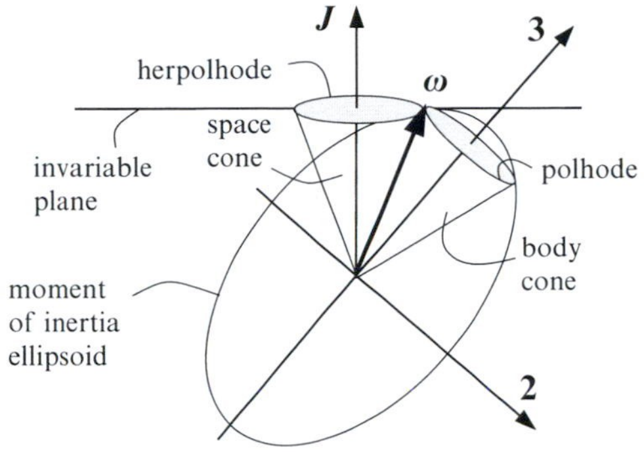

prolate symmetric top

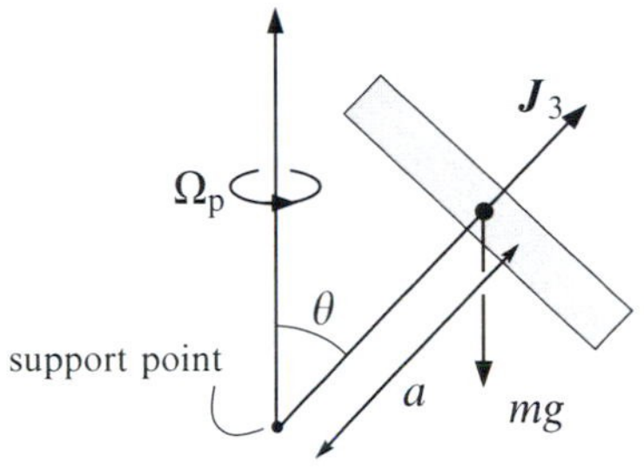

gyroscope

Euler's equations[a]	$G_1 = I_1\dot{\omega}_1 + (I_3 - I_2)\omega_2\omega_3$	(3.184)	G_i external couple (=0 for free rotation) I_i principal moments of inertia ω_i angular velocity of rotation
	$G_2 = I_2\dot{\omega}_2 + (I_1 - I_3)\omega_3\omega_1$	(3.185)	
	$G_3 = I_3\dot{\omega}_3 + (I_2 - I_1)\omega_1\omega_2$	(3.186)	
Free symmetric top[b] ($I_3 < I_2 = I_1$)	$\Omega_b = \frac{I_1 - I_3}{I_1}\omega_3$	(3.187)	Ω_b body frequency Ω_s space frequency J total angular momentum
	$\Omega_s = \frac{J}{I_1}$	(3.188)	
Free asymmetric top[c]	$\Omega_b^2 = \frac{(I_1 - I_3)(I_2 - I_3)}{I_1 I_2}\omega_3^2$	(3.189)	
Steady gyroscopic precession	$\Omega_p^2 I_1' \cos\theta - \Omega_p J_3 + mga = 0$	(3.190)	Ω_p precession angular velocity θ angle from vertical J_3 angular momentum around symmetry axis m mass g gravitational acceleration a distance of centre of mass from support point I_1' moment of inertia about support point
	$\Omega_p \simeq \begin{cases} Mga/J_3 & \text{(slow)} \\ J_3/(I_1'\cos\theta) & \text{(fast)} \end{cases}$	(3.191)	
Gyroscopic stability	$J_3^2 \geq 4I_1' mga\cos\theta$	(3.192)	
Gyroscopic limit ("sleeping top")	$J_3^2 \gg I_1' mga$	(3.193)	
Nutation rate	$\Omega_n = J_3/I_1'$	(3.194)	Ω_n nutation angular velocity
Gyroscope released from rest	$\Omega_p = \frac{mga}{J_3}(1 - \cos\Omega_n t)$	(3.195)	t time

[a]Components are with respect to the principal axes, rotating with the body.
[b]The body frequency is the angular velocity (with respect to principal axes) of **ω** around the **3**-axis. The space frequency is the angular velocity of the **3**-axis around **J**, i.e., the angular velocity at which the body cone moves around the space cone.
[c]**J** close to **3**-axis. If $\Omega_b^2 < 0$, the body tumbles.

3.6 Oscillating systems

Free oscillations

Differential equation	$\frac{d^2x}{dt^2}+2\gamma\frac{dx}{dt}+\omega_0^2x=0$	(3.196)	x t γ ω_0	oscillating variable time damping factor (per unit mass) undamped angular frequency
Underdamped solution ($\gamma<\omega_0$)	$x=Ae^{-\gamma t}\cos(\omega t+\phi)$ where $\omega=(\omega_0^2-\gamma^2)^{1/2}$	(3.197) (3.198)	A ϕ ω	amplitude constant phase constant angular eigenfrequency
Critically damped solution ($\gamma=\omega_0$)	$x=e^{-\gamma t}(A_1+A_2t)$	(3.199)	A_i	amplitude constants
Overdamped solution ($\gamma>\omega_0$)	$x=e^{-\gamma t}(A_1e^{qt}+A_2e^{-qt})$ where $q=(\gamma^2-\omega_0^2)^{1/2}$	(3.200) (3.201)		
Logarithmic decrement[a]	$\Delta=\ln\frac{a_n}{a_{n+1}}=\frac{2\pi\gamma}{\omega}$	(3.202)	Δ a_n	logarithmic decrement nth displacement maximum
Quality factor	$Q=\frac{\omega_0}{2\gamma}\quad\left[\simeq\frac{\pi}{\Delta}\quad\text{if}\quad Q\gg 1\right]$	(3.203)	Q	quality factor

[a]The *decrement* is usually the ratio of successive displacement *maxima* but is sometimes taken as the ratio of successive displacement *extrema*, reducing Δ by a factor of 2. Logarithms are sometimes taken to base 10, introducing a further factor of $\log_{10}e$.

Forced oscillations

Differential equation	$\frac{d^2x}{dt^2}+2\gamma\frac{dx}{dt}+\omega_0^2x=F_0e^{i\omega_f t}$	(3.204)	x t γ	oscillating variable time damping factor (per unit mass)
Steady-state solution[a]	$x=Ae^{i(\omega_f t-\phi)}$, where $A=F_0[(\omega_0^2-\omega_f^2)^2+(2\gamma\omega_f)^2]^{-1/2}$ $\simeq\frac{F_0/(2\omega_0)}{[(\omega_0-\omega_f)^2+\gamma^2]^{1/2}}\quad(\gamma\ll\omega_f)$ $\tan\phi=\frac{2\gamma\omega_f}{\omega_0^2-\omega_f^2}$	(3.205) (3.206) (3.207) (3.208)	ω_0 F_0 ω_f A ϕ	undamped angular frequency force amplitude (per unit mass) forcing angular frequency amplitude phase lag of response behind driving force
Amplitude resonance[b]	$\omega_{ar}^2=\omega_0^2-2\gamma^2$	(3.209)	ω_{ar}	amplitude resonant forcing angular frequency
Velocity resonance[c]	$\omega_{vr}=\omega_0$	(3.210)	ω_{vr}	velocity resonant forcing angular frequency
Quality factor	$Q=\frac{\omega_0}{2\gamma}$	(3.211)	Q	quality factor
Impedance	$Z=2\gamma+i\frac{\omega_f^2-\omega_0^2}{\omega_f}$	(3.212)	Z	impedance (per unit mass)

[a]Excluding the free oscillation terms.
[b]Forcing frequency for maximum displacement.
[c]Forcing frequency for maximum velocity. Note $\phi=\pi/2$ at this frequency.

3.7 Generalised dynamics

Lagrangian dynamics

Action	$S=\int_{t_1}^{t_2} L(\boldsymbol{q},\dot{\boldsymbol{q}},t)\,\mathrm{d}t$	(3.213)	S / $\boldsymbol{q}$ / $\dot{\boldsymbol{q}}$	action ($\delta S=0$ for the motion) / generalised coordinates / generalised velocities
Euler–Lagrange equation	$\frac{\mathrm{d}}{\mathrm{d}t}\left(\frac{\partial L}{\partial \dot{q}_i}\right)-\frac{\partial L}{\partial q_i}=0$	(3.214)	L / t / m	Lagrangian / time / mass
Lagrangian of particle in external field	$L=\frac{1}{2}mv^2-U(\boldsymbol{r},t)$	(3.215)	$\boldsymbol{v}$ / $\boldsymbol{r}$ / U / T	velocity / position vector / potential energy / kinetic energy
	$=T-U$	(3.216)		
Relativistic Lagrangian of a charged particle	$L=-\frac{m_0c^2}{\gamma}-e(\phi-\boldsymbol{A}\cdot\boldsymbol{v})$	(3.217)	m_0 / γ / $+e$ / ϕ / $\boldsymbol{A}$	(rest) mass / Lorentz factor / positive charge / electric potential / magnetic vector potential
Generalised momenta	$p_i=\frac{\partial L}{\partial \dot{q}_i}$	(3.218)	p_i	generalised momenta

Hamiltonian dynamics

Hamiltonian	$H=\sum_i p_i\dot{q}_i-L$	(3.219)	L / p_i / $\dot{q}_i$	Lagrangian / generalised momenta / generalised velocities
Hamilton's equations	$\dot{q}_i=\frac{\partial H}{\partial p_i};\qquad \dot{p}_i=-\frac{\partial H}{\partial q_i}$	(3.220)	H / q_i	Hamiltonian / generalised coordinates
Hamiltonian of particle in external field	$H=\frac{1}{2}mv^2+U(\boldsymbol{r},t)$	(3.221)	v / $\boldsymbol{r}$ / U / T	particle speed / position vector / potential energy / kinetic energy
	$=T+U$	(3.222)		
Relativistic Hamiltonian of a charged particle	$H=(m_0^2c^4+\lvert\boldsymbol{p}-e\boldsymbol{A}\rvert^2c^2)^{1/2}+e\phi$	(3.223)	m_0 / c / $+e$ / ϕ / $\boldsymbol{A}$	(rest) mass / speed of light / positive charge / electric potential / vector potential
Poisson brackets	$[f,g]=\sum_i\left(\frac{\partial f}{\partial q_i}\frac{\partial g}{\partial p_i}-\frac{\partial f}{\partial p_i}\frac{\partial g}{\partial q_i}\right)$	(3.224)	$\boldsymbol{p}$ / t	particle momentum / time
	$[q_i,g]=\frac{\partial g}{\partial p_i},\qquad [p_i,g]=-\frac{\partial g}{\partial q_i}$	(3.225)	f,g / $[\cdot,\cdot]$	arbitrary functions / Poisson bracket (also see *Commutators* on page 26)
	$[H,g]=0\quad\text{if}\quad \frac{\partial g}{\partial t}=0,\quad \frac{\mathrm{d}g}{\mathrm{d}t}=0$	(3.226)		
Hamilton–Jacobi equation	$\frac{\partial S}{\partial t}+H\left(q_i,\frac{\partial S}{\partial q_i},t\right)=0$	(3.227)	S	action

3.8 Elasticity

Elasticity definitions (simple)[a]

Stress	$\tau = F/A$	(3.228)	τ F A	stress applied force cross-sectional area
Strain	$e = \delta l / l$	(3.229)	e δl l	strain change in length length
Young modulus (Hooke's law)	$E = \tau / e = \text{constant}$	(3.230)	E	Young modulus
Poisson ratio[b]	$\sigma = -\dfrac{\delta w / w}{\delta l / l}$	(3.231)	σ δw w	Poisson ratio change in width width

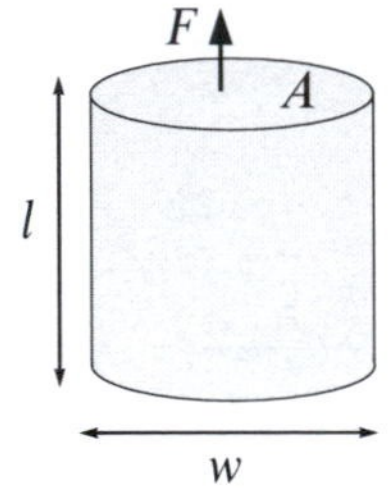

[a]These apply to a thin wire under longitudinal stress.
[b]Solids obeying Hooke's law are restricted by thermodynamics to $-1 \le \sigma \le 1/2$, but none are known with $\sigma < 0$. Non-Hookean materials can show $\sigma > 1/2$.

Elasticity definitions (general)

Stress tensor[a]	$\tau_{ij} = \dfrac{\text{force} \parallel i \text{ direction}}{\text{area} \perp j \text{ direction}}$	(3.232)	τ_{ij}	stress tensor ($\tau_{ij} = \tau_{ji}$)
Strain tensor	$e_{kl} = \dfrac{1}{2}\left(\dfrac{\partial u_k}{\partial x_l} + \dfrac{\partial u_l}{\partial x_k}\right)$	(3.233)	e_{kl} u_k x_k	strain tensor ($e_{kl} = e_{lk}$) displacement $\parallel$ to x_k coordinate system
Elastic modulus	$\tau_{ij} = \lambda_{ijkl} e_{kl}$	(3.234)	λ_{ijkl}	elastic modulus
Elastic energy[b]	$U = \dfrac{1}{2}\lambda_{ijkl} e_{ij} e_{kl}$	(3.235)	U	potential energy
Volume strain (dilatation)	$e_{\mathrm{v}} = \dfrac{\delta V}{V} = e_{11} + e_{22} + e_{33}$	(3.236)	e_{v} δV V	volume strain change in volume volume
Shear strain	$e_{kl} = \underbrace{(e_{kl} - \frac{1}{3} e_{\mathrm{v}} \delta_{kl})}_{\text{pure shear}} + \underbrace{\frac{1}{3} e_{\mathrm{v}} \delta_{kl}}_{\text{dilatation}}$	(3.237)	δ_{kl}	Kronecker delta
Hydrostatic compression	$\tau_{ij} = -p\delta_{ij}$	(3.238)	p	hydrostatic pressure

[a]τ_{ii} are normal stresses, $\tau_{ij}\,(i \neq j)$ are torsional stresses.
[b]As usual, products are implicitly summed over repeated indices.

Isotropic elastic solids

Lamé coefficients	$\mu=\dfrac{E}{2(1+\sigma)}$ (3.239) $\lambda=\dfrac{E\sigma}{(1+\sigma)(1-2\sigma)}$ (3.240)		μ,λ Lamé coefficients E Young modulus σ Poisson ratio
Longitudinal modulus[a]	$M_{\mathrm{l}}=\dfrac{E(1-\sigma)}{(1+\sigma)(1-2\sigma)}=\lambda+2\mu$	(3.241)	M_{l} longitudinal elastic modulus
Diagonalised equations[b]	$e_{ii}=\dfrac{1}{E}[\tau_{ii}-\sigma(\tau_{jj}+\tau_{kk})]$ (3.242) $\tau_{ii}=M_{\mathrm{l}}\left[e_{ii}+\dfrac{\sigma}{1-\sigma}(e_{jj}+e_{kk})\right]$ (3.243) $\mathbf{t}=2\mu\mathbf{e}+\lambda\mathbf{1}\,\mathrm{tr}(\mathbf{e})$ (3.244)		e_{ii} strain in i direction τ_{ii} stress in i direction $\mathbf{e}$ strain tensor $\mathbf{t}$ stress tensor $\mathbf{1}$ unit matrix $\mathrm{tr}(\cdot)$ trace
Bulk modulus (compression modulus)	$K=\dfrac{E}{3(1-2\sigma)}=\lambda+\dfrac{2}{3}\mu$ (3.245) $\dfrac{1}{K_T}=-\dfrac{1}{V}\dfrac{\partial V}{\partial p}\Big\vert_T$ (3.246) $-p=Ke_{\mathrm{v}}$ (3.247)		K bulk modulus K_T isothermal bulk modulus V volume p pressure T temperature
Shear modulus (rigidity modulus)	$\mu=\dfrac{E}{2(1+\sigma)}$ (3.248) $\tau_{\mathrm{T}}=\mu\theta_{\mathrm{sh}}$ (3.249)		e_{v} volume strain μ shear modulus τ_{T} transverse stress θ_{sh} shear strain
Young modulus	$E=\dfrac{9\mu K}{\mu+3K}$	(3.250)	
Poisson ratio	$\sigma=\dfrac{3K-2\mu}{2(3K+\mu)}$	(3.251)	

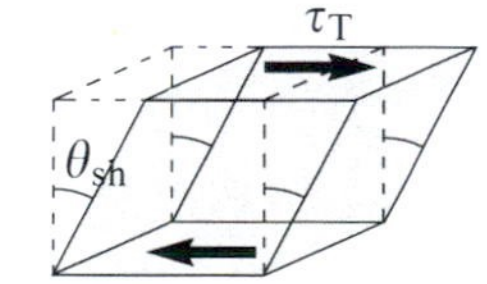

[a]In an extended medium.
[b]Axes aligned along eigenvectors of the stress and strain tensors.

Torsion

Torsional rigidity (for a homogeneous rod)	$G=C\dfrac{\phi}{l}$	(3.252)	G twisting couple C torsional rigidity l rod length ϕ twist angle in length l
Thin circular cylinder	$C=2\pi a^3\mu t$	(3.253)	a radius t wall thickness μ shear modulus
Thick circular cylinder	$C=\dfrac{1}{2}\mu\pi(a_2^4-a_1^4)$	(3.254)	a_1 inner radius a_2 outer radius
Arbitrary thin-walled tube	$C=\dfrac{4A^2\mu t}{P}$	(3.255)	A cross-sectional area P perimeter
Long flat ribbon	$C=\dfrac{1}{3}\mu wt^3$	(3.256)	w cross-sectional width

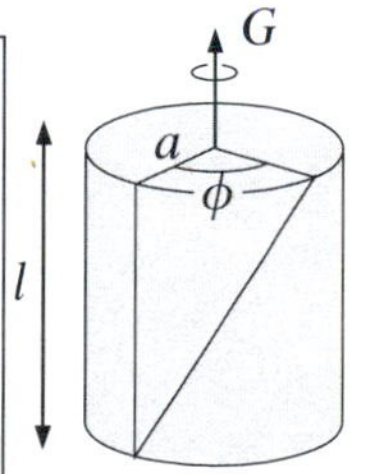

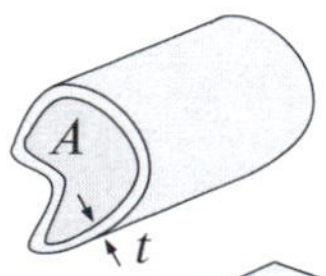

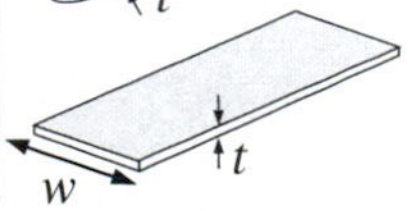

Bending beams[a]

Bending moment	$G_b = \frac{E}{R_c}\int \xi^2\,\mathrm{d}s$ (3.257) $= \frac{EI}{R_c}$ (3.258)	G_b bending moment E Young modulus R_c radius of curvature $\mathrm{d}s$ area element ξ distance to neutral surface from $\mathrm{d}s$ I moment of area	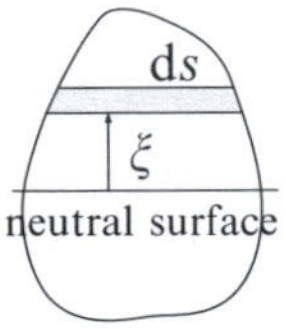 (cross section)
Light beam, horizontal at $x=0$, weight at $x=l$	$y = \frac{W}{2EI}\left(l - \frac{x}{3}\right)x^2$ (3.259)	y displacement from horizontal W end-weight l beam length x distance along beam	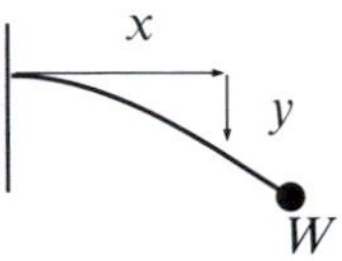
Heavy beam	$EI\frac{\mathrm{d}^4y}{\mathrm{d}x^4} = w(x)$ (3.260)	w beam weight per unit length	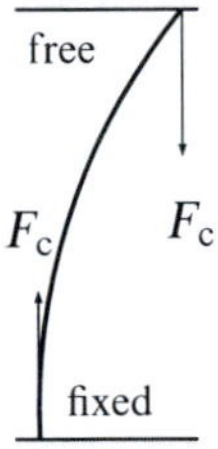
Euler strut failure	$F_c = \begin{cases} \pi^2 EI/l^2 & \text{(free ends)} \\ 4\pi^2 EI/l^2 & \text{(fixed ends)} \\ \pi^2 EI/(4l^2) & \text{(1 free end)} \end{cases}$ (3.261)	F_c critical compression force l strut length	

[a]The radius of curvature is approximated by $1/R_c \simeq \mathrm{d}^2y/\mathrm{d}x^2$.

Elastic wave velocities[a]

In an infinite isotropic solid[b]	$v_t = (\mu/\rho)^{1/2}$ (3.262) $v_l = (M_l/\rho)^{1/2}$ (3.263) $\frac{v_l}{v_t} = \left(\frac{2-2\sigma}{1-2\sigma}\right)^{1/2}$ (3.264)	v_t speed of transverse wave v_l speed of longitudinal wave μ shear modulus ρ density M_l longitudinal modulus $\left(= \frac{E(1-\sigma)}{(1+\sigma)(1-2\sigma)}\right)$
In a fluid	$v_l = (K/\rho)^{1/2}$ (3.265)	K bulk modulus
On a thin plate (wave travelling along x, plate thin in z)	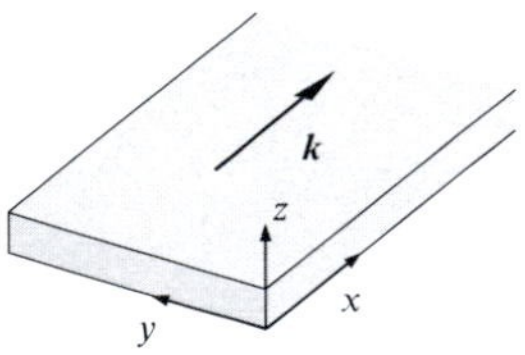 $v_l^{(x)} = \left[\frac{E}{\rho(1-\sigma^2)}\right]^{1/2}$ (3.266) $v_t^{(y)} = (\mu/\rho)^{1/2}$ (3.267) $v_t^{(z)} = k\left[\frac{Et^2}{12\rho(1-\sigma^2)}\right]^{1/2}$ (3.268)	$v_l^{(i)}$ speed of longitudinal wave (displacement $\parallel i$) $v_t^{(i)}$ speed of transverse wave (displacement $\parallel i$) E Young modulus σ Poisson ratio k wavenumber $(=2\pi/\lambda)$ t plate thickness (in z, $t \ll \lambda$)
In a thin circular rod	$v_l = (E/\rho)^{1/2}$ (3.269) $v_\phi = (\mu/\rho)^{1/2}$ (3.270) $v_t = \frac{ka}{2}\left(\frac{E}{\rho}\right)^{1/2}$ (3.271)	v_ϕ torsional wave velocity a rod radius $(\ll \lambda)$

[a]Waves that produce "bending" are generally dispersive. Wave (phase) speeds are quoted throughout.
[b]Transverse waves are also known as shear waves, or S-waves. Longitudinal waves are also known as pressure waves, or P-waves.

Waves in strings and springs[a]

In a spring	$v_l = (\kappa l/\rho_l)^{1/2}$	(3.272)	v_l speed of longitudinal wave κ spring constant[b] l spring length ρ_l mass per unit length[c]
On a stretched string	$v_t = (T/\rho_l)^{1/2}$	(3.273)	v_t speed of transverse wave T tension
On a stretched sheet	$v_t = (\tau/\rho_A)^{1/2}$	(3.274)	τ tension per unit width ρ_A mass per unit area

[a]Wave amplitude assumed $\ll$ wavelength.
[b]In the sense κ = force/extension.
[c]Measured along the axis of the spring.

3

Propagation of elastic waves

Acoustic impedance	$Z = \dfrac{\text{force}}{\text{response velocity}} = \dfrac{F}{\dot{u}}$ $= (E'\rho)^{1/2}$	(3.275) (3.276)	Z impedance F stress force u strain displacement
Wave velocity/impedance relation	if $v = \left(\dfrac{E'}{\rho}\right)^{1/2}$ then $Z = (E'\rho)^{1/2} = \rho v$	(3.277) (3.278)	E' elastic modulus ρ density v wave phase velocity
Mean energy density (nondispersive waves)	$\mathcal{U} = \dfrac{1}{2}E'k^2u_0^2$ $= \dfrac{1}{2}\rho\omega^2u_0^2$ $P = \mathcal{U}v$	(3.279) (3.280) (3.281)	$\mathcal{U}$ energy density k wavenumber ω angular frequency u_0 maximum displacement P mean energy flux
Normal coefficients[a]	$r = \dfrac{u_r}{u_i} = -\dfrac{\tau_r}{\tau_i} = \dfrac{Z_1 - Z_2}{Z_1 + Z_2}$ $t = \dfrac{2Z_1}{Z_1 + Z_2}$	(3.282) (3.283)	r reflection coefficient t transmission coefficient τ stress
Snell's law[b]	$\dfrac{\sin\theta_i}{v_i} = \dfrac{\sin\theta_r}{v_r} = \dfrac{\sin\theta_t}{v_t}$	(3.284)	θ_i angle of incidence θ_r angle of reflection θ_t angle of refraction

[a]For stress and strain amplitudes. Because these reflection and transmission coefficients are usually defined in terms of displacement, u, rather than stress, there are differences between these coefficients and their equivalents defined in electromagnetism [see Equation (7.179) and page 154].
[b]Angles defined from the normal to the interface. An incident plane pressure wave will generally excite both shear and pressure waves in reflection and transmission. Use the velocity appropriate for the wave type.

3.9 Fluid dynamics

Ideal fluids[a]

Continuity[b]	$\frac{\partial \rho}{\partial t}+\nabla\cdot(\rho\boldsymbol{v})=0$	(3.285)	ρ $\boldsymbol{v}$ t	density fluid velocity field time
Kelvin circulation	$\Gamma=\oint \boldsymbol{v}\cdot \mathrm{d}\boldsymbol{l}=\text{constant}$ $=\int_S \boldsymbol{\omega}\cdot \mathrm{d}\boldsymbol{s}$	(3.286) (3.287)	Γ $\mathrm{d}\boldsymbol{l}$ $\mathrm{d}\boldsymbol{s}$ $\boldsymbol{\omega}$	circulation loop element element of surface bounded by loop vorticity ($=\nabla\times\boldsymbol{v}$)
Euler's equation[c]	$\frac{\partial \boldsymbol{v}}{\partial t}+(\boldsymbol{v}\cdot\nabla)\boldsymbol{v}=-\frac{\nabla p}{\rho}+\boldsymbol{g}$ or $\frac{\partial}{\partial t}(\nabla\times\boldsymbol{v})=\nabla\times[\boldsymbol{v}\times(\nabla\times\boldsymbol{v})]$	(3.288) (3.289)	p $\boldsymbol{g}$ $(\boldsymbol{v}\cdot\nabla)$	pressure gravitational field strength advective operator
Bernoulli's equation (incompressible flow)	$\frac{1}{2}\rho v^2+p+\rho g z=\text{constant}$	(3.290)	z	altitude
Bernoulli's equation (compressible adiabatic flow)[d]	$\frac{1}{2}v^2+\frac{\gamma}{\gamma-1}\frac{p}{\rho}+gz=\text{constant}$ $=\frac{1}{2}v^2+c_pT+gz$	(3.291) (3.292)	γ c_p T	ratio of specific heat capacities (c_p/c_V) specific heat capacity at constant pressure temperature
Hydrostatics	$\nabla p=\rho\boldsymbol{g}$	(3.293)		
Adiabatic lapse rate (ideal gas)	$\frac{\mathrm{d}T}{\mathrm{d}z}=-\frac{g}{c_p}$	(3.294)		

[a]No thermal conductivity or viscosity.
[b]True generally.
[c]The second form of Euler's equation applies to incompressible flow only.
[d]Equation (3.292) is true only for an ideal gas.

Potential flow[a]

Velocity potential	$\boldsymbol{v}=\nabla\phi$ $\nabla^2\phi=0$	(3.295) (3.296)	$\boldsymbol{v}$ ϕ	velocity velocity potential
Vorticity condition	$\boldsymbol{\omega}=\nabla\times\boldsymbol{v}=0$	(3.297)	$\boldsymbol{\omega}$ $\boldsymbol{F}$	vorticity drag force on moving sphere
Drag force on a sphere[b]	$\boldsymbol{F}=-\frac{2}{3}\pi\rho a^3\dot{\boldsymbol{u}}=-\frac{1}{2}M_{\mathrm{d}}\dot{\boldsymbol{u}}$	(3.298)	a $\dot{\boldsymbol{u}}$ ρ M_{d}	sphere radius sphere acceleration fluid density displaced fluid mass

[a]For incompressible fluids.
[b]The effect of this drag force is to give the sphere an additional effective mass equal to half the mass of fluid displaced.

Viscous flow (incompressible)[a]

Fluid stress	$\tau_{ij} = -p\delta_{ij} + \eta\left(\frac{\partial v_i}{\partial x_j} + \frac{\partial v_j}{\partial x_i}\right)$	(3.299)	τ_{ij} p η v_i δ_{ij}	fluid stress tensor hydrostatic pressure shear viscosity velocity along i axis Kronecker delta
Navier–Stokes equation[b]	$\frac{\partial \boldsymbol{v}}{\partial t} + (\boldsymbol{v}\cdot\nabla)\boldsymbol{v} = -\frac{\nabla p}{\rho} - \frac{\eta}{\rho}\nabla\times\boldsymbol{\omega} + \boldsymbol{g}$	(3.300)	$\boldsymbol{v}$ $\boldsymbol{\omega}$ $\boldsymbol{g}$ ρ	fluid velocity field vorticity gravitational acceleration density
	$= -\frac{\nabla p}{\rho} + \frac{\eta}{\rho}\nabla^2\boldsymbol{v} + \boldsymbol{g}$	(3.301)		
Kinematic viscosity	$\nu = \eta/\rho$	(3.302)	ν	kinematic viscosity

[a] I.e., $\nabla\cdot\boldsymbol{v} = 0$, $\eta \neq 0$.
[b] Neglecting bulk (second) viscosity.

Laminar viscous flow

Between parallel plates	$v_z(y) = \frac{1}{2\eta}y(h-y)\frac{\partial p}{\partial z}$	(3.303)	v_z z y η p	flow velocity direction of flow distance from plate shear viscosity pressure
Along a circular pipe[a]	$v_z(r) = \frac{1}{4\eta}(a^2 - r^2)\frac{\partial p}{\partial z}$	(3.304)	r	distance from pipe axis
	$Q = \frac{\mathrm{d}V}{\mathrm{d}t} = \frac{\pi a^4}{8\eta}\frac{\partial p}{\partial z}$	(3.305)	a V	pipe radius volume
Circulating between concentric rotating cylinders[b]	$G_z = \frac{4\pi\eta a_1^2 a_2^2}{a_2^2 - a_1^2}(\omega_2 - \omega_1)$	(3.306)	G_z ω_i	axial couple between cylinders per unit length angular velocity of ith cylinder
Along an annular pipe	$Q = \frac{\pi}{8\eta}\frac{\partial p}{\partial z}\left[a_2^4 - a_1^4 - \frac{(a_2^2 - a_1^2)^2}{\ln(a_2/a_1)}\right]$	(3.307)	a_1 a_2 Q	inner radius outer radius volume discharge rate

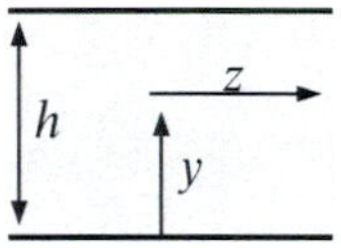

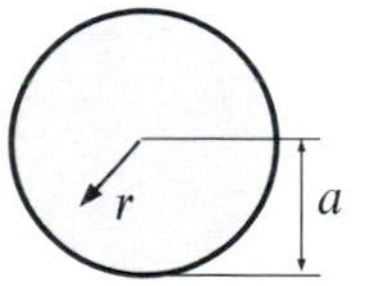

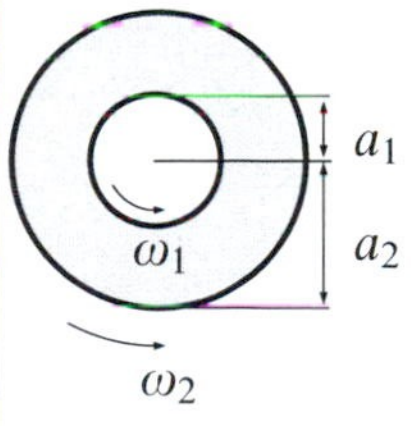

[a] Poiseuille flow.
[b] Couette flow.

Drag[a]

On a sphere (Stokes's law)	$F = 6\pi a\eta v$	(3.308)	F a	drag force radius
On a disk, broadside to flow	$F = 16a\eta v$	(3.309)	v η	velocity shear viscosity
On a disk, edge on to flow	$F = 32a\eta v/3$	(3.310)		

[a] For Reynolds numbers $\ll 1$.

Characteristic numbers

Reynolds number	$\mathsf{Re}=\frac{\rho UL}{\eta}=\frac{\text{inertial force}}{\text{viscous force}}$	(3.311)	Re ρ U L η	Reynolds number density characteristic velocity characteristic scale-length shear viscosity
Froude number[a]	$\mathsf{F}=\frac{U^2}{Lg}=\frac{\text{inertial force}}{\text{gravitational force}}$	(3.312)	F g	Froude number gravitational acceleration
Strouhal number[b]	$\mathsf{S}=\frac{U\tau}{L}=\frac{\text{evolution scale}}{\text{physical scale}}$	(3.313)	S τ	Strouhal number characteristic timescale
Prandtl number	$\mathsf{P}=\frac{\eta c_p}{\lambda}=\frac{\text{momentum transport}}{\text{heat transport}}$	(3.314)	P c_p λ	Prandtl number Specific heat capacity at constant pressure thermal conductivity
Mach number	$\mathsf{M}=\frac{U}{c}=\frac{\text{speed}}{\text{sound speed}}$	(3.315)	M c	Mach number sound speed
Rossby number	$\mathsf{Ro}=\frac{U}{\Omega L}=\frac{\text{inertial force}}{\text{Coriolis force}}$	(3.316)	Ro Ω	Rossby number angular velocity

[a]Sometimes the square root of this expression. L is usually the fluid depth.
[b]Sometimes the reciprocal of this expression.

Fluid waves

Sound waves	$v_{\mathrm{p}}=\left(\frac{K}{\rho}\right)^{1/2}=\left(\frac{\mathrm{d}p}{\mathrm{d}\rho}\right)^{1/2}$	(3.317)	v_{p} K p ρ	wave (phase) speed bulk modulus pressure density
In an ideal gas (adiabatic conditions)[a]	$v_{\mathrm{p}}=\left(\frac{\gamma RT}{\mu}\right)^{1/2}=\left(\frac{\gamma p}{\rho}\right)^{1/2}$	(3.318)	γ R T μ	ratio of heat capacities molar gas constant (absolute) temperature mean molecular mass
Gravity waves on a liquid surface[b]	$\omega^2=gk\tanh kh$	(3.319)	v_{g} h λ k g ω	group speed of wave liquid depth wavelength wavenumber gravitational acceleration angular frequency
	$v_{\mathrm{g}}\simeq\begin{cases}\frac{1}{2}\left(\frac{g}{k}\right)^{1/2} & (h\gg\lambda)\\ (gh)^{1/2} & (h\ll\lambda)\end{cases}$	(3.320)		
Capillary waves (ripples)[c]	$\omega^2=\frac{\sigma k^3}{\rho}$	(3.321)	σ	surface tension
Capillary–gravity waves ($h\gg\lambda$)	$\omega^2=gk+\frac{\sigma k^3}{\rho}$	(3.322)		

[a]If the waves are isothermal rather than adiabatic then $v_{\mathrm{p}}=(p/\rho)^{1/2}$.
[b]Amplitude $\ll$ wavelength.
[c]In the limit $k^2\gg g\rho/\sigma$.

Doppler effect[a]

Source at rest, observer moving at u	$\frac{\nu'}{\nu} = 1 - \frac{\lvert \boldsymbol{u} \rvert}{v_{\rm p}} \cos\theta$	(3.323)	ν', ν'' observed frequency ν emitted frequency $v_{\rm p}$ wave (phase) speed in fluid
Observer at rest, source moving at u	$\frac{\nu''}{\nu} = \frac{1}{1 - \frac{\lvert \boldsymbol{u} \rvert}{v_{\rm p}} \cos\theta}$	(3.324)	$\boldsymbol{u}$ velocity θ angle between wavevector, $\boldsymbol{k}$, and $\boldsymbol{u}$

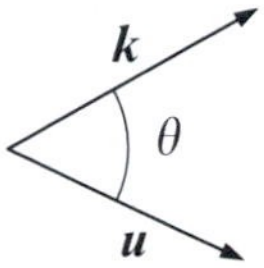

[a]For plane waves in a stationary fluid.

Wave speeds

Phase speed	$v_{\rm p} = \frac{\omega}{k} = \nu\lambda$	(3.325)	$v_{\rm p}$ phase speed ν frequency ω angular frequency ($=2\pi\nu$) λ wavelength k wavenumber ($=2\pi/\lambda$)
Group speed	$v_{\rm g} = \frac{{\rm d}\omega}{{\rm d}k}$	(3.326)	$v_{\rm g}$ group speed
	$= v_{\rm p} - \lambda \frac{{\rm d}v_{\rm p}}{{\rm d}\lambda}$	(3.327)	

Shocks

Mach wedge[a]	$\sin\theta_{\rm w} = \frac{v_{\rm p}}{v_{\rm b}}$	(3.328)	$\theta_{\rm w}$ wedge semi-angle $v_{\rm p}$ wave (phase) speed $v_{\rm b}$ body speed
Kelvin wedge[b]	$\lambda_{\rm K} = \frac{4\pi v_{\rm b}^2}{3g}$	(3.329)	$\lambda_{\rm K}$ characteristic wavelength g gravitational acceleration
	$\theta_{\rm w} = \arcsin(1/3) = 19^\circ.5$	(3.330)	
Spherical adiabatic shock[c]	$r \simeq \left(\frac{Et^2}{\rho_0}\right)^{1/5}$	(3.331)	r shock radius E energy release t time ρ_0 density of undisturbed medium
Rankine–Hugoniot shock relations[d]	$\frac{p_2}{p_1} = \frac{2\gamma {\rm M}_1^2 - (\gamma - 1)}{\gamma + 1}$	(3.332)	1 upstream values 2 downstream values p pressure
	$\frac{v_1}{v_2} = \frac{\rho_2}{\rho_1} = \frac{\gamma + 1}{(\gamma - 1) + 2/{\rm M}_1^2}$	(3.333)	v velocity T temperature ρ density
	$\frac{T_2}{T_1} = \frac{[2\gamma {\rm M}_1^2 - (\gamma - 1)][2 + (\gamma - 1){\rm M}_1^2]}{(\gamma + 1)^2 {\rm M}_1^2}$	(3.334)	γ ratio of specific heats M Mach number

[a]Approximating the wake generated by supersonic motion of a body in a nondispersive medium.
[b]For gravity waves, e.g., in the wake of a boat. Note that the wedge semi-angle is independent of $v_{\rm b}$.
[c]Sedov–Taylor relation.
[d]Solutions for a steady, normal shock, in the frame moving with the shock front. If $\gamma = 5/3$ then $v_1/v_2 \leq 4$.

Surface tension

Definition	$\sigma_{\mathrm{lv}}=\dfrac{\text{surface energy}}{\text{area}}$	(3.335)
	$=\dfrac{\text{surface tension}}{\text{length}}$	(3.336)
Laplace's formula[a]	$\Delta p=\sigma_{\mathrm{lv}}\left(\dfrac{1}{R_1}+\dfrac{1}{R_2}\right)$	(3.337)
Capillary constant	$c_{\mathrm{c}}=\left(\dfrac{2\sigma_{\mathrm{lv}}}{g\rho}\right)^{1/2}$	(3.338)
Capillary rise (circular tube)	$h=\dfrac{2\sigma_{\mathrm{lv}}\cos\theta}{\rho g a}$	(3.339)
Contact angle	$\cos\theta=\dfrac{\sigma_{\mathrm{wv}}-\sigma_{\mathrm{wl}}}{\sigma_{\mathrm{lv}}}$	(3.340)

σ_{lv}	surface tension (liquid/vapour interface)
Δp	pressure difference over surface
R_i	principal radii of curvature
c_{c}	capillary constant
ρ	liquid density
g	gravitational acceleration
h	rise height
θ	contact angle
a	tube radius
σ_{wv}	wall/vapour surface tension
σ_{wl}	wall/liquid surface tension

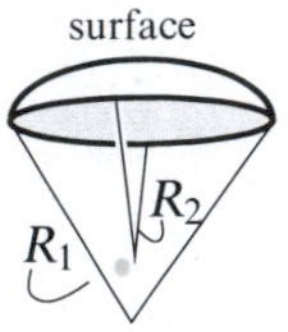

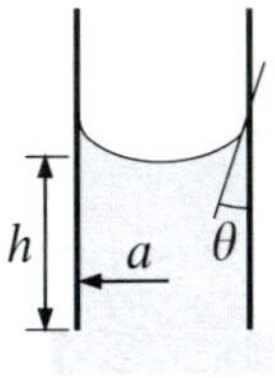

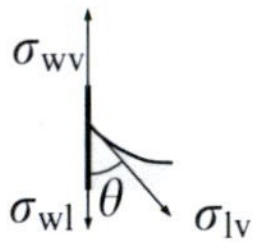

[a] For a spherical bubble in a liquid $\Delta p=2\sigma_{\mathrm{lv}}/R$. For a soap bubble (two surfaces) $\Delta p=4\sigma_{\mathrm{lv}}/R$.

Chapter 4 Quantum physics

4.1 Introduction

Quantum ideas occupy such a pivotal position in physics that different notations and algebras appropriate to each field have been developed. In the spirit of this book, only those formulas that are commonly present in undergraduate courses and that can be simply presented in tabular form are included here. For example, much of the detail of atomic spectroscopy and of specific perturbation analyses has been omitted, as have ideas from the somewhat specialised field of quantum electrodynamics. Traditionally, quantum physics is understood through standard "toy" problems, such as the potential step and the one-dimensional harmonic oscillator, and these are reproduced here. Operators are distinguished from observables using the "hat" notation, so that the momentum observable, p_x, has the operator $\hat{p_x} = -\mathrm{i}\hbar\partial/\partial x$.

For clarity, many relations that can be generalised to three dimensions in an obvious way have been stated in their one-dimensional form, and wavefunctions are implicitly taken as normalised functions of space and time unless otherwise stated. With the exception of the last panel, all equations should be taken as nonrelativistic, so that "total energy" is the sum of potential and kinetic energies, excluding the rest mass energy.

4

4.2 Quantum definitions

Quantum uncertainty relations

De Broglie relation	$p = \frac{h}{\lambda}$	(4.1)	$p, \boldsymbol{p}$	particle momentum
			h	Planck constant
	$\boldsymbol{p} = \hbar \boldsymbol{k}$	(4.2)	$\hbar$	$h/(2\pi)$
			λ	de Broglie wavelength
			$\boldsymbol{k}$	de Broglie wavevector
Planck–Einstein relation	$E = h\nu = \hbar\omega$	(4.3)	E	energy
			ν	frequency
			ω	angular frequency ($=2\pi\nu$)
Dispersion[a]	$(\Delta a)^2 = \langle (a - \langle a \rangle)^2 \rangle$	(4.4)	a,b	observables[b]
	$= \langle a^2 \rangle - \langle a \rangle^2$	(4.5)	$\langle \cdot \rangle$	expectation value
			$(\Delta a)^2$	dispersion of a
General uncertainty relation	$(\Delta a)^2 (\Delta b)^2 \geq \frac{1}{4} \langle \mathrm{i}[\hat{a}, \hat{b}] \rangle^2$	(4.6)	$\hat{a}$	operator for observable a
			$[\cdot,\cdot]$	commutator (see page 26)
Momentum–position uncertainty relation[c]	$\Delta p \Delta x \geq \frac{\hbar}{2}$	(4.7)	x	particle position
Energy–time uncertainty relation	$\Delta E \Delta t \geq \frac{\hbar}{2}$	(4.8)	t	time
Number–phase uncertainty relation	$\Delta n \Delta \phi \geq \frac{1}{2}$	(4.9)	n	number of photons
			ϕ	wave phase

[a]Dispersion in quantum physics corresponds to variance in statistics.
[b]An observable is a directly measurable parameter of a system.
[c]Also known as the "Heisenberg uncertainty relation."

Wavefunctions

Probability density	$\mathrm{pr}(x,t)\,\mathrm{d}x = \lvert \psi(x,t) \rvert^2\,\mathrm{d}x$	(4.10)	pr	probability density
			ψ	wavefunction
Probability density current[a]	$j(x) = \frac{\hbar}{2\mathrm{i}m}\left(\psi^* \frac{\partial \psi}{\partial x} - \psi \frac{\partial \psi^*}{\partial x}\right)$	(4.11)	$j, \boldsymbol{j}$	probability density current
			$\hbar$	(Planck constant)/(2π)
	$\boldsymbol{j} = \frac{\hbar}{2\mathrm{i}m}\left[\psi^*(\boldsymbol{r})\nabla\psi(\boldsymbol{r}) - \psi(\boldsymbol{r})\nabla\psi^*(\boldsymbol{r})\right]$	(4.12)	x	position coordinate
			$\hat{\boldsymbol{p}}$	momentum operator
	$= \frac{1}{m}\Re(\psi^* \hat{\boldsymbol{p}} \psi)$	(4.13)	m	particle mass
			$\Re$	real part of
			t	time
Continuity equation	$\nabla \cdot \boldsymbol{j} = -\frac{\partial}{\partial t}(\psi\psi^*)$	(4.14)		
Schrödinger equation	$\hat{H}\psi = \mathrm{i}\hbar \frac{\partial \psi}{\partial t}$	(4.15)	H	Hamiltonian
Particle stationary states[b]	$-\frac{\hbar^2}{2m}\frac{\partial^2 \psi(x)}{\partial x^2} + V(x)\psi(x) = E\psi(x)$	(4.16)	V	potential energy
			E	total energy

[a]For particles. In three dimensions, suitable units would be particles $\mathrm{m}^{-2}\mathrm{s}^{-1}$.
[b]Time-independent Schrödinger equation for a particle, in one dimension.

Operators

Hermitian conjugate operator	$\int (\hat{a}\phi)^* \psi \, \mathrm{d}x = \int \phi^* \hat{a} \psi \, \mathrm{d}x$ (4.17)		$\hat{a}$ Hermitian conjugate operator ψ, ϕ normalisable functions
Position operator	$\hat{x^n} = x^n$ (4.18)		* complex conjugate x, y position coordinates
Momentum operator	$\hat{p_x^n} = \frac{\hbar^n}{\mathrm{i}^n} \frac{\partial^n}{\partial x^n}$ (4.19)		n arbitrary integer ≥ 1 p_x momentum coordinate
Kinetic energy operator	$\hat{T} = -\frac{\hbar^2}{2m} \frac{\partial^2}{\partial x^2}$ (4.20)		T kinetic energy $\hbar$ (Planck constant)/(2π) m particle mass
Hamiltonian operator	$\hat{H} = -\frac{\hbar^2}{2m} \frac{\partial^2}{\partial x^2} + V(x)$ (4.21)		H Hamiltonian V potential energy
Angular momentum operators	$\hat{L}_z = \hat{x}\hat{p}_y - \hat{y}\hat{p}_x$ (4.22) $\hat{L}^2 = \hat{L}_x^2 + \hat{L}_y^2 + \hat{L}_z^2$ (4.23)		L_z angular momentum along z axis (sim. x and y) L total angular momentum
Parity operator	$\hat{P}\psi(\boldsymbol{r}) = \psi(-\boldsymbol{r})$ (4.24)		$\hat{P}$ parity operator $\boldsymbol{r}$ position vector

Expectation value

Expectation value[a]	$\langle a \rangle = \langle \hat{a} \rangle = \int \Psi^* \hat{a} \Psi \, \mathrm{d}x$ (4.25) $= \langle \Psi \| \hat{a} \| \Psi \rangle$ (4.26)		$\langle a \rangle$ expectation value of a $\hat{a}$ operator for a Ψ (spatial) wavefunction x (spatial) coordinate
Time dependence	$\frac{\mathrm{d}}{\mathrm{d}t}\langle \hat{a} \rangle = \frac{\mathrm{i}}{\hbar}\langle [\hat{H}, \hat{a}] \rangle + \left\langle \frac{\partial \hat{a}}{\partial t} \right\rangle$ (4.27)		t time $\hbar$ (Planck constant)/(2π)
Relation to eigenfunctions	if $\hat{a}\psi_n = a_n \psi_n$ and $\Psi = \sum c_n \psi_n$ then $\langle a \rangle = \sum \|c_n\|^2 a_n$ (4.28)		ψ_n eigenfunctions of $\hat{a}$ a_n eigenvalues n dummy index c_n probability amplitudes
Ehrenfest's theorem	$m \frac{\mathrm{d}}{\mathrm{d}t}\langle \boldsymbol{r} \rangle = \langle \boldsymbol{p} \rangle$ (4.29) $\frac{\mathrm{d}}{\mathrm{d}t}\langle \boldsymbol{p} \rangle = -\langle \nabla V \rangle$ (4.30)		m particle mass $\boldsymbol{r}$ position vector $\boldsymbol{p}$ momentum V potential energy

[a]Equation (4.26) uses the Dirac "bra-ket" notation for integrals involving operators. The presence of vertical bars distinguishes this use of angled brackets from that on the left-hand side of the equations. Note that $\langle a \rangle$ and $\langle \hat{a} \rangle$ are taken as equivalent.

Dirac notation

Matrix element[a]	$a_{nm} = \int \psi_n^* \hat{a} \psi_m \, dx$ $= \langle n\|\hat{a}\|m\rangle$	(4.31) (4.32)	n,m a_{nm} ψ_n $\hat{a}$ x	eigenvector indices matrix element basis states operator spatial coordinate
Bra vector	bra state vector $= \langle n\|$	(4.33)	$\langle \cdot \|$	bra
Ket vector	ket state vector $= \|m\rangle$	(4.34)	$\| \cdot \rangle$	ket
Scalar product	$\langle n\|m\rangle = \int \psi_n^* \psi_m \, dx$	(4.35)		
Expectation	if $\Psi = \sum_n c_n \psi_n$ then $\langle a \rangle = \sum_m \sum_n c_n^* c_m a_{nm}$	(4.36) (4.37)	Ψ c_n	wavefunction probability amplitudes

[a]The Dirac bracket, $\langle n|\hat{a}|m\rangle$, can also be written $\langle \psi_n|\hat{a}|\psi_m\rangle$.

4.3 Wave mechanics

Potential step[a]

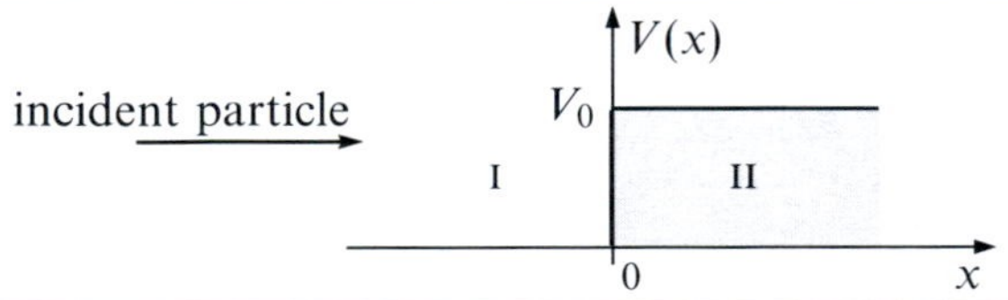

Potential function	$V(x) = \begin{cases} 0 & (x<0) \\ V_0 & (x \geq 0) \end{cases}$	(4.38)	V V_0 $\hbar$	particle potential energy step height (Planck constant)/(2π)
Wavenumbers	$\hbar^2 k^2 = 2mE \quad (x<0)$ $\hbar^2 q^2 = 2m(E - V_0) \quad (x>0)$	(4.39) (4.40)	k,q m E	particle wavenumbers particle mass total particle energy
Amplitude reflection coefficient	$r = \dfrac{k-q}{k+q}$	(4.41)	r	amplitude reflection coefficient
Amplitude transmission coefficient	$t = \dfrac{2k}{k+q}$	(4.42)	t	amplitude transmission coefficient
Probability currents[b]	$j_{\mathrm{I}} = \dfrac{\hbar k}{m}(1 - \|r\|^2)$ $j_{\mathrm{II}} = \dfrac{\hbar q}{m}\|t\|^2$	(4.43) (4.44)	j_{I} j_{II}	particle flux in zone I particle flux in zone II

[a]One-dimensional interaction with an incident particle of total energy $E = \mathrm{KE} + V$. If $E < V_0$ then q is imaginary and $|r|^2 = 1$. $1/|q|$ is then a measure of the tunnelling depth.
[b]Particle flux with the sign of increasing x.

Potential well[a]

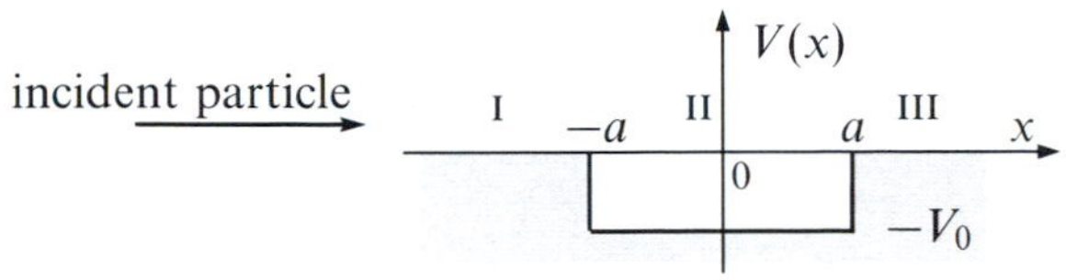

Potential function	$V(x)=\begin{cases}0 & (\lvert x\rvert > a)\\ -V_0 & (\lvert x\rvert \le a)\end{cases}$ (4.45)		V particle potential energy V_0 well depth $\hbar$ (Planck constant)/(2π) $2a$ well width
Wavenumbers	$\hbar^2k^2=2mE \quad (\lvert x\rvert > a)$ (4.46) $\hbar^2q^2=2m(E+V_0) \quad (\lvert x\rvert < a)$ (4.47)		k,q particle wavenumbers m particle mass E total particle energy
Amplitude reflection coefficient	$r=\frac{\mathrm{i}e^{-2ika}(q^2-k^2)\sin 2qa}{2kq\cos 2qa-\mathrm{i}(q^2+k^2)\sin 2qa}$ (4.48)		r amplitude reflection coefficient
Amplitude transmission coefficient	$t=\frac{2kqe^{-2ika}}{2kq\cos 2qa-\mathrm{i}(q^2+k^2)\sin 2qa}$ (4.49)		t amplitude transmission coefficient
Probability currents[b]	$j_{\mathrm{I}}=\frac{\hbar k}{m}(1-\lvert r\rvert^2)$ (4.50) $j_{\mathrm{III}}=\frac{\hbar k}{m}\lvert t\rvert^2$ (4.51)		j_{I} particle flux in zone I j_{III} particle flux in zone III
Ramsauer effect[c]	$E_n=-V_0+\frac{n^2\hbar^2\pi^2}{8ma^2}$ (4.52)		n integer >0 E_n Ramsauer energy
Bound states $(V_0<E<0)$[d]	$\tan qa=\begin{cases}\lvert k\rvert/q & \text{even parity}\\ -q/\lvert k\rvert & \text{odd parity}\end{cases}$ (4.53) $q^2-\lvert k\rvert^2=2mV_0/\hbar^2$ (4.54)		

[a]One-dimensional interaction with an incident particle of total energy $E=\mathrm{KE}+V>0$.
[b]Particle flux in the sense of increasing x.
[c]Incident energy for which $2qa=n\pi$, $\lvert r\rvert=0$, and $\lvert t\rvert=1$.
[d]When $E<0$, k is purely imaginary. $\lvert k\rvert$ and q are obtained by solving these implicit equations.

Barrier tunnelling[a]

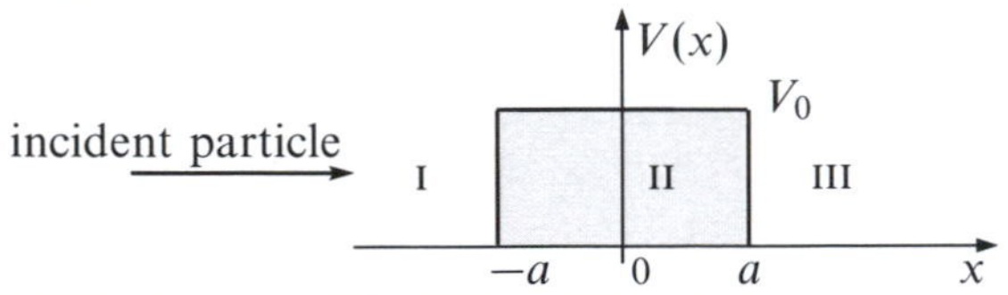

Potential function	$V(x)=\begin{cases}0 & (\lvert x\rvert>a)\\ V_0 & (\lvert x\rvert\le a)\end{cases}$ (4.55)	V V_0 $\hbar$ $2a$	particle potential energy well depth (Planck constant)/(2π) barrier width
Wavenumber and tunnelling constant	$\hbar^2k^2=2mE \quad (\lvert x\rvert>a)$ (4.56) $\hbar^2\kappa^2=2m(V_0-E) \quad (\lvert x\rvert<a)$ (4.57)	k κ m E	incident wavenumber tunnelling constant particle mass total energy ($<V_0$)
Amplitude reflection coefficient	$r=\frac{-\mathrm{i}e^{-2\mathrm{i}ka}(k^2+\kappa^2)\sinh 2\kappa a}{2k\kappa\cosh 2\kappa a-\mathrm{i}(k^2-\kappa^2)\sinh 2\kappa a}$ (4.58)	r	amplitude reflection coefficient
Amplitude transmission coefficient	$t=\frac{2k\kappa e^{-2\mathrm{i}ka}}{2k\kappa\cosh 2\kappa a-\mathrm{i}(k^2-\kappa^2)\sinh 2\kappa a}$ (4.59)	t	amplitude transmission coefficient
Tunnelling probability	$\lvert t\rvert^2=\frac{4k^2\kappa^2}{(k^2+\kappa^2)^2\sinh^2 2\kappa a+4k^2\kappa^2}$ (4.60) $\simeq\frac{16k^2\kappa^2}{(k^2+\kappa^2)^2}\exp(-4\kappa a) \quad (\lvert t\rvert^2\ll 1)$ (4.61)	$\lvert t\rvert^2$	tunnelling probability
Probability currents[b]	$j_{\mathrm{I}}=\frac{\hbar k}{m}(1-\lvert r\rvert^2)$ (4.62) $j_{\mathrm{III}}=\frac{\hbar k}{m}\lvert t\rvert^2$ (4.63)	j_{I} j_{III}	particle flux in zone I particle flux in zone III

[a]By a particle of total energy $E=\mathrm{KE}+V$, through a one-dimensional rectangular potential barrier height $V_0>E$.
[b]Particle flux in the sense of increasing x.

Particle in a rectangular box[a]

Eigen-functions	$\Psi_{lmn}=\left(\frac{8}{abc}\right)^{1/2}\sin\frac{l\pi x}{a}\sin\frac{m\pi y}{b}\sin\frac{n\pi z}{c}$ (4.64)	Ψ_{lmn} a,b,c l,m,n	eigenfunctions box dimensions integers ≥ 1
Energy levels	$E_{lmn}=\frac{h^2}{8M}\left(\frac{l^2}{a^2}+\frac{m^2}{b^2}+\frac{n^2}{c^2}\right)$ (4.65)	E_{lmn} h M	energy Planck constant particle mass
Density of states	$\rho(E)\,\mathrm{d}E=\frac{4\pi}{h^3}(2M^3E)^{1/2}\,\mathrm{d}E$ (4.66)	$\rho(E)$	density of states (per unit volume)

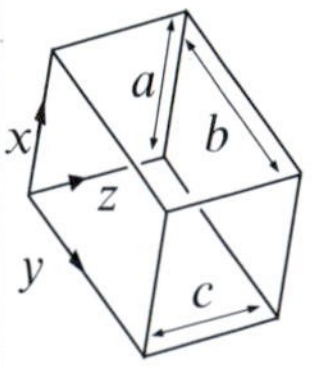

[a]Spinless particle in a rectangular box bounded by the planes $x=0$, $y=0$, $z=0$, $x=a$, $y=b$, and $z=c$. The potential is zero inside and infinite outside the box.

Harmonic oscillator

Schrödinger equation	$$-\frac{\hbar^2}{2m}\frac{\partial^2\psi_n}{\partial x^2}+\frac{1}{2}m\omega^2x^2\psi_n=E_n\psi_n$$	(4.67)	$\hbar$ (Planck constant)/(2π) m mass ψ_n nth eigenfunction x displacement
Energy levels[a]	$$E_n=\left(n+\frac{1}{2}\right)\hbar\omega$$	(4.68)	n integer ≥ 0 ω angular frequency E_n total energy in nth state
Eigen-functions	$$\psi_n=\frac{H_n(x/a)\exp[-x^2/(2a^2)]}{(n!2^n a\pi^{1/2})^{1/2}}$$ where $a=\left(\frac{\hbar}{m\omega}\right)^{1/2}$	(4.69)	H_n Hermite polynomials
Hermite polynomials	$H_0(y)=1,\quad H_1(y)=2y,\quad H_2(y)=4y^2-2$ $H_{n+1}(y)=2yH_n(y)-2nH_{n-1}(y)$	(4.70)	y dummy variable

[a] E_0 is the zero-point energy of the oscillator.

4.4 Hydrogenic atoms

Bohr model[a]

Quantisation condition	$$\mu r_n^2\Omega=n\hbar$$	(4.71)	r_n nth orbit radius Ω orbital angular speed n principal quantum number (>0)
Bohr radius	$$a_0=\frac{\epsilon_0 h^2}{\pi m_e e^2}=\frac{\alpha}{4\pi R_\infty}\simeq 52.9\,\text{pm}$$	(4.72)	a_0 Bohr radius μ reduced mass ($\simeq m_e$) $-e$ electronic charge
Orbit radius	$$r_n=\frac{n^2}{Z}a_0\frac{m_e}{\mu}$$	(4.73)	Z atomic number h Planck constant $\hbar$ $h/(2\pi)$
Total energy	$$E_n=-\frac{\mu e^4 Z^2}{8\epsilon_0^2 h^2 n^2}=-R_\infty hc\frac{\mu}{m_e}\frac{Z^2}{n^2}$$	(4.74)	E_n total energy of nth orbit ϵ_0 permittivity of free space m_e electron mass
Fine structure constant	$$\alpha=\frac{\mu_0 ce^2}{2h}=\frac{e^2}{4\pi\epsilon_0\hbar c}\simeq\frac{1}{137}$$	(4.75)	α fine structure constant μ_0 permeability of free space
Hartree energy	$$E_H=\frac{\hbar^2}{m_e a_0^2}\simeq 4.36\times 10^{-18}\,\text{J}$$	(4.76)	E_H Hartree energy
Rydberg constant	$$R_\infty=\frac{m_e c\alpha^2}{2h}=\frac{m_e e^4}{8h^3\epsilon_0^2 c}=\frac{E_H}{2hc}$$	(4.77)	R_∞ Rydberg constant c speed of light
Rydberg's formula[b]	$$\frac{1}{\lambda_{mn}}=R_\infty\frac{\mu}{m_e}Z^2\left(\frac{1}{n^2}-\frac{1}{m^2}\right)$$	(4.78)	λ_{mn} photon wavelength m integer $>n$

[a] Because the Bohr model is strictly a two-body problem, the equations use reduced mass, $\mu=m_e m_{nuc}/(m_e+m_{nuc})\simeq m_e$, where m_{nuc} is the nuclear mass, throughout. The orbit radius is therefore the electron–nucleus distance.
[b] Wavelength of the spectral line corresponding to electron transitions between orbits m and n.

Hydrogenlike atoms – Schrödinger solution[a]

Schrödinger equation

$$-\frac{\hbar^2}{2\mu}\nabla^2\Psi_{nlm}-\frac{Ze^2}{4\pi\epsilon_0 r}\Psi_{nlm}=E_n\Psi_{nlm} \qquad \text{with} \quad \mu=\frac{m_e m_{nuc}}{m_e+m_{nuc}} \tag{4.79}$$

Eigenfunctions

$$\Psi_{nlm}(r,\theta,\phi)=\left[\frac{(n-l-1)!}{2n(n+l)!}\right]^{1/2}\left(\frac{2}{an}\right)^{3/2}x^l e^{-x/2}L_{n-l-1}^{2l+1}(x)Y_l^m(\theta,\phi) \tag{4.80}$$

$$\text{with} \quad a=\frac{m_e}{\mu}\frac{a_0}{Z}, \qquad x=\frac{2r}{an}, \qquad \text{and} \quad L_{n-l-1}^{2l+1}(x)=\sum_{k=0}^{n-l-1}\frac{(l+n)!(-x)^k}{(2l+1+k)!(n-l-1-k)!k!}$$

Total energy

$$E_n=-\frac{\mu e^4 Z^2}{8\epsilon_0^2 h^2 n^2} \tag{4.81}$$

Radial expectation values

$$\langle r\rangle=\frac{a}{2}[3n^2-l(l+1)] \tag{4.82}$$

$$\langle r^2\rangle=\frac{a^2n^2}{2}[5n^2+1-3l(l+1)] \tag{4.83}$$

$$\langle 1/r\rangle=\frac{1}{an^2} \tag{4.84}$$

$$\langle 1/r^2\rangle=\frac{2}{(2l+1)n^3a^2} \tag{4.85}$$

Allowed quantum numbers and selection rules[b]

$$n=1,2,3,\ldots \tag{4.86}$$

$$l=0,1,2,\ldots,(n-1) \tag{4.87}$$

$$m=0,\pm1,\pm2,\ldots,\pm l \tag{4.88}$$

$$\Delta n\neq 0 \tag{4.89}$$

$$\Delta l=\pm1 \tag{4.90}$$

$$\Delta m=0 \quad \text{or} \quad \pm1 \tag{4.91}$$

E_n	total energy
ϵ_0	permittivity of free space
h	Planck constant
m_e	mass of electron
$\hbar$	$h/2\pi$
μ	reduced mass ($\simeq m_e$)
m_{nuc}	mass of nucleus
Ψ_{nlm}	eigenfunctions
Ze	charge of nucleus
$-e$	electronic charge
L_p^q	associated Laguerre polynomials[c]
a	classical orbit radius, $n=1$
r	electron–nucleus separation
Y_l^m	spherical harmonics
a_0	Bohr radius $=\frac{\epsilon_0 h^2}{\pi m_e e^2}$

$$\Psi_{100}=\frac{a^{-3/2}}{\pi^{1/2}}e^{-r/a} \qquad \Psi_{200}=\frac{a^{-3/2}}{4(2\pi)^{1/2}}\left(2-\frac{r}{a}\right)e^{-r/2a}$$

$$\Psi_{210}=\frac{a^{-3/2}}{4(2\pi)^{1/2}}\frac{r}{a}e^{-r/2a}\cos\theta \qquad \Psi_{21\pm1}=\mp\frac{a^{-3/2}}{8\pi^{1/2}}\frac{r}{a}e^{-r/2a}\sin\theta\, e^{\pm i\phi}$$

$$\Psi_{300}=\frac{a^{-3/2}}{81(3\pi)^{1/2}}\left(27-18\frac{r}{a}+2\frac{r^2}{a^2}\right)e^{-r/3a} \qquad \Psi_{310}=\frac{2^{1/2}a^{-3/2}}{81\pi^{1/2}}\left(6-\frac{r}{a}\right)\frac{r}{a}e^{-r/3a}\cos\theta$$

$$\Psi_{31\pm1}=\mp\frac{a^{-3/2}}{81\pi^{1/2}}\left(6-\frac{r}{a}\right)\frac{r}{a}e^{-r/3a}\sin\theta\, e^{\pm i\phi} \qquad \Psi_{320}=\frac{a^{-3/2}}{81(6\pi)^{1/2}}\frac{r^2}{a^2}e^{-r/3a}(3\cos^2\theta-1)$$

$$\Psi_{32\pm1}=\mp\frac{a^{-3/2}}{81\pi^{1/2}}\frac{r^2}{a^2}e^{-r/3a}\sin\theta\cos\theta\, e^{\pm i\phi} \qquad \Psi_{32\pm2}=\frac{a^{-3/2}}{162\pi^{1/2}}\frac{r^2}{a^2}e^{-r/3a}\sin^2\theta\, e^{\pm 2i\phi}$$

[a] For a single bound electron in a perfect nuclear Coulomb potential (nonrelativistic and spin-free).
[b] For dipole transitions between orbitals.
[c] The sign and indexing definitions for this function vary. This form is appropriate to Equation (4.80).

Orbital angular dependence

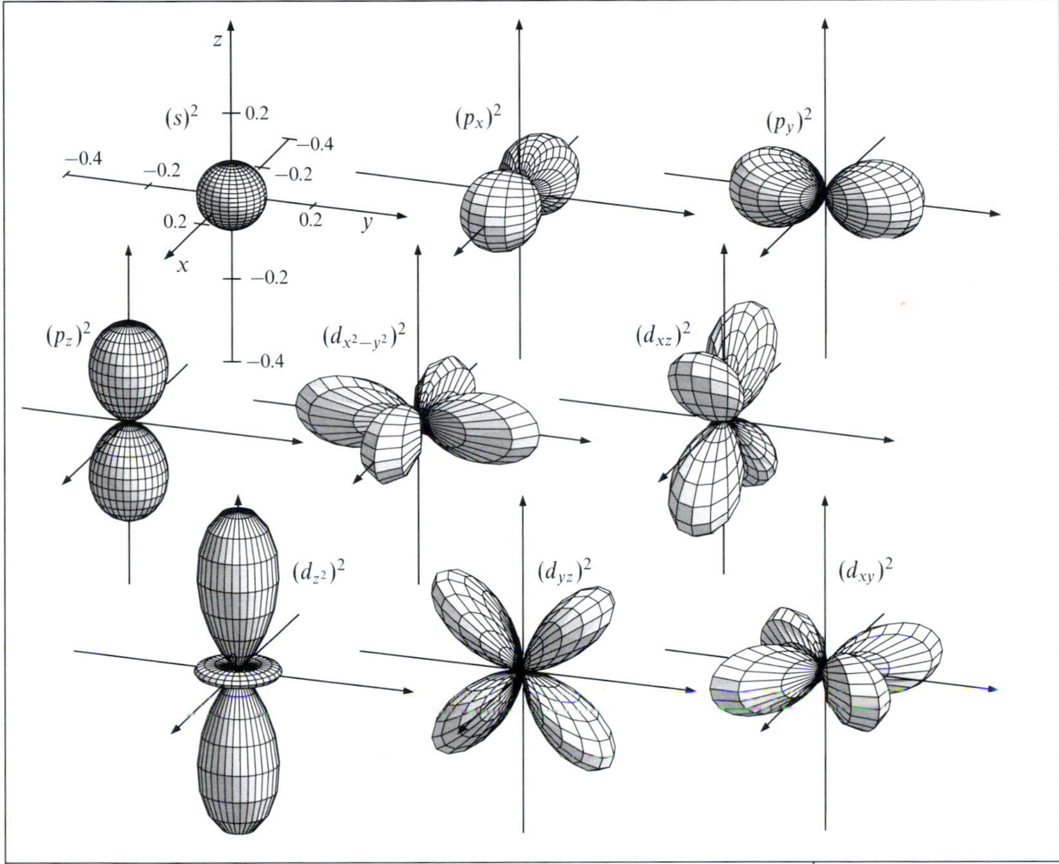

s orbital ($l=0$)	$s = Y_0^0 = \text{constant}$	(4.92)	Y_l^m spherical harmonics[a]
p orbitals ($l=1$)	$p_x = \frac{-1}{2^{1/2}}(Y_1^1 - Y_1^{-1}) \propto \cos\phi \sin\theta$	(4.93)	θ,ϕ spherical polar coordinates
	$p_y = \frac{\mathbf{i}}{2^{1/2}}(Y_1^1 + Y_1^{-1}) \propto \sin\phi \sin\theta$	(4.94)	
	$p_z = Y_1^0 \propto \cos\theta$	(4.95)	
d orbitals ($l=2$)	$d_{x^2-y^2} = \frac{1}{2^{1/2}}(Y_2^2 + Y_2^{-2}) \propto \sin^2\theta \cos 2\phi$	(4.96)	
	$d_{xz} = \frac{-1}{2^{1/2}}(Y_2^1 - Y_2^{-1}) \propto \sin\theta \cos\theta \cos\phi$	(4.97)	
	$d_{z^2} = Y_2^0 \propto (3\cos^2\theta - 1)$	(4.98)	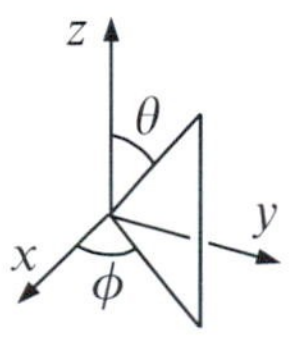
	$d_{yz} = \frac{\mathbf{i}}{2^{1/2}}(Y_2^1 + Y_2^{-1}) \propto \sin\theta \cos\theta \sin\phi$	(4.99)	
	$d_{xy} = \frac{-\mathbf{i}}{2^{1/2}}(Y_2^2 - Y_2^{-2}) \propto \sin^2\theta \sin 2\phi$	(4.100)	

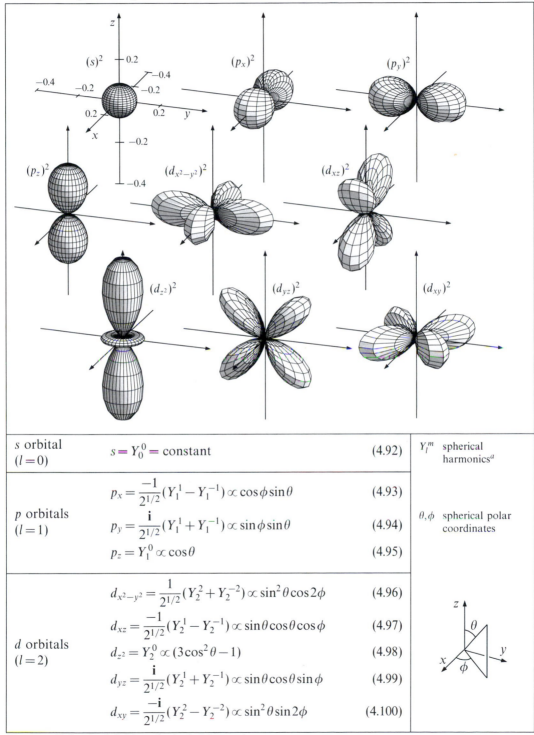

[a]See page 49 for the definition of spherical harmonics.

4.5 Angular momentum

Orbital angular momentum

Angular momentum operators	$\hat{\boldsymbol{L}}=\boldsymbol{r}\times\hat{\boldsymbol{p}}$	(4.101)	L angular momentum
	$\hat{L}_z=\frac{\hbar}{\mathrm{i}}\left(x\frac{\partial}{\partial y}-y\frac{\partial}{\partial x}\right)$	(4.102)	p linear momentum
	$=\frac{\hbar}{\mathrm{i}}\frac{\partial}{\partial\phi}$	(4.103)	r position vector
	$\hat{L}^2=\hat{L}_x^{\,2}+\hat{L}_y^{\,2}+\hat{L}_z^{\,2}$	(4.104)	xyz Cartesian coordinates
	$=-\hbar^2\left[\frac{1}{\sin\theta}\frac{\partial}{\partial\theta}\left(\sin\theta\frac{\partial}{\partial\theta}\right)+\frac{1}{\sin^2\theta}\frac{\partial^2}{\partial\phi^2}\right]$	(4.105)	$r\theta\phi$ spherical polar coordinates; $\hbar$ (Planck constant)/(2π)
Ladder operators	$\hat{L}_\pm=\hat{L}_x\pm\mathrm{i}\hat{L}_y$	(4.106)	$\hat{L}_\pm$ ladder operators
	$=\hbar e^{\pm\mathrm{i}\phi}\left(\mathrm{i}\cot\theta\frac{\partial}{\partial\phi}\pm\frac{\partial}{\partial\theta}\right)$	(4.107)	$Y_l^{m_l}$ spherical harmonics
	$\hat{L}_\pm Y_l^{m_l}=\hbar[l(l+1)-m_l(m_l\pm1)]^{1/2}Y_l^{m_l\pm1}$	(4.108)	l,m_l integers
Eigenfunctions and eigenvalues	$\hat{L}^2Y_l^{m_l}=l(l+1)\hbar^2Y_l^{m_l}\qquad(l\geq0)$	(4.109)	
	$\hat{L}_zY_l^{m_l}=m_l\hbar Y_l^{m_l}\qquad(\lvert m_l\rvert\leq l)$	(4.110)	
	$\hat{L}_z[\hat{L}_\pm Y_l^{m_l}(\theta,\phi)]=(m_l\pm1)\hbar\hat{L}_\pm Y_l^{m_l}(\theta,\phi)$	(4.111)	
	l-multiplicity $=(2l+1)$	(4.112)	

Angular momentum commutation relations[a]

Conservation of angular momentum[b]	$[\hat{H},\hat{L}_z]=0$	(4.113)	L angular momentum; p momentum; H Hamiltonian; $\hat{L}_\pm$ ladder operators

$[\hat{L}_z,x]=\mathrm{i}\hbar y$	(4.114)	$[\hat{L}_x,\hat{L}_y]=\mathrm{i}\hbar\hat{L}_z$	(4.120)
$[\hat{L}_z,y]=-\mathrm{i}\hbar x$	(4.115)	$[\hat{L}_z,\hat{L}_x]=\mathrm{i}\hbar\hat{L}_y$	(4.121)
$[\hat{L}_z,z]=0$	(4.116)	$[\hat{L}_y,\hat{L}_z]=\mathrm{i}\hbar\hat{L}_x$	(4.122)
$[\hat{L}_z,\hat{p}_x]=\mathrm{i}\hbar\hat{p}_y$	(4.117)	$[\hat{L}_+,\hat{L}_z]=-\hbar\hat{L}_+$	(4.123)
$[\hat{L}_z,\hat{p}_y]=-\mathrm{i}\hbar\hat{p}_x$	(4.118)	$[\hat{L}_-,\hat{L}_z]=\hbar\hat{L}_-$	(4.124)
$[\hat{L}_z,\hat{p}_z]=0$	(4.119)	$[\hat{L}_+,\hat{L}_-]=2\hbar\hat{L}_z$	(4.125)
		$[\hat{L}^2,\hat{L}_\pm]=0$	(4.126)

$$[\hat{L}^2,\hat{L}_x]=[\hat{L}^2,\hat{L}_y]=[\hat{L}^2,\hat{L}_z]=0 \qquad (4.127)$$

[a]The commutation of a and b is defined as $[a,b]=ab-ba$ (see page 26). Similar expressions hold for S and J.
[b]For motion under a central force.

Clebsch–Gordan coefficients[a]

$$\langle j,-m_j|l_1,-m_1;l_2,-m_2\rangle=(-1)^{l_1+l_2-j}\langle j,m_j|l_1,m_1;l_2,m_2\rangle$$

Key: $l_1 \times l_2$; header row gives m_j and the values of j (j, j, …); each row gives m_1, m_2 and the coefficients $\langle j,m_j|l_1,m_1;l_2,m_2\rangle$.

m_1	m_2	j	j	…
m_1	m_2	coefficients		
m_1	m_2	$\langle j,m_j\|l_1,m_1;l_2,m_2\rangle$		
⋮	⋮	⋮		

$1/2 \times 1/2$

$m_j=+1$

m_1	m_2	1
+1/2	+1/2	1

$m_j=0$

m_1	m_2	1	0
+1/2	−1/2	1/2	1/2
−1/2	+1/2	1/2	−1/2

$1 \times 1/2$

$m_j=+3/2$

m_1	m_2	3/2
+1	+1/2	1

$m_j=+1/2$

m_1	m_2	3/2	1/2
+1	−1/2	1/3	2/3
0	+1/2	2/3	−1/3

$3/2 \times 1/2$

$m_j=+2$

m_1	m_2	2
+3/2	+1/2	1

$m_j=+1$

m_1	m_2	2	1
+3/2	−1/2	1/4	3/4
+1/2	+1/2	3/4	−1/4

$m_j=0$

m_1	m_2	2	1
+1/2	−1/2	1/2	1/2
−1/2	+1/2	1/2	−1/2

$2 \times 1/2$

$m_j=+5/2$

m_1	m_2	5/2
+2	+1/2	1

$m_j=+3/2$

m_1	m_2	5/2	3/2
+2	−1/2	1/5	4/5
+1	+1/2	4/5	−1/5

$m_j=+1/2$

m_1	m_2	5/2	3/2
+1	−1/2	2/5	3/5
0	+1/2	3/5	−2/5

1×1

$m_j=+2$

m_1	m_2	2
+1	+1	1

$m_j=+1$

m_1	m_2	2	1
+1	0	1/2	1/2
0	+1	1/2	−1/2

$m_j=0$

m_1	m_2	2	1	0
+1	−1	1/6	1/2	1/3
0	0	2/3	0	−1/3
−1	+1	1/6	−1/2	1/3

$3/2 \times 1$

$m_j=+5/2$

m_1	m_2	5/2
+3/2	+1	1

$m_j=+3/2$

m_1	m_2	5/2	3/2
+3/2	0	2/5	3/5
+1/2	+1	3/5	−2/5

$m_j=+1/2$

m_1	m_2	5/2	3/2	1/2
+3/2	−1	1/10	2/5	1/2
1/2	0	3/5	1/15	−1/3
−1/2	+1	3/10	−8/15	1/6

2×1

$m_j=+3$

m_1	m_2	3
+2	+1	1

$m_j=+2$

m_1	m_2	3	2
+2	0	1/3	2/3
+1	+1	2/3	−1/3

$m_j=+1$

m_1	m_2	3	2	1
+2	−1	1/15	1/3	3/5
+1	0	8/15	1/6	−3/10
0	+1	6/15	−1/2	1/10

$m_j=0$

m_1	m_2	3	2	1
+1	−1	1/5	1/2	3/10
0	0	3/5	0	−2/5
−1	+1	1/5	−1/2	3/10

$3/2 \times 3/2$

$m_j=+3$

m_1	m_2	3
+3/2	+3/2	1

$m_j=+2$

m_1	m_2	3	2
+3/2	+1/2	1/2	1/2
+1/2	+3/2	1/2	−1/2

$m_j=+1$

m_1	m_2	3	2	1
+3/2	−1/2	1/5	1/2	3/10
+1/2	+1/2	3/5	0	−2/5
−1/2	+3/2	1/5	−1/2	3/10

$m_j=0$

m_1	m_2	3	2	1	0
+3/2	−3/2	1/20	1/4	9/20	1/4
+1/2	−1/2	9/20	1/4	−1/20	−1/4
−1/2	+1/2	9/20	−1/4	−1/20	1/4
−3/2	+3/2	1/20	−1/4	9/20	−1/4

$2 \times 3/2$

$m_j=+7/2$

m_1	m_2	7/2
+2	+3/2	1

$m_j=+5/2$

m_1	m_2	7/2	5/2
+2	+1/2	3/7	4/7
+1	+3/2	4/7	−3/7

$m_j=+3/2$

m_1	m_2	7/2	5/2	3/2
+2	−1/2	1/7	16/35	2/5
+1	+1/2	4/7	1/35	−2/5
0	+3/2	2/7	−18/35	1/5

$m_j=+1/2$

m_1	m_2	7/2	5/2	3/2	1/2
+2	−3/2	1/35	6/35	2/5	2/5
+1	−1/2	12/35	5/14	0	−3/10
0	+1/2	18/35	−3/35	−1/5	1/5
−1	+3/2	4/35	−27/70	2/5	−1/10

2×2

$m_j=+4$

m_1	m_2	4
+2	+2	1

$m_j=+3$

m_1	m_2	4	3
+2	+1	1/2	1/2
+1	+2	1/2	−1/2

$m_j=+2$

m_1	m_2	4	3	2
+2	0	3/14	1/2	2/7
+1	+1	4/7	0	−3/7
0	+2	3/14	−1/2	2/7

$m_j=+1$

m_1	m_2	4	3	2	1
+2	−1	1/14	3/10	3/7	1/5
+1	0	3/7	1/5	−1/14	−3/10
0	+1	3/7	−1/5	−1/14	3/10
−1	+2	1/14	−3/10	3/7	−1/5

$m_j=0$

m_1	m_2	4	3	2	1	0
+2	−2	1/70	1/10	2/7	2/5	1/5
+1	−1	8/35	2/5	1/14	−1/10	−1/5
0	0	18/35	0	−2/7	0	1/5
−1	+1	8/35	−2/5	1/14	1/10	−1/5
−2	+2	1/70	−1/10	2/7	−2/5	1/5

[a]Or "Wigner coefficients," using the Condon–Shortley sign convention. Note that a square root is assumed over all coefficient digits, so that "−3/10" corresponds to $-\sqrt{3/10}$. Also for clarity, only values of $m_j \geq 0$ are listed here. The coefficients for $m_j<0$ can be obtained from the symmetry relation $\langle j,-m_j|l_1,-m_1;l_2,-m_2\rangle=(-1)^{l_1+l_2-j}\langle j,m_j|l_1,m_1;l_2,m_2\rangle$.

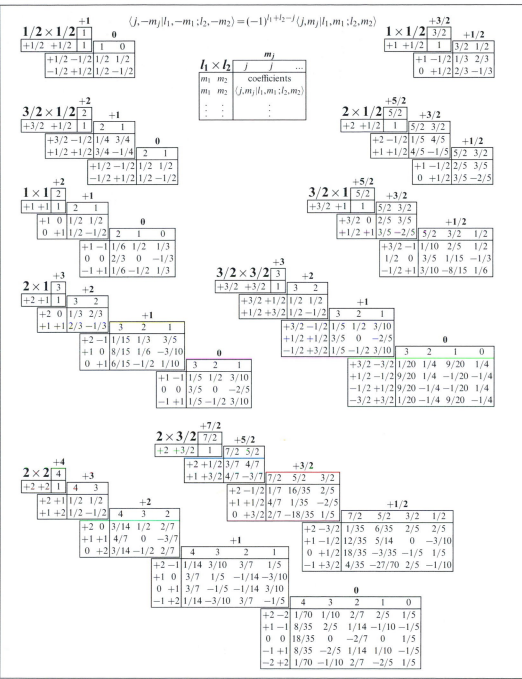

Angular momentum addition[a]

Total angular momentum	$\boldsymbol{J} = \boldsymbol{L} + \boldsymbol{S}$	(4.128)	$\boldsymbol{J}, J$ total angular momentum
	$\hat{J}_z = \hat{L}_z + \hat{S}_z$	(4.129)	$\boldsymbol{L}, L$ orbital angular momentum
	$\hat{J}^2 = \hat{L}^2 + \hat{S}^2 + 2\widehat{\boldsymbol{L}\cdot\boldsymbol{S}}$	(4.130)	$\boldsymbol{S}, S$ spin angular momentum
	$\hat{J}_z \psi_{j,m_j} = m_j \hbar \psi_{j,m_j}$	(4.131)	ψ eigenfunctions
	$\hat{J}^2 \psi_{j,m_j} = j(j+1)\hbar^2 \psi_{j,m_j}$	(4.132)	m_j magnetic quantum number $\lvert m_j \rvert \leq j$
	j-multiplicity $= (2l+1)(2s+1)$	(4.133)	j $(l+s) \geq j \geq \lvert l-s \rvert$
Mutually commuting sets	$\{L^2, S^2, J^2, J_z, \boldsymbol{L}\cdot\boldsymbol{S}\}$	(4.134)	$\{\}$ set of mutually commuting observables
	$\{L^2, S^2, L_z, S_z, J_z\}$	(4.135)	
Clebsch–Gordan coefficients[b]	$\lvert j, m_j \rangle = \sum_{\substack{m_l, m_s \\ m_s + m_l = m_j}} \langle j, m_j \lvert l, m_l; s, m_s \rangle \lvert l, m_l \rangle \lvert s, m_s \rangle$	(4.136)	$\lvert \cdot \rangle$ eigenstates; $\langle \cdot \lvert \cdot \rangle$ Clebsch–Gordan coefficients

[a]Summing spin and orbital angular momenta as examples, eigenstates $\lvert s, m_s \rangle$ and $\lvert l, m_l \rangle$.
[b]Or "Wigner coefficients." Assuming no L–S interaction.

Magnetic moments

Bohr magneton	$\mu_B = \frac{e\hbar}{2m_e}$	(4.137)	μ_B Bohr magneton; $-e$ electronic charge; $\hbar$ (Planck constant)/(2π); m_e electron mass
Gyromagnetic ratio[a]	$\gamma = \frac{\text{orbital magnetic moment}}{\text{orbital angular momentum}}$	(4.138)	γ gyromagnetic ratio
Electron orbital gyromagnetic ratio	$\gamma_e = \frac{-\mu_B}{\hbar}$	(4.139)	γ_e electron gyromagnetic ratio
	$= \frac{-e}{2m_e}$	(4.140)	
Spin magnetic moment of an electron[b]	$\mu_{e,z} = -g_e \mu_B m_s$	(4.141)	$\mu_{e,z}$ z component of spin magnetic moment
	$= \pm g_e \gamma_e \frac{\hbar}{2}$	(4.142)	g_e electron g-factor ($\simeq 2.002$)
	$= \pm \frac{g_e e\hbar}{4m_e}$	(4.143)	m_s spin quantum number ($\pm 1/2$)
Landé g-factor[c]	$\mu_J = g_J \sqrt{J(J+1)}\mu_B$	(4.144)	μ_J total magnetic moment
	$\mu_{J,z} = -g_J \mu_B m_J$	(4.145)	$\mu_{J,z}$ z component of μ_J; m_J magnetic quantum number
	$g_J = 1 + \frac{J(J+1) + S(S+1) - L(L+1)}{2J(J+1)}$	(4.146)	J, L, S total, orbital, and spin quantum numbers; g_J Landé g-factor

[a]Or "magnetogyric ratio."
[b]The electron g-factor equals exactly 2 in Dirac theory. The modification $g_e = 2 + \alpha/\pi + \ldots$, where α is the fine structure constant, comes from quantum electrodynamics.
[c]Relating the spin + orbital angular momenta of an electron to its total magnetic moment, assuming $g_e = 2$.

Quantum paramagnetism

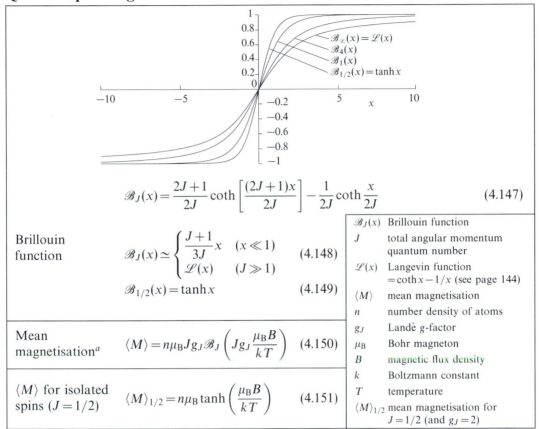

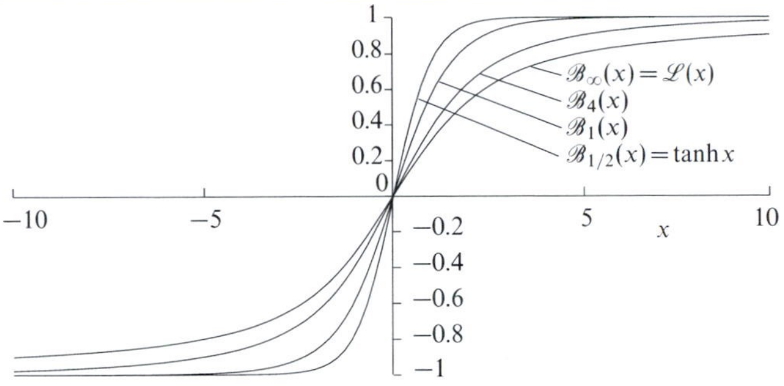

$$\mathscr{B}_J(x) = \frac{2J+1}{2J} \coth\left[\frac{(2J+1)x}{2J}\right] - \frac{1}{2J}\coth\frac{x}{2J} \qquad (4.147)$$

Brillouin function

$$\mathscr{B}_J(x) \simeq \begin{cases} \dfrac{J+1}{3J}x & (x \ll 1) \\ \mathscr{L}(x) & (J \gg 1) \end{cases} \qquad (4.148)$$

$$\mathscr{B}_{1/2}(x) = \tanh x \qquad (4.149)$$

Mean magnetisation[a]

$$\langle M \rangle = n\mu_B J g_J \mathscr{B}_J\left(J g_J \frac{\mu_B B}{kT}\right) \qquad (4.150)$$

$\langle M \rangle$ for isolated spins ($J = 1/2$)

$$\langle M \rangle_{1/2} = n\mu_B \tanh\left(\frac{\mu_B B}{kT}\right) \qquad (4.151)$$

$\mathscr{B}_J(x)$	Brillouin function
J	total angular momentum quantum number
$\mathscr{L}(x)$	Langevin function $= \coth x - 1/x$ (see page 144)
$\langle M \rangle$	mean magnetisation
n	number density of atoms
g_J	Landé g-factor
μ_B	Bohr magneton
B	magnetic flux density
k	Boltzmann constant
T	temperature
$\langle M \rangle_{1/2}$	mean magnetisation for $J = 1/2$ (and $g_J = 2$)

[a] Of an ensemble of atoms in thermal equilibrium at temperature T, each with total angular momentum quantum number J.

4.6 Perturbation theory

Time-independent perturbation theory

Unperturbed states	$\hat{H}_0\psi_n = E_n\psi_n$ (ψ_n nondegenerate)	(4.152)	$\hat{H}_0$ ψ_n E_n n	unperturbed Hamiltonian eigenfunctions of $\hat{H}_0$ eigenvalues of $\hat{H}_0$ integer ≥ 0
Perturbed Hamiltonian	$\hat{H} = \hat{H}_0 + \hat{H}'$	(4.153)	$\hat{H}$ $\hat{H}'$	perturbed Hamiltonian perturbation ($\ll \hat{H}_0$)
Perturbed eigenvalues[a]	$E'_k = E_k + \langle\psi_k\|\hat{H}'\|\psi_k\rangle + \sum_{n\neq k}\frac{\|\langle\psi_k\|\hat{H}'\|\psi_n\rangle\|^2}{E_k - E_n} + \ldots$	(4.154)	E'_k $\langle\|\|\rangle$	perturbed eigenvalue ($\simeq E_k$) Dirac bracket
Perturbed eigen-functions[b]	$\psi'_k = \psi_k + \sum_{n\neq k}\frac{\langle\psi_k\|\hat{H}'\|\psi_n\rangle}{E_k - E_n}\psi_n + \ldots$	(4.155)	ψ'_k	perturbed eigenfunction ($\simeq \psi_k$)

[a]To second order.
[b]To first order.

Time-dependent perturbation theory

Unperturbed stationary states	$\hat{H}_0\psi_n = E_n\psi_n$	(4.156)	$\hat{H}_0$ ψ_n E_n n	unperturbed Hamiltonian eigenfunctions of $\hat{H}_0$ eigenvalues of $\hat{H}_0$ integer ≥ 0
Perturbed Hamiltonian	$\hat{H}(t) = \hat{H}_0 + \hat{H}'(t)$	(4.157)	$\hat{H}$ $\hat{H}'(t)$ t	perturbed Hamiltonian perturbation ($\ll \hat{H}_0$) time
Schrödinger equation	$[\hat{H}_0 + \hat{H}'(t)]\Psi(t) = \mathrm{i}\hbar\frac{\partial\Psi(t)}{\partial t}$ $\Psi(t=0) = \psi_0$	(4.158) (4.159)	Ψ ψ_0 $\hbar$	wavefunction initial state (Planck constant)/(2π)
Perturbed wave-function[a]	$\Psi(t) = \sum_n c_n(t)\psi_n \exp(-\mathrm{i}E_n t/\hbar)$ where	(4.160)	c_n	probability amplitudes
	$c_n = \frac{-\mathrm{i}}{\hbar}\int_0^t \langle\psi_n\|\hat{H}'(t')\|\psi_0\rangle \exp[\mathrm{i}(E_n - E_0)t'/\hbar]\,\mathrm{d}t'$	(4.161)		
Fermi's golden rule	$\Gamma_{i\to f} = \frac{2\pi}{\hbar}\|\langle\psi_f\|\hat{H}'\|\psi_i\rangle\|^2\rho(E_f)$	(4.162)	$\Gamma_{i\to f}$ $\rho(E_f)$	transition probability per unit time from state i to state f density of final states

[a]To first order.

4.7 High energy and nuclear physics

Nuclear decay

Nuclear decay law	$N(t)=N(0)\mathrm{e}^{-\lambda t}$	(4.163)	$N(t)$ number of nuclei remaining after time t t time
Half-life and mean life	$T_{1/2}=\dfrac{\ln 2}{\lambda}$	(4.164)	λ decay constant $T_{1/2}$ half-life
	$\langle T\rangle = 1/\lambda$	(4.165)	$\langle T\rangle$ mean lifetime
Successive decays $1\rightarrow 2\rightarrow 3$ (species 3 stable)	$N_1(t)=N_1(0)\mathrm{e}^{-\lambda_1 t}$	(4.166)	N_1 population of species 1 N_2 population of species 2 N_3 population of species 3
	$N_2(t)=N_2(0)\mathrm{e}^{-\lambda_2 t}+\dfrac{N_1(0)\lambda_1(\mathrm{e}^{-\lambda_1 t}-\mathrm{e}^{-\lambda_2 t})}{\lambda_2-\lambda_1}$	(4.167)	λ_1 decay constant $1\rightarrow 2$
	$N_3(t)=N_3(0)+N_2(0)(1-\mathrm{e}^{-\lambda_2 t})+N_1(0)\left(1+\dfrac{\lambda_1\mathrm{e}^{-\lambda_2 t}-\lambda_2\mathrm{e}^{-\lambda_1 t}}{\lambda_2-\lambda_1}\right)$	(4.168)	λ_2 decay constant $2\rightarrow 3$
Geiger's law[a]	$v^3=a(R-x)$	(4.169)	v velocity of α particle x distance from source a constant
Geiger–Nuttall rule	$\log\lambda=b+c\log R$	(4.170)	R range b,c constants for each series α, β, and γ

[a]For α particles in air (empirical).

Nuclear binding energy

Liquid drop model[a]	$$B=a_\mathrm{v}A-a_\mathrm{s}A^{2/3}-a_\mathrm{c}\frac{Z^2}{A^{1/3}}-a_\mathrm{a}\frac{(N-Z)^2}{A}+\delta(A)$$	(4.171)	N number of neutrons A mass number $(=N+Z)$ B semi-empirical binding energy Z number of protons a_v volume term ($\sim 15.8\,\mathrm{MeV}$)
	$$\delta(A)\simeq\begin{cases}+a_\mathrm{p}A^{-3/4} & Z,N \text{ both even}\\ -a_\mathrm{p}A^{-3/4} & Z,N \text{ both odd}\\ 0 & \text{otherwise}\end{cases}$$	(4.172)	a_s surface term ($\sim 18.0\,\mathrm{MeV}$) a_c Coulomb term ($\sim 0.72\,\mathrm{MeV}$) a_a asymmetry term ($\sim 23.5\,\mathrm{MeV}$) a_p pairing term ($\sim 33.5\,\mathrm{MeV}$)
Semi-empirical mass formula	$M(Z,A)=ZM_\mathrm{H}+Nm_\mathrm{n}-B$	(4.173)	$M(Z,A)$ atomic mass M_H mass of hydrogen atom m_n neutron mass

[a]Coefficient values are empirical and approximate.

Nuclear collisions

Breit–Wigner formula[a]	$\sigma(E)=\frac{\pi}{k^2}g\frac{\Gamma_{ab}\Gamma_c}{(E-E_0)^2+\Gamma^2/4}$	(4.174)	$\sigma(E)$	cross-section for $a+b\rightarrow c$
	$g=\frac{2J+1}{(2s_a+1)(2s_b+1)}$	(4.175)	k	incoming wavenumber
			g	spin factor
			E	total energy (PE + KE)
			E_0	resonant energy
			Γ	width of resonant state R
Total width	$\Gamma=\Gamma_{ab}+\Gamma_c$	(4.176)	Γ_{ab}	partial width into $a+b$
			Γ_c	partial width into c
Resonance lifetime	$\tau=\frac{\hbar}{\Gamma}$	(4.177)	τ	resonance lifetime
			J	total angular momentum quantum number of R
			$s_{a,b}$	spins of a and b
Born scattering formula[b]	$\frac{d\sigma}{d\Omega}=\left\lvert\frac{2\mu}{\hbar^2}\int_0^\infty\frac{\sin Kr}{Kr}V(r)r^2\,dr\right\rvert^2$	(4.178)	$\frac{d\sigma}{d\Omega}$	differential collision cross-section
			μ	reduced mass
			K	$=\lvert\boldsymbol{k}_{in}-\boldsymbol{k}_{out}\rvert$ (see footnote)
			r	radial distance
			$V(r)$	potential energy of interaction
Mott scattering formula[c]	$\frac{d\sigma}{d\Omega}=\left(\frac{\alpha}{4E}\right)^2\left[\csc^4\frac{\chi}{2}+\sec^4\frac{\chi}{2}+\frac{A\cos\left(\frac{\alpha}{\hbar v}\ln\tan^2\frac{\chi}{2}\right)}{\sin^2\frac{\chi}{2}\cos\frac{\chi}{2}}\right]$	(4.179)	$\hbar$	(Planck constant)/2π
			α/r	scattering potential energy
			χ	scattering angle
	$\frac{d\sigma}{d\Omega}\simeq\left(\frac{\alpha}{2E}\right)^2\frac{4-3\sin^2\chi}{\sin^4\chi}\quad(A=-1,\ \alpha\ll v\hbar)$	(4.180)	v	closing velocity
			A	$=2$ for spin-zero particles, $=-1$ for spin-half particles

[a]For the reaction $a+b\leftrightarrow R\rightarrow c$ in the centre of mass frame.
[b]For a central field. The Born approximation holds when the potential energy of scattering, V, is much less than the total kinetic energy. K is the magnitude of the change in the particle's wavevector due to scattering.
[c]For identical particles undergoing Coulomb scattering in the centre of mass frame. Nonidentical particles obey the Rutherford scattering formula (page 72).

Relativistic wave equations[a]

Klein–Gordon equation (massive, spin zero particles)	$(\nabla^2-m^2)\psi=\frac{\partial^2\psi}{\partial t^2}$	(4.181)	ψ	wavefunction
			m	particle mass
			t	time
Weyl equations (massless, spin 1/2 particles)	$\frac{\partial\boldsymbol{\psi}}{\partial t}=\pm\left(\boldsymbol{\sigma}_x\frac{\partial\boldsymbol{\psi}}{\partial x}+\boldsymbol{\sigma}_y\frac{\partial\boldsymbol{\psi}}{\partial y}+\boldsymbol{\sigma}_z\frac{\partial\boldsymbol{\psi}}{\partial z}\right)$	(4.182)	$\boldsymbol{\psi}$	spinor wavefunction
			$\boldsymbol{\sigma}_i$	Pauli spin matrices (see page 26)
Dirac equation (massive, spin 1/2 particles)	$(\mathrm{i}\gamma^\mu\partial_\mu-m)\boldsymbol{\psi}=0$	(4.183)	i	$\mathrm{i}^2=-1$
	where $\partial_\mu=\left(\frac{\partial}{\partial t},\frac{\partial}{\partial x},\frac{\partial}{\partial y},\frac{\partial}{\partial z}\right)$	(4.184)	γ^μ	Dirac matrices: $\gamma^0=\begin{pmatrix}\mathbf{1}_2 & 0\\ 0 & -\mathbf{1}_2\end{pmatrix}$, $\gamma^i=\begin{pmatrix}0 & \boldsymbol{\sigma}_i\\ -\boldsymbol{\sigma}_i & 0\end{pmatrix}$
	$(\gamma^0)^2=\mathbf{1}_4;\quad(\gamma^1)^2=(\gamma^2)^2=(\gamma^3)^2=-\mathbf{1}_4$	(4.185)	$\mathbf{1}_n$	$n\times n$ unit matrix

[a]Written in natural units, with $c=\hbar=1$.

Chapter 5 Thermodynamics

5.1 Introduction

The term *thermodynamics* is used here loosely and includes classical thermodynamics, statistical thermodynamics, thermal physics, and radiation processes. Notation in these subjects can be confusing and the conventions used here are those found in the majority of modern treatments. In particular:

- The internal energy of a system is defined in terms of the heat supplied *to* the system plus the work done *on* the system, that is, $\mathrm{d}U = \text{đ}Q + \text{đ}W$.
- The lowercase symbol p is used for pressure. Probability density functions are denoted by $\mathrm{pr}(x)$ and microstate probabilities by p_i.
- With the exception of *specific intensity*, quantities are taken as specific if they refer to unit mass and are distinguished from the extensive equivalent by using lowercase. Hence *specific volume*, v, equals V/m, where V is the volume of gas and m its mass. Also, the *specific heat capacity* of a gas at constant pressure is $c_p = C_p/m$, where C_p is the heat capacity of mass m of gas. Molar values take a subscript "m" (e.g., V_m for molar volume) and remain in upper case.
- The component held constant during a partial differentiation is shown after a vertical bar; hence $\left.\frac{\partial V}{\partial p}\right|_T$ is the partial differential of volume with respect to pressure, holding temperature constant.

The thermal properties of solids are dealt with more explicitly in the section on solid state physics (page 123). Note that in solid state literature *specific heat capacity* is often taken to mean heat capacity per unit volume.

5.2 Classical thermodynamics

Thermodynamic laws

Thermodynamic temperature[a]	$T \propto \lim_{p\to 0}(pV)$	(5.1)	T	thermodynamic temperature
			V	volume of a fixed mass of gas
			p	gas pressure
Kelvin temperature scale	$T/\mathrm{K} = 273.16 \dfrac{\lim_{p\to 0}(pV)_T}{\lim_{p\to 0}(pV)_{\mathrm{tr}}}$	(5.2)	K	kelvin unit
			tr	temperature of the triple point of water
First law[b]	$\mathrm{d}U = \text{đ}Q + \text{đ}W$	(5.3)	$\mathrm{d}U$	change in internal energy
			$\text{đ}W$	work done on system
			$\text{đ}Q$	heat supplied to system
Entropy[c]	$\mathrm{d}S = \dfrac{\text{đ}Q_{\mathrm{rev}}}{T} \geq \dfrac{\text{đ}Q}{T}$	(5.4)	S	experimental entropy
			T	temperature
			rev	reversible change

[a] As determined with a gas thermometer. The idea of temperature is associated with the zeroth law of thermodynamics: *If two systems are in thermal equilibrium with a third, they are also in thermal equilibrium with each other.*

[b] The đ notation represents a differential change in a quantity that is not a function of state of the system.

[c] Associated with the second law of thermodynamics: *No process is possible with the sole effect of completely converting heat into work* (Kelvin statement).

Thermodynamic work[a]

Hydrostatic pressure	$\text{đ}W = -p\,\mathrm{d}V$	(5.5)	p	(hydrostatic) pressure
			$\mathrm{d}V$	volume change
Surface tension	$\text{đ}W = \gamma\,\mathrm{d}A$	(5.6)	$\text{đ}W$	work done on the system
			γ	surface tension
			$\mathrm{d}A$	change in area
Electric field	$\text{đ}W = \boldsymbol{E}\cdot\mathrm{d}\boldsymbol{p}$	(5.7)	$\boldsymbol{E}$	electric field
			$\mathrm{d}\boldsymbol{p}$	induced electric dipole moment
Magnetic field	$\text{đ}W = \boldsymbol{B}\cdot\mathrm{d}\boldsymbol{m}$	(5.8)	$\boldsymbol{B}$	magnetic flux density
			$\mathrm{d}\boldsymbol{m}$	induced magnetic dipole moment
Electric current	$\text{đ}W = \Delta\phi\,\mathrm{d}q$	(5.9)	$\Delta\phi$	potential difference
			$\mathrm{d}q$	charge moved

[a] The sources of electric and magnetic fields are taken as being outside the thermodynamic system on which they are working.

Cycle efficiencies (thermodynamic)[a]

Heat engine	$\eta = \dfrac{\text{work extracted}}{\text{heat input}} \le \dfrac{T_h - T_l}{T_h}$	(5.10)	η efficiency T_h higher temperature T_l lower temperature
Refrigerator	$\eta = \dfrac{\text{heat extracted}}{\text{work done}} \le \dfrac{T_l}{T_h - T_l}$	(5.11)	
Heat pump	$\eta = \dfrac{\text{heat supplied}}{\text{work done}} \le \dfrac{T_h}{T_h - T_l}$	(5.12)	
Otto cycle[b]	$\eta = \dfrac{\text{work extracted}}{\text{heat input}} = 1 - \left(\dfrac{V_2}{V_1}\right)^{\gamma - 1}$	(5.13)	$\frac{V_1}{V_2}$ compression ratio γ ratio of heat capacities (assumed constant)

[a]The equalities are for reversible cycles, such as Carnot cycles, operating between temperatures T_h and T_l.
[b]Idealised reversible "petrol" (heat) engine.

Heat capacities

Constant volume	$C_V = \left.\dfrac{đQ}{dT}\right\|_V = \left.\dfrac{\partial U}{\partial T}\right\|_V = T\left.\dfrac{\partial S}{\partial T}\right\|_V$	(5.14)	C_V heat capacity, V constant Q heat T temperature V volume U internal energy S entropy
Constant pressure	$C_p = \left.\dfrac{đQ}{dT}\right\|_p = \left.\dfrac{\partial H}{\partial T}\right\|_p = T\left.\dfrac{\partial S}{\partial T}\right\|_p$	(5.15)	C_p heat capacity, p constant p pressure H enthalpy
Difference in heat capacities	$C_p - C_V = \left(\left.\dfrac{\partial U}{\partial V}\right\|_T + p\right)\left.\dfrac{\partial V}{\partial T}\right\|_p$	(5.16)	β_p isobaric expansivity κ_T isothermal compressibility
	$= \dfrac{VT\beta_p^2}{\kappa_T}$	(5.17)	
Ratio of heat capacities	$\gamma = \dfrac{C_p}{C_V} = \dfrac{\kappa_T}{\kappa_S}$	(5.18)	γ ratio of heat capacities κ_S adiabatic compressibility

Thermodynamic coefficients

Isobaric expansivity[a]	$\beta_p = \dfrac{1}{V}\left.\dfrac{\partial V}{\partial T}\right\|_p$	(5.19)	β_p isobaric expansivity V volume T temperature
Isothermal compressibility	$\kappa_T = -\dfrac{1}{V}\left.\dfrac{\partial V}{\partial p}\right\|_T$	(5.20)	κ_T isothermal compressibility p pressure
Adiabatic compressibility	$\kappa_S = -\dfrac{1}{V}\left.\dfrac{\partial V}{\partial p}\right\|_S$	(5.21)	κ_S adiabatic compressibility
Isothermal bulk modulus	$K_T = \dfrac{1}{\kappa_T} = -V\left.\dfrac{\partial p}{\partial V}\right\|_T$	(5.22)	K_T isothermal bulk modulus
Adiabatic bulk modulus	$K_S = \dfrac{1}{\kappa_S} = -V\left.\dfrac{\partial p}{\partial V}\right\|_S$	(5.23)	K_S adiabatic bulk modulus

[a]Also called "cubic expansivity" or "volume expansivity." The linear expansivity is $\alpha_p = \beta_p/3$.

5

Expansion processes

Joule expansion[a]	$\eta = \left.\frac{\partial T}{\partial V}\right\|_U = -\frac{T^2}{C_V}\left.\frac{\partial(p/T)}{\partial T}\right\|_V$	(5.24)	η Joule coefficient T temperature p pressure U internal energy C_V heat capacity, V constant
	$= -\frac{1}{C_V}\left(T\left.\frac{\partial p}{\partial T}\right\|_V - p\right)$	(5.25)	
Joule–Kelvin expansion[b]	$\mu = \left.\frac{\partial T}{\partial p}\right\|_H = \frac{T^2}{C_p}\left.\frac{\partial(V/T)}{\partial T}\right\|_p$	(5.26)	μ Joule–Kelvin coefficient V volume H enthalpy C_p heat capacity, p constant
	$= \frac{1}{C_p}\left(T\left.\frac{\partial V}{\partial T}\right\|_p - V\right)$	(5.27)	

[a] Expansion with no change in internal energy.
[b] Expansion with no change in enthalpy. Also known as a "Joule–Thomson expansion" or "throttling" process.

Thermodynamic potentials[a]

Internal energy	$\mathrm{d}U = T\,\mathrm{d}S - p\,\mathrm{d}V + \mu\,\mathrm{d}N$	(5.28)	U internal energy T temperature S entropy μ chemical potential N number of particles
Enthalpy	$H = U + pV$	(5.29)	H enthalpy p pressure V volume
	$\mathrm{d}H = T\,\mathrm{d}S + V\,\mathrm{d}p + \mu\,\mathrm{d}N$	(5.30)	
Helmholtz free energy[b]	$F = U - TS$	(5.31)	F Helmholtz free energy
	$\mathrm{d}F = -S\,\mathrm{d}T - p\,\mathrm{d}V + \mu\,\mathrm{d}N$	(5.32)	
Gibbs free energy[c]	$G = U - TS + pV$	(5.33)	G Gibbs free energy
	$= F + pV = H - TS$	(5.34)	
	$\mathrm{d}G = -S\,\mathrm{d}T + V\,\mathrm{d}p + \mu\,\mathrm{d}N$	(5.35)	
Grand potential	$\Phi = F - \mu N$	(5.36)	Φ grand potential
	$\mathrm{d}\Phi = -S\,\mathrm{d}T - p\,\mathrm{d}V - N\,\mathrm{d}\mu$	(5.37)	
Gibbs–Duhem relation	$-S\,\mathrm{d}T + V\,\mathrm{d}p - N\,\mathrm{d}\mu = 0$	(5.38)	
Availability	$A = U - T_0 S + p_0 V$	(5.39)	A availability T_0 temperature of surroundings p_0 pressure of surroundings
	$\mathrm{d}A = (T - T_0)\,\mathrm{d}S - (p - p_0)\,\mathrm{d}V$	(5.40)	

[a] $\mathrm{d}N = 0$ for a closed system.
[b] Sometimes called the "work function."
[c] Sometimes called the "thermodynamic potential."

Maxwell's relations

Maxwell 1	$\left.\frac{\partial T}{\partial V}\right\|_S = -\left.\frac{\partial p}{\partial S}\right\|_V \quad \left(= \frac{\partial^2 U}{\partial S \partial V}\right)$	(5.41)	U T V	internal energy temperature volume
Maxwell 2	$\left.\frac{\partial T}{\partial p}\right\|_S = \left.\frac{\partial V}{\partial S}\right\|_p \quad \left(= \frac{\partial^2 H}{\partial p \partial S}\right)$	(5.42)	H S p	enthalpy entropy pressure
Maxwell 3	$\left.\frac{\partial p}{\partial T}\right\|_V = \left.\frac{\partial S}{\partial V}\right\|_T \quad \left(= \frac{\partial^2 F}{\partial T \partial V}\right)$	(5.43)	F	Helmholtz free energy
Maxwell 4	$\left.\frac{\partial V}{\partial T}\right\|_p = -\left.\frac{\partial S}{\partial p}\right\|_T \quad \left(= \frac{\partial^2 G}{\partial p \partial T}\right)$	(5.44)	G	Gibbs free energy

Gibbs–Helmholtz equations

$U = -T^2 \left.\frac{\partial(F/T)}{\partial T}\right\|_V$	(5.45)	F U	Helmholtz free energy internal energy
$G = -V^2 \left.\frac{\partial(F/V)}{\partial V}\right\|_T$	(5.46)	G H	Gibbs free energy enthalpy
$H = -T^2 \left.\frac{\partial(G/T)}{\partial T}\right\|_p$	(5.47)	T p V	temperature pressure volume

Phase transitions

Heat absorbed	$L = T(S_2 - S_1)$	(5.48)	L T S	(latent) heat absorbed ($1 \rightarrow 2$) temperature of phase change entropy
Clausius–Clapeyron equation[a]	$\frac{\mathrm{d}p}{\mathrm{d}T} = \frac{S_2 - S_1}{V_2 - V_1}$ $= \frac{L}{T(V_2 - V_1)}$	(5.49) (5.50)	p V 1,2	pressure volume phase states
Coexistence curve[b]	$p(T) \propto \exp\left(\frac{-L}{RT}\right)$	(5.51)	R	molar gas constant
Ehrenfest's equation[c]	$\frac{\mathrm{d}p}{\mathrm{d}T} = \frac{\beta_{p2} - \beta_{p1}}{\kappa_{T2} - \kappa_{T1}}$ $= \frac{1}{VT} \frac{C_{p2} - C_{p1}}{\beta_{p2} - \beta_{p1}}$	(5.52) (5.53)	β_p κ_T C_p	isobaric expansivity isothermal compressibility heat capacity (p constant)
Gibbs's phase rule	$\mathsf{P} + \mathsf{F} = \mathsf{C} + 2$	(5.54)	P F C	number of phases in equilibrium number of degrees of freedom number of components

[a] Phase boundary gradient for a first-order transition. Equation (5.50) is sometimes called the "Clapeyron equation."
[b] For $V_2 \gg V_1$, e.g., if phase 1 is a liquid and phase 2 a vapour.
[c] For a second-order phase transition.

5

5.3 Gas laws

Ideal gas

Joule's law	$U=U(T)$	(5.55)	U internal energy T temperature
Boyle's law	$pV\|_T=\text{constant}$	(5.56)	p pressure V volume
Equation of state (Ideal gas law)	$pV=nRT$	(5.57)	n number of moles R molar gas constant
Adiabatic equations	$pV^{\gamma}=\text{constant}$ $TV^{(\gamma-1)}=\text{constant}$ $T^{\gamma}p^{(1-\gamma)}=\text{constant}$ $\Delta W=\frac{1}{\gamma-1}(p_2V_2-p_1V_1)$	(5.58) (5.59) (5.60) (5.61)	γ ratio of heat capacities (C_p/C_V) ΔW work done on system
Internal energy	$U=\frac{nRT}{\gamma-1}$	(5.62)	
Reversible isothermal expansion	$\Delta Q=nRT\ln(V_2/V_1)$	(5.63)	ΔQ heat supplied to system 1,2 initial and final states
Joule expansion[a]	$\Delta S=nR\ln(V_2/V_1)$	(5.64)	ΔS change in entropy of the system

[a]Since $\Delta Q=0$ for a Joule expansion, ΔS is due entirely to irreversibility. Because entropy is a function of state it has the same value as for the reversible isothermal expansion, where $\Delta S=\Delta Q/T$.

Virial expansion

Virial expansion	$pV=RT\left(1+\frac{B_2(T)}{V}+\frac{B_3(T)}{V^2}+\cdots\right)$	(5.65)	p pressure V volume R molar gas constant T temperature B_i virial coefficients
Boyle temperature	$B_2(T_{\mathrm{B}})=0$	(5.66)	T_{B} Boyle temperature

Van der Waals gas

Equation of state	$\left(p+\frac{a}{V_{\mathrm{m}}^2}\right)(V_{\mathrm{m}}-b)=RT$	(5.67)	p pressure V_{m} molar volume R molar gas constant T temperature a,b van der Waals' constants
Critical point	$T_{\mathrm{c}}=8a/(27Rb)$ $p_{\mathrm{c}}=a/(27b^2)$ $V_{\mathrm{mc}}=3b$	(5.68) (5.69) (5.70)	T_{c} critical temperature p_{c} critical pressure V_{mc} critical molar volume
Reduced equation of state	$\left(p_{\mathrm{r}}+\frac{3}{V_{\mathrm{r}}^2}\right)(3V_{\mathrm{r}}-1)=8T_{\mathrm{r}}$	(5.71)	$p_{\mathrm{r}}=p/p_{\mathrm{c}}$ $V_{\mathrm{r}}=V_{\mathrm{m}}/V_{\mathrm{mc}}$ $T_{\mathrm{r}}=T/T_{\mathrm{c}}$

Dieterici gas

Equation of state	$p=\frac{RT}{V_{\mathrm{m}}-b'}\exp\left(\frac{-a'}{RTV_{\mathrm{m}}}\right)$	(5.72)	p pressure V_{m} molar volume R molar gas constant T temperature a',b' Dieterici's constants
Critical point	$T_{\mathrm{c}}=a'/(4Rb')$ $p_{\mathrm{c}}=a'/(4b'^2\mathrm{e}^2)$ $V_{\mathrm{mc}}=2b'$	(5.73) (5.74) (5.75)	T_{c} critical temperature p_{c} critical pressure V_{mc} critical molar volume e $=2.71828\ldots$
Reduced equation of state	$p_{\mathrm{r}}=\frac{T_{\mathrm{r}}}{2V_{\mathrm{r}}-1}\exp\left(2-\frac{2}{V_{\mathrm{r}}T_{\mathrm{r}}}\right)$	(5.76)	$p_{\mathrm{r}}=p/p_{\mathrm{c}}$ $V_{\mathrm{r}}=V_{\mathrm{m}}/V_{\mathrm{mc}}$ $T_{\mathrm{r}}=T/T_{\mathrm{c}}$

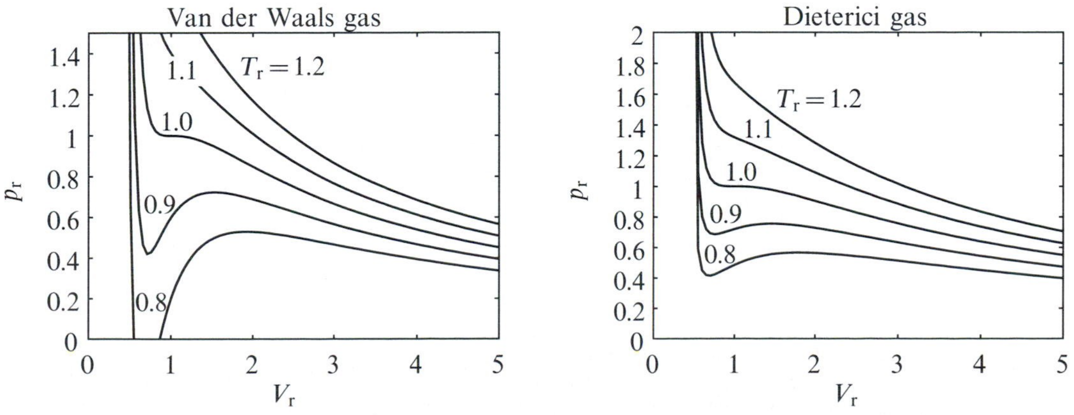

5.4 Kinetic theory

Monatomic gas

Pressure	$p=\frac{1}{3}nm\langle c^2\rangle$	(5.77)	p pressure n number density $=N/V$ m particle mass $\langle c^2\rangle$ mean squared particle velocity
Equation of state of an ideal gas	$pV=NkT$	(5.78)	V volume k Boltzmann constant N number of particles T temperature
Internal energy	$U=\frac{3}{2}NkT=\frac{N}{2}m\langle c^2\rangle$	(5.79)	U internal energy
Heat capacities	$C_V=\frac{3}{2}Nk$	(5.80)	C_V heat capacity, constant V
	$C_p=C_V+Nk=\frac{5}{2}Nk$	(5.81)	C_p heat capacity, constant p
	$\gamma=\frac{C_p}{C_V}=\frac{5}{3}$	(5.82)	γ ratio of heat capacities
Entropy (Sackur–Tetrode equation)[a]	$S=Nk\ln\left[\left(\frac{mkT}{2\pi\hbar^2}\right)^{3/2}e^{5/2}\frac{V}{N}\right]$	(5.83)	S entropy $\hbar$ $=$(Planck constant)/(2π) e $=2.71828\ldots$

[a]For the uncondensed gas. The factor $\left(\frac{mkT}{2\pi\hbar^2}\right)^{3/2}$ is the quantum concentration of the particles, n_Q. Their thermal de Broglie wavelength, λ_T, approximately equals $n_Q^{-1/3}$.

Maxwell–Boltzmann distribution[a]

Particle speed distribution	$\mathrm{pr}(c)\,\mathrm{d}c=\left(\frac{m}{2\pi kT}\right)^{3/2}\exp\left(\frac{-mc^2}{2kT}\right)4\pi c^2\,\mathrm{d}c$	(5.84)	pr probability density m particle mass k Boltzmann constant T temperature c particle speed
Particle energy distribution	$\mathrm{pr}(E)\,\mathrm{d}E=\frac{2E^{1/2}}{\pi^{1/2}(kT)^{3/2}}\exp\left(\frac{-E}{kT}\right)\mathrm{d}E$	(5.85)	E particle kinetic energy $(=mc^2/2)$
Mean speed	$\langle c\rangle=\left(\frac{8kT}{\pi m}\right)^{1/2}$	(5.86)	$\langle c\rangle$ mean speed
rms speed	$c_{\mathrm{rms}}=\left(\frac{3kT}{m}\right)^{1/2}=\left(\frac{3\pi}{8}\right)^{1/2}\langle c\rangle$	(5.87)	c_{rms} root mean squared speed
Most probable speed	$\hat{c}=\left(\frac{2kT}{m}\right)^{1/2}=\left(\frac{\pi}{4}\right)^{1/2}\langle c\rangle$	(5.88)	$\hat{c}$ most probable speed

[a]Probability density functions normalised so that $\int_0^\infty \mathrm{pr}(x)\,\mathrm{d}x=1$.

Transport properties

Mean free path[a]	$l = \dfrac{1}{\sqrt{2}\pi d^2 n}$ (5.89)	l mean free path d molecular diameter n particle number density
Survival equation[b]	$\mathrm{pr}(x) = \exp(-x/l)$ (5.90)	pr probability x linear distance
Flux through a plane[c]	$J = \dfrac{1}{4} n \langle c \rangle$ (5.91)	J molecular flux $\langle c \rangle$ mean molecular speed
Self-diffusion (Fick's law of diffusion)[d]	$\boldsymbol{J} = -D \nabla n$ (5.92) where $D \simeq \dfrac{2}{3} l \langle c \rangle$ (5.93)	D diffusion coefficient
Thermal conductivity[d]	$\boldsymbol{H} = -\lambda \nabla T$ (5.94) $\nabla^2 T = \dfrac{1}{D} \dfrac{\partial T}{\partial t}$ (5.95) for monatomic gas $\lambda \simeq \dfrac{5}{4} \rho l \langle c \rangle c_V$ (5.96)	H heat flux per unit area λ thermal conductivity T temperature ρ density c_V specific heat capacity, V constant
Viscosity[d]	$\eta \simeq \dfrac{1}{2} \rho l \langle c \rangle$ (5.97)	η dynamic viscosity x displacement of sphere in x direction after time t
Brownian motion (of a sphere)	$\langle x^2 \rangle = \dfrac{kTt}{3\pi\eta a}$ (5.98)	k Boltzmann constant t time interval a sphere radius
Free molecular flow (Knudsen flow)[e]	$\dfrac{\mathrm{d}M}{\mathrm{d}t} = \dfrac{4R_\mathrm{p}^3}{3L} \left(\dfrac{2\pi m}{k} \right)^{1/2} \left(\dfrac{p_1}{T_1^{1/2}} - \dfrac{p_2}{T_2^{1/2}} \right)$ (5.99)	$\frac{\mathrm{d}M}{\mathrm{d}t}$ mass flow rate R_p pipe radius L pipe length m particle mass p pressure

[a]For a perfect gas of hard, spherical particles with a Maxwell–Boltzmann speed distribution.
[b]Probability of travelling distance x without a collision.
[c]From the side where the number density is n, assuming an isotropic velocity distribution. Also known as "collision number."
[d]Simplistic kinetic theory yields numerical coefficients of 1/3 for D, λ and η.
[e]Through a pipe from end 1 to end 2, assuming $R_\mathrm{p} \ll l$ (i.e., at very low pressure).

Gas equipartition

Classical equipartition[a]	$E_\mathrm{q} = \dfrac{1}{2} kT$ (5.100)	E_q energy per quadratic degree of freedom k Boltzmann constant T temperature
Ideal gas heat capacities	$C_V = \dfrac{1}{2} f N k = \dfrac{1}{2} f n R$ (5.101) $C_p = Nk \left(1 + \dfrac{f}{2} \right)$ (5.102) $\gamma = \dfrac{C_p}{C_V} = 1 + \dfrac{2}{f}$ (5.103)	C_V heat capacity, V constant C_p heat capacity, p constant N number of molecules f number of degrees of freedom n number of moles R molar gas constant γ ratio of heat capacities

[a]System in thermal equilibrium at temperature T.

5.5 Statistical thermodynamics

Statistical entropy

Boltzmann formula[a]	$S = k \ln W$ $\simeq k \ln g(E)$	(5.104) (5.105)	S k W $g(E)$	entropy Boltzmann constant number of accessible microstates density of microstates with energy E
Gibbs entropy[b]	$S = -k \sum_i p_i \ln p_i$	(5.106)	$\sum_i$ p_i	sum over microstates probability that the system is in microstate i
N two-level systems	$W = \dfrac{N!}{(N-n)!n!}$	(5.107)	N n	number of systems number in upper state
N harmonic oscillators	$W = \dfrac{(Q+N-1)!}{Q!(N-1)!}$	(5.108)	Q	total number of energy quanta available

[a]Sometimes called "configurational entropy." Equation (5.105) is true only for large systems.
[b]Sometimes called "canonical entropy."

Ensemble probabilities

Microcanonical ensemble[a]	$p_i = \dfrac{1}{W}$	(5.109)	p_i W	probability that the system is in microstate i number of accessible microstates
Partition function[b]	$Z = \sum_i e^{-\beta E_i}$	(5.110)	Z $\sum_i$ β E_i	partition function sum over microstates $=1/(kT)$ energy of microstate i
Canonical ensemble (Boltzmann distribution)[c]	$p_i = \dfrac{1}{Z} e^{-\beta E_i}$	(5.111)	k T	Boltzmann constant temperature
Grand partition function	$\Xi = \sum_i e^{-\beta(E_i - \mu N_i)}$	(5.112)	Ξ μ N_i	grand partition function chemical potential number of particles in microstate i
Grand canonical ensemble (Gibbs distribution)[d]	$p_i = \dfrac{1}{\Xi} e^{-\beta(E_i - \mu N_i)}$	(5.113)		

[a]Energy fixed.
[b]Also called "sum over states."
[c]Temperature fixed.
[d]Temperature fixed. Exchange of both heat and particles with a reservoir.

Macroscopic thermodynamic variables

Helmholtz free energy	$F=-kT\ln Z$	(5.114)	F Helmholtz free energy k Boltzmann constant T temperature Z partition function
Grand potential	$\Phi=-kT\ln\Xi$	(5.115)	Φ grand potential Ξ grand partition function
Internal energy	$U=F+TS=-\left.\dfrac{\partial \ln Z}{\partial\beta}\right\|_{V,N}$	(5.116)	U internal energy β $=1/(kT)$
Entropy	$S=-\left.\dfrac{\partial F}{\partial T}\right\|_{V,N}=\left.\dfrac{\partial(kT\ln Z)}{\partial T}\right\|_{V,N}$	(5.117)	S entropy N number of particles
Pressure	$p=-\left.\dfrac{\partial F}{\partial V}\right\|_{T,N}=\left.\dfrac{\partial(kT\ln Z)}{\partial V}\right\|_{T,N}$	(5.118)	p pressure
Chemical potential	$\mu=\left.\dfrac{\partial F}{\partial N}\right\|_{V,T}=-\left.\dfrac{\partial(kT\ln Z)}{\partial N}\right\|_{V,T}$	(5.119)	μ chemical potential

Identical particles

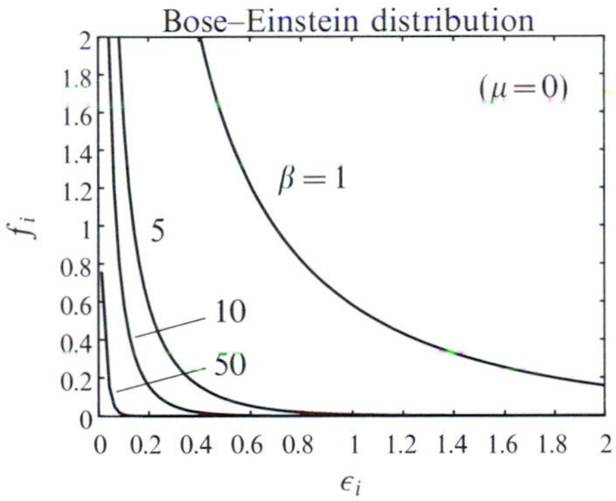

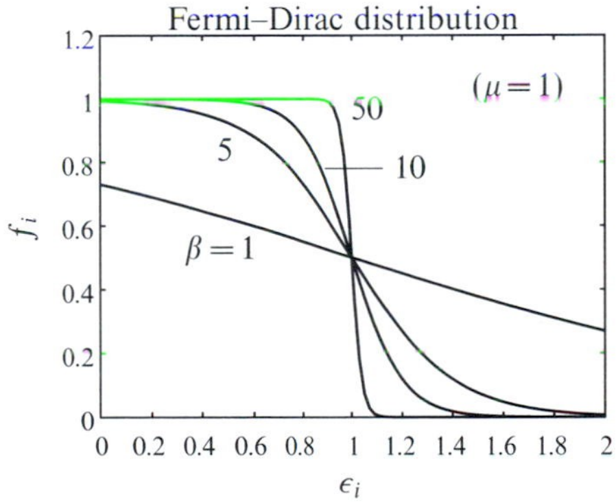

Bose–Einstein distribution[a]	$f_i=\dfrac{1}{e^{\beta(\epsilon_i-\mu)}-1}$	(5.120)	f_i mean occupation number of ith state β $=1/(kT)$
Fermi–Dirac distribution[b]	$f_i=\dfrac{1}{e^{\beta(\epsilon_i-\mu)}+1}$	(5.121)	ϵ_i energy quantum for ith state μ chemical potential
Fermi energy[c]	$\epsilon_F=\dfrac{\hbar^2}{2m}\left(\dfrac{6\pi^2 n}{g}\right)^{2/3}$	(5.122)	ϵ_F Fermi energy $\hbar$ (Planck constant)/(2π) n particle number density m particle mass
Bose condensation temperature	$T_c=\dfrac{2\pi\hbar^2}{mk}\left[\dfrac{n}{g\zeta(3/2)}\right]^{2/3}$	(5.123)	g spin degeneracy $(=2s+1)$ ζ Riemann zeta function $\zeta(3/2)\simeq 2.612$ T_c Bose condensation temperature

[a] For bosons. $f_i\geq 0$.
[b] For fermions. $0\leq f_i\leq 1$.
[c] For noninteracting particles. At low temperatures, $\mu\simeq\epsilon_F$.

Population densities[a]

Boltzmann excitation equation	$\frac{n_{mj}}{n_{lj}} = \frac{g_{mj}}{g_{lj}} \exp\left[\frac{-(\chi_{mj}-\chi_{lj})}{kT}\right]$ (5.124) $= \frac{g_{mj}}{g_{lj}} \exp\left(\frac{-h\nu_{lm}}{kT}\right)$ (5.125)	n_{ij} g_{ij} χ_{ij}	number density of atoms in excitation level i of ionisation state j ($j=0$ if not ionised) level degeneracy excitation energy relative to the ground state
Partition function	$Z_j(T) = \sum_i g_{ij} \exp\left(\frac{-\chi_{ij}}{kT}\right)$ (5.126) $\frac{n_{ij}}{N_j} = \frac{g_{ij}}{Z_j(T)} \exp\left(\frac{-\chi_{ij}}{kT}\right)$ (5.127)	ν_{ij} h k T	photon transition frequency Planck constant Boltzmann constant temperature
Saha equation (general)	$n_{ij} = n_{0,j+1} n_e \frac{g_{ij}}{g_{0,j+1}} \frac{h^3}{2} (2\pi m_e kT)^{-3/2} \exp\left(\frac{\chi_{Ij}-\chi_{ij}}{kT}\right)$ (5.128)	Z_j N_j	partition function for ionisation state j total number density in ionisation state j
Saha equation (ion populations)	$\frac{N_j}{N_{j+1}} = n_e \frac{Z_j(T)}{Z_{j+1}(T)} \frac{h^3}{2} (2\pi m_e kT)^{-3/2} \exp\left(\frac{\chi_{Ij}}{kT}\right)$ (5.129)	n_e m_e χ_{Ij}	electron number density electron mass ionisation energy of atom in ionisation state j

[a]All equations apply only under conditions of local thermodynamic equilibrium (LTE). In atoms with no magnetic splitting, the degeneracy of a level with total angular momentum quantum number J is $g_{ij} = 2J+1$.

5.6 Fluctuations and noise

Thermodynamic fluctuations[a]

Fluctuation probability	$\mathrm{pr}(x) \propto \exp[S(x)/k]$ (5.130) $\propto \exp\left[\frac{-A(x)}{kT}\right]$ (5.131)	pr x S A	probability density unconstrained variable entropy availability
General variance	$\mathrm{var}[x] = kT \left[\frac{\partial^2 A(x)}{\partial x^2}\right]^{-1}$ (5.132)	var[·] k T	mean square deviation Boltzmann constant temperature
Temperature fluctuations	$\mathrm{var}[T] = kT \left.\frac{\partial T}{\partial S}\right\|_V = \frac{kT^2}{C_V}$ (5.133)	V C_V	volume heat capacity, V constant
Volume fluctuations	$\mathrm{var}[V] = -kT \left.\frac{\partial V}{\partial p}\right\|_T = \kappa_T V kT$ (5.134)	p κ_T	pressure isothermal compressibility
Entropy fluctuations	$\mathrm{var}[S] = kT \left.\frac{\partial S}{\partial T}\right\|_p = kC_p$ (5.135)	C_p	heat capacity, p constant
Pressure fluctuations	$\mathrm{var}[p] = -kT \left.\frac{\partial p}{\partial V}\right\|_S = \frac{K_S kT}{V}$ (5.136)	K_S	adiabatic bulk modulus
Density fluctuations	$\mathrm{var}[n] = \frac{n^2}{V^2} \mathrm{var}[V] = \frac{n^2}{V} \kappa_T kT$ (5.137)	n	number density

[a]In part of a large system, whose mean temperature is fixed. Quantum effects are assumed negligible.

Noise

Nyquist's noise theorem	$dw = kT\cdot\beta\epsilon(e^{\beta\epsilon}-1)^{-1}\,d\nu$	(5.138)	w	exchangeable noise power
	$= kT_N\,d\nu$	(5.139)	k	Boltzmann constant
	$\simeq kT\,d\nu \qquad (h\nu \ll kT)$	(5.140)	T	temperature
			T_N	noise temperature
			$\beta\epsilon$	$=h\nu/(kT)$
			ν	frequency
			h	Planck constant
Johnson (thermal) noise voltage[a]	$v_{rms} = (4kT_N R\Delta\nu)^{1/2}$	(5.141)	v_{rms}	rms noise voltage
			R	resistance
			$\Delta\nu$	bandwidth
Shot noise (electrical)	$I_{rms} = (2eI_0\Delta\nu)^{1/2}$	(5.142)	I_{rms}	rms noise current
			$-e$	electronic charge
			I_0	mean current
Noise figure[b]	$f_{dB} = 10\log_{10}\left(1+\frac{T_N}{T_0}\right)$	(5.143)	f_{dB}	noise figure (decibels)
			T_0	ambient temperature (usually taken as 290 K)
Relative power	$G = 10\log_{10}\left(\frac{P_2}{P_1}\right)$	(5.144)	G	decibel gain of P_2 over P_1
			P_1, P_2	power levels

[a]Thermal voltage over an open-circuit resistance.
[b]Noise figure can also be defined as $f = 1 + T_N/T_0$, when it is also called "noise factor."

5

5.7 Radiation processes

Radiometry[a]

Radiant energy[b]	$$Q_e = \iiint L_e \cos\theta \, dA \, d\Omega \, dt \quad \text{J} \qquad (5.145)$$	Q_e radiant energy L_e radiance (generally a function of position and direction) θ angle between dir. of $d\Omega$ and normal to dA Ω solid angle
Radiant flux ("radiant power")	$$\Phi_e = \frac{\partial Q_e}{\partial t} \quad \text{W} \qquad (5.146)$$ $$= \iint L_e \cos\theta \, dA \, d\Omega \qquad (5.147)$$	A area t time Φ_e radiant flux
Radiant energy density[c]	$$W_e = \frac{\partial Q_e}{\partial V} \quad \text{J m}^{-3} \qquad (5.148)$$	W_e radiant energy density dV differential volume of propagation medium
Radiant exitance[d]	$$M_e = \frac{\partial \Phi_e}{\partial A} \quad \text{W m}^{-2} \qquad (5.149)$$ $$= \int L_e \cos\theta \, d\Omega \qquad (5.150)$$	M_e radiant exitance
Irradiance[e]	$$E_e = \frac{\partial \Phi_e}{\partial A} \quad \text{W m}^{-2} \qquad (5.151)$$ $$= \int L_e \cos\theta \, d\Omega \qquad (5.152)$$	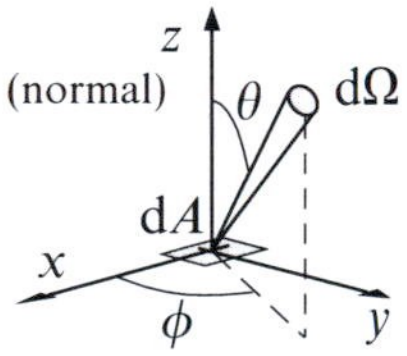
Radiant intensity	$$I_e = \frac{\partial \Phi_e}{\partial \Omega} \quad \text{W sr}^{-1} \qquad (5.153)$$ $$= \int L_e \cos\theta \, dA \qquad (5.154)$$	E_e irradiance I_e radiant intensity
Radiance	$$L_e = \frac{1}{\cos\theta} \frac{\partial^2 \Phi_e}{dA \, d\Omega} \quad \text{W m}^{-2} \text{sr}^{-1} \qquad (5.155)$$ $$= \frac{1}{\cos\theta} \frac{\partial I_e}{\partial A} \qquad (5.156)$$	

[a]Radiometry is concerned with the treatment of light as energy.
[b]Sometimes called "total energy." Note that we assume opaque radiant surfaces, so that $0 \le \theta \le \pi/2$.
[c]The instantaneous amount of radiant energy contained in a unit volume of propagation medium.
[d]Power per unit area leaving a surface. For a perfectly diffusing surface, $M_e = \pi L_e$.
[e]Power per unit area incident on a surface.

Photometry[a]

Luminous energy ("total light")	$Q_{\mathrm{v}}=\iiint L_{\mathrm{v}}\cos\theta\,\mathrm{d}A\,\mathrm{d}\Omega\,\mathrm{d}t$	lm s	(5.157)	Q_{v} luminous energy L_{v} luminance (generally a function of position and direction) θ angle between dir. of dΩ and normal to dA Ω solid angle
Luminous flux	$\Phi_{\mathrm{v}}=\dfrac{\partial Q_{\mathrm{v}}}{\partial t}$	lumen (lm)	(5.158)	A area t time Φ_{v} luminous flux
	$=\iint L_{\mathrm{v}}\cos\theta\,\mathrm{d}A\,\mathrm{d}\Omega$		(5.159)	
Luminous density[b]	$W_{\mathrm{v}}=\dfrac{\partial Q_{\mathrm{v}}}{\partial V}$	$\mathrm{lm\,s\,m^{-3}}$	(5.160)	W_{v} luminous density V volume
Luminous exitance[c]	$M_{\mathrm{v}}=\dfrac{\partial \Phi_{\mathrm{v}}}{\partial A}$	lx $(\mathrm{lm\,m^{-2}})$	(5.161)	M_{v} luminous exitance
	$=\int L_{\mathrm{v}}\cos\theta\,\mathrm{d}\Omega$		(5.162)	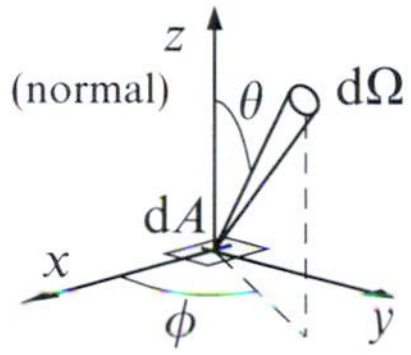
Illuminance ("illumination")[d]	$E_{\mathrm{v}}=\dfrac{\partial \Phi_{\mathrm{v}}}{\partial A}$	$\mathrm{lm\,m^{-2}}$	(5.163)	
	$=\int L_{\mathrm{v}}\cos\theta\,\mathrm{d}\Omega$		(5.164)	
Luminous intensity[e]	$I_{\mathrm{v}}=\dfrac{\partial \Phi_{\mathrm{v}}}{\partial \Omega}$	cd	(5.165)	E_{v} illuminance I_{v} luminous intensity
	$=\int L_{\mathrm{v}}\cos\theta\,\mathrm{d}A$		(5.166)	
Luminance ("photometric brightness")	$L_{\mathrm{v}}=\dfrac{1}{\cos\theta}\dfrac{\partial^2 \Phi_{\mathrm{v}}}{\mathrm{d}A\,\mathrm{d}\Omega}$	$\mathrm{cd\,m^{-2}}$	(5.167)	
	$=\dfrac{1}{\cos\theta}\dfrac{\partial I_{\mathrm{v}}}{\partial A}$		(5.168)	
Luminous efficacy	$K=\dfrac{\Phi_{\mathrm{v}}}{\Phi_{\mathrm{e}}}=\dfrac{L_{\mathrm{v}}}{L_{\mathrm{e}}}=\dfrac{I_{\mathrm{v}}}{I_{\mathrm{e}}}$	$\mathrm{lm\,W^{-1}}$	(5.169)	K luminous efficacy L_{e} radiance Φ_{e} radiant flux I_{e} radiant intensity
Luminous efficiency	$V(\lambda)=\dfrac{K(\lambda)}{K_{\mathrm{max}}}$		(5.170)	V luminous efficiency λ wavelength K_{max} spectral maximum of $K(\lambda)$

[a] Photometry is concerned with the treatment of light as seen by the human eye.
[b] The instantaneous amount of luminous energy contained in a unit volume of propagating medium.
[c] Luminous emitted flux per unit area.
[d] Luminous incident flux per unit area. The derived SI unit is the lux (lx). $1\,\mathrm{lx}=1\,\mathrm{lm\,m^{-2}}$.
[e] The SI unit of luminous intensity is the candela (cd). $1\,\mathrm{cd}=1\,\mathrm{lm\,sr^{-1}}$.

Radiative transfer[a]

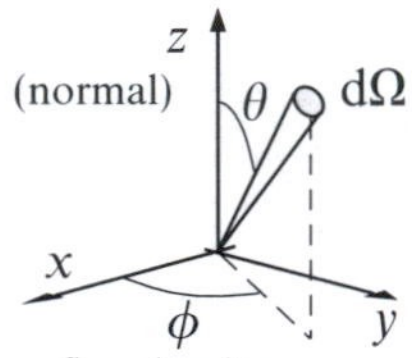

Flux density (through a plane)	$F_\nu = \int I_\nu(\theta,\phi)\cos\theta\,\mathrm{d}\Omega \quad \mathrm{W\,m^{-2}\,Hz^{-1}}$	(5.171)
Mean intensity[b]	$J_\nu = \frac{1}{4\pi}\int I_\nu(\theta,\phi)\,\mathrm{d}\Omega \quad \mathrm{W\,m^{-2}\,Hz^{-1}}$	(5.172)
Spectral energy density[c]	$u_\nu = \frac{1}{c}\int I_\nu(\theta,\phi)\,\mathrm{d}\Omega \quad \mathrm{J\,m^{-3}\,Hz^{-1}}$	(5.173)
Specific emission coefficient	$j_\nu = \frac{\epsilon_\nu}{\rho} \quad \mathrm{W\,kg^{-1}\,Hz^{-1}\,sr^{-1}}$	(5.174)
Gas linear absorption coefficient ($\alpha_\nu \ll 1$)	$\alpha_\nu = n\sigma_\nu = \frac{1}{l_\nu} \quad \mathrm{m^{-1}}$	(5.175)
Opacity[d]	$\kappa_\nu = \frac{\alpha_\nu}{\rho} \quad \mathrm{kg^{-1}\,m^2}$	(5.176)
Optical depth	$\tau_\nu = \int \kappa_\nu \rho\,\mathrm{d}s$	(5.177)
Transfer equation[e]	$\frac{1}{\rho}\frac{\mathrm{d}I_\nu}{\mathrm{d}s} = -\kappa_\nu I_\nu + j_\nu$	(5.178)
	or $\frac{\mathrm{d}I_\nu}{\mathrm{d}s} = -\alpha_\nu I_\nu + \epsilon_\nu$	(5.179)
Kirchhoff's law[f]	$S_\nu \equiv \frac{j_\nu}{\kappa_\nu} = \frac{\epsilon_\nu}{\alpha_\nu}$	(5.180)
Emission from a homogeneous medium	$I_\nu = S_\nu(1-\mathrm{e}^{-\tau_\nu})$	(5.181)

- F_ν flux density
- I_ν specific intensity ($\mathrm{W\,m^{-2}\,Hz^{-1}\,sr^{-1}}$)
- J_ν mean intensity
- u_ν spectral energy density
- Ω solid angle
- θ angle between normal and direction of Ω
- j_ν specific emission coefficient
- ϵ_ν emission coefficient ($\mathrm{W\,m^{-3}\,Hz^{-1}\,sr^{-1}}$)
- ρ density
- α_ν linear absorption coefficient
- n particle number density
- σ_ν particle cross section
- l_ν mean free path
- κ_ν opacity
- τ_ν optical depth, or optical thickness
- $\mathrm{d}s$ line element
- S_ν source function

[a]The definitions of these quantities vary in the literature. Those presented here are common in meteorology and astrophysics. Note particularly that the ambiguous term *specific* is taken to mean "per unit frequency interval" in the case of specific intensity and "per unit mass per unit frequency interval" in the case of specific emission coefficient.
[b]In radio astronomy, flux density is usually taken as $S = 4\pi J_\nu$.
[c]Assuming a refractive index of 1.
[d]Or "mass absorption coefficient."
[e]Or "Schwarzschild's equation."
[f]Under conditions of local thermal equilibrium (LTE), the source function, S_ν, equals the Planck function, $B_\nu(T)$ [see Equation (5.182)].

Blackbody radiation

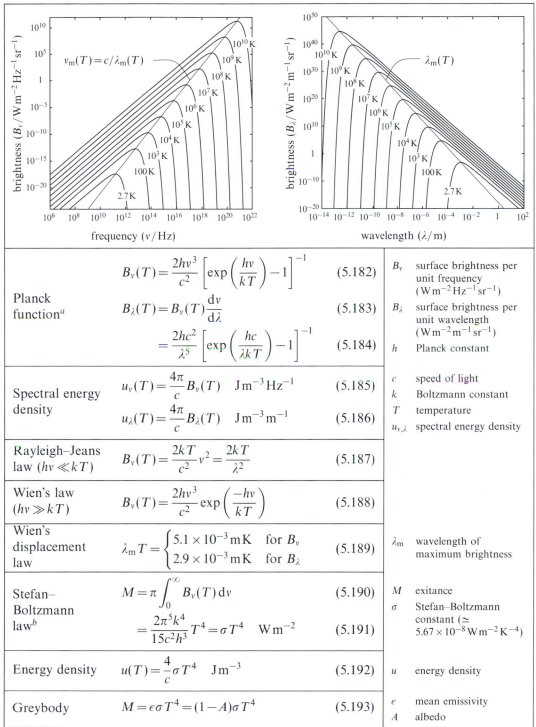

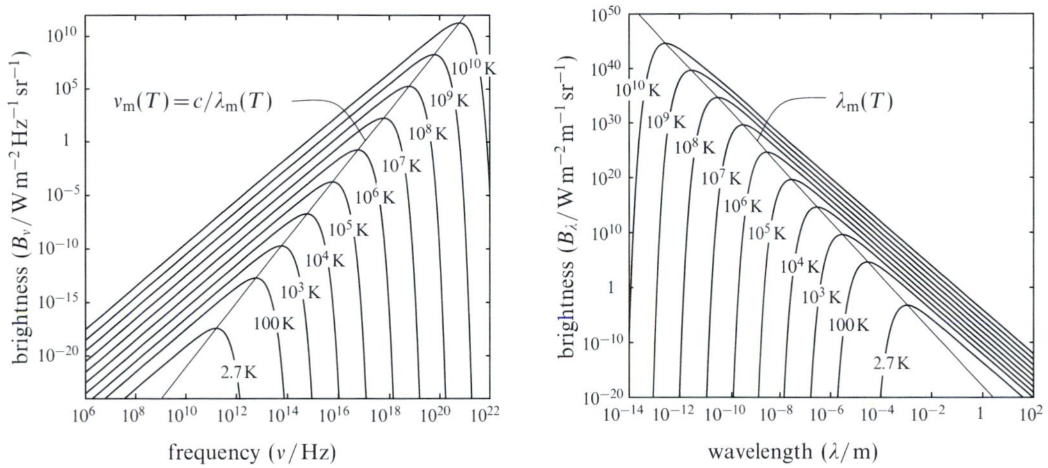

Planck function[a]	$B_\nu(T)=\frac{2h\nu^3}{c^2}\left[\exp\left(\frac{h\nu}{kT}\right)-1\right]^{-1}$	(5.182)	B_ν surface brightness per unit frequency (W m^{-2} Hz^{-1} sr^{-1})
	$B_\lambda(T)=B_\nu(T)\frac{d\nu}{d\lambda}$	(5.183)	B_λ surface brightness per unit wavelength (W m^{-2} m^{-1} sr^{-1})
	$=\frac{2hc^2}{\lambda^5}\left[\exp\left(\frac{hc}{\lambda kT}\right)-1\right]^{-1}$	(5.184)	h Planck constant
Spectral energy density	$u_\nu(T)=\frac{4\pi}{c}B_\nu(T)$ J m^{-3} Hz^{-1}	(5.185)	c speed of light; k Boltzmann constant
	$u_\lambda(T)=\frac{4\pi}{c}B_\lambda(T)$ J m^{-3} m^{-1}	(5.186)	T temperature; $u_{\nu,\lambda}$ spectral energy density
Rayleigh–Jeans law ($h\nu \ll kT$)	$B_\nu(T)=\frac{2kT}{c^2}\nu^2=\frac{2kT}{\lambda^2}$	(5.187)	
Wien's law ($h\nu \gg kT$)	$B_\nu(T)=\frac{2h\nu^3}{c^2}\exp\left(\frac{-h\nu}{kT}\right)$	(5.188)	
Wien's displacement law	$\lambda_m T=\begin{cases}5.1\times10^{-3}\,\text{m K} & \text{for } B_\nu\\ 2.9\times10^{-3}\,\text{m K} & \text{for } B_\lambda\end{cases}$	(5.189)	λ_m wavelength of maximum brightness
Stefan–Boltzmann law[b]	$M=\pi\int_0^\infty B_\nu(T)\,d\nu$	(5.190)	M exitance
	$=\frac{2\pi^5k^4}{15c^2h^3}T^4=\sigma T^4$ W m^{-2}	(5.191)	σ Stefan–Boltzmann constant ($\simeq 5.67\times10^{-8}$ W m^{-2} K^{-4})
Energy density	$u(T)=\frac{4}{c}\sigma T^4$ J m^{-3}	(5.192)	u energy density
Greybody	$M=\epsilon\sigma T^4=(1-A)\sigma T^4$	(5.193)	ϵ mean emissivity; A albedo

[a]With respect to the projected area of the surface. Surface brightness is also known simply as "brightness." "Specific intensity" is used for reception.

[b]Sometimes "Stefan's law." Exitance is the total radiated energy from unit area of the body per unit time.

Chapter 6 Solid state physics

6.1 Introduction

This section covers a few selected topics in solid state physics. There is no attempt to do more than scratch the surface of this vast field, although the basics of many undergraduate texts on the subject are covered. In addition a period table of elements, together with some of their physical properties, is displayed on the next two pages.

Periodic table (overleaf) Data for the periodic table of elements are taken from *Pure Appl. Chem.*, **71**, 1593–1607 (1999), from the 16th edition of Kaye and Laby *Tables of Physical and Chemical Constants* (Longman, 1995) and from the 74th edition of the CRC *Handbook of Chemistry and Physics* (CRC Press, 1993). Note that melting and boiling points have been converted to kelvins by adding 273.15 to the Celsius values listed in Kaye and Laby. The standard atomic masses reflect the relative isotopic abundances in samples found naturally on Earth, and the number of significant figures reflect the variations between samples. Elements with atomic masses shown in square brackets have no stable nuclides, and the values reflect the mass numbers of the longest-lived isotopes. Crystallographic data are based on the most common forms of the elements (the α-form, unless stated otherwise) stable under standard conditions. Densities are for the solid state. For full details and footnotes for each element, the reader is advised to consult the original texts.

Elements 110, 111, 112 and 114 are known to exist but their names are not yet permanent.

6.2 Periodic table

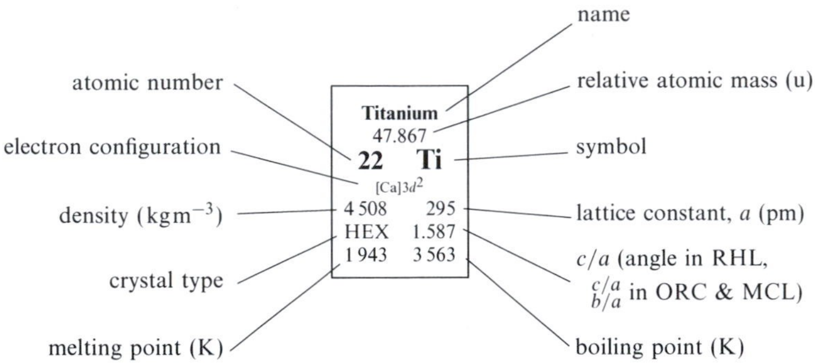

	1	2	3	4	5	6	7	8	9
1	**Hydrogen** 1.007 94 **1 H** $1s^1$ 89 (β) 378 HEX 1.632 13.80 20.28								
2	**Lithium** 6.941 **3 Li** [He]$2s^1$ 533 (β) 351 BCC 453.65 1 613	**Beryllium** 9.012 182 **4 Be** [He]$2s^2$ 1 846 229 HEX 1.568 1 560 2 745							
3	**Sodium** 22.989 770 **11 Na** [Ne]$3s^1$ 966 429 BCC 370.8 1 153	**Magnesium** 24.305 0 **12 Mg** [Ne]$3s^2$ 1 738 321 HEX 1.624 923 1 363							
4	**Potassium** 39.098 3 **19 K** [Ar]$4s^1$ 862 532 BCC 336.5 1 033	**Calcium** 40.078 **20 Ca** [Ar]$4s^2$ 1 530 559 FCC 1 113 1 757	**Scandium** 44.955 910 **21 Sc** [Ca]$3d^1$ 2 992 331 HEX 1.592 1 813 3 103	**Titanium** 47.867 **22 Ti** [Ca]$3d^2$ 4 508 295 HEX 1.587 1 943 3 563	**Vanadium** 50.941 5 **23 V** [Ca]$3d^3$ 6 090 302 BCC 2 193 3 673	**Chromium** 51.996 1 **24 Cr** [Ar]$3d^5 4s^1$ 7 194 388 BCC 2 180 2 943	**Manganese** 54.938 049 **25 Mn** [Ca]$3d^5$ 7 473 891 FCC 1 523 2 333	**Iron** 55.845 **26 Fe** [Ca]$3d^6$ 7 873 287 BCC 1 813 3 133	**Cobalt** 58.933 200 **27 Co** [Ca]$3d^7$ 8 800 (ϵ) 251 HEX 1.623 1 768 3 203
5	**Rubidium** 85.467 8 **37 Rb** [Kr]$5s^1$ 1 533 571 BCC 312.4 963.1	**Strontium** 87.62 **38 Sr** [Kr]$5s^2$ 2 583 608 FCC 1 050 1 653	**Yttrium** 88.905 85 **39 Y** [Sr]$4d^1$ 4 475 365 HEX 1.571 1 798 3 613	**Zirconium** 91.224 **40 Zr** [Sr]$4d^2$ 6 507 323 HEX 1.593 2 123 4 673	**Niobium** 92.906 38 **41 Nb** [Kr]$4d^4 5s^1$ 8 578 330 BCC 2 750 4 973	**Molybdenum** 95.94 **42 Mo** [Kr]$4d^5 5s^1$ 10 222 315 BCC 2 896 4 913	**Technetium** [98] **43 Tc** [Sr]$4d^5$ 11 496 274 HEX 1.604 2 433 4 533	**Ruthenium** 101.07 **44 Ru** [Kr]$4d^7 5s^1$ 12 360 270 HEX 1.582 2 603 4 423	**Rhodium** 102.905 50 **45 Rh** [Kr]$4d^8 5s^1$ 12 420 380 FCC 2 236 3 973
6	**Caesium** 132.905 45 **55 Cs** [Xe]$6s^1$ 1 900 614 BCC 301.6 943.2	**Barium** 137.327 **56 Ba** [Xe]$6s^2$ 3 594 502 BCC 1 001 2 173	**Lanthanides** **57 – 71**	**Hafnium** 178.49 **72 Hf** [Yb]$5d^2$ 13 276 319 HEX 1.581 2 503 4 873	**Tantalum** 180.947 9 **73 Ta** [Yb]$5d^3$ 16 670 330 BCC 3 293 5 833	**Tungsten** 183.84 **74 W** [Yb]$5d^4$ 19 254 316 BCC 3 695 5 823	**Rhenium** 186.207 **75 Re** [Yb]$5d^5$ 21 023 276 HEX 1.615 3 459 5 873	**Osmium** 190.23 **76 Os** [Yb]$5d^6$ 22 580 273 HEX 1.606 3 303 5 273	**Iridium** 192.217 **77 Ir** [Yb]$5d^7$ 22 550 384 FCC 2 720 4 703
7	**Francium** [223] **87 Fr** [Rn]$7s^1$ 300 923	**Radium** [226] **88 Ra** [Rn]$7s^2$ 5 000 515 BCC 973 1 773	**Actinides** **89 – 103**	**Rutherfordium** [261] **104 Rf** [Ra]$5f^{14} 6d^2$	**Dubnium** [262] **105 Db** [Ra]$5f^{14} 6d^3$?	**Seaborgium** [263] **106 Sg** [Ra]$5f^{14} 6d^4$?	**Bohrium** [264] **107 Bh** [Ra]$5f^{14} 6d^5$?	**Hassium** [265] **108 Hs** [Ra]$5f^{14} 6d^6$?	**Meitnerium** [268] **109 Mt** [Ra]$5f^{14} 6d^7$?

Lanthanides	**Lanthanum** 138.905 5 **57 La** [Ba]$5d^1$ 6 174 377 HEX 3.23 1 193 3 733	**Cerium** 140.116 **58 Ce** [Ba]$4f^1 5d^1$ 6 711 (γ) 516 FCC 1 073 3 693	**Praseodymium** 140.907 65 **59 Pr** [Ba]$4f^3$ 6 779 367 HEX 3.222 1 204 3 783	**Neodymium** 144.24 **60 Nd** [Ba]$4f^4$ 7 000 366 HEX 3.225 1 289 3 343	**Promethium** [145] **61 Pm** [Ba]$4f^5$ 7 220 365 HEX 3.19 1 415 3 573	**Samarium** 150.36 **62 Sm** [Ba]$4f^6$ 7 536 363 HEX 7.221 1 443 2 063
Actinides	**Actinium** [227] **89 Ac** [Ra]$6d^1$ 10 060 531 FCC 1 323 3 473	**Thorium** 232.038 1 **90 Th** [Ra]$6d^2$ 11 725 508 FCC 2 023 5 063	**Protactinium** 231.035 88 **91 Pa** [Rn]$5f^2 6d^1 7s^2$ 15 370 392 TET 0.825 1 843 4 273	**Uranium** 238.028 9 **92 U** [Rn]$5f^3 6d^1 7s^2$ 19 050 285 ORC $\frac{1.736}{2.056}$ 1 405.3 4 403	**Neptunium** [237] **93 Np** [Rn]$5f^4 6d^1 7s^2$ 20 450 666 ORC $\frac{0.733}{0.709}$ 913 4 173	**Plutonium** [244] **94 Pu** [Rn]$5f^6 7s^2$ 19 816 618 MCL $\frac{1.773}{0.780}$ 913 3 503

BCC	body-centred cubic
CUB	simple cubic
DIA	diamond
FCC	face-centred cubic
HEX	hexagonal
MCL	monoclinic
ORC	orthorhombic
RHL	rhombohedral
TET	tetragonal
(t-pt)	triple point

10	11	12	13	14	15	16	17	18
								Helium 4.002 602 **2 He** $1s^2$ 120 356 HEX 1.631 3-5 4.22
			Boron 10.811 **5 B** [Be]$2p^1$ 2 466 1017 RHL 65°7′ 2 348 4 273	**Carbon** 12.0107 **6 C** [Be]$2p^2$ 2 266 357 DIA 4 763 (t-pt)	**Nitrogen** 14.006 74 **7 N** [Be]$2p^3$ 1 035 (β) 405 HEX 1.631 63 77.35	**Oxygen** 15.999 4 **8 O** [Be]$2p^4$ 1 460 (γ) 683 CUB 54.36 90.19	**Fluorine** 18.998 403 2 **9 F** [Be]$2p^5$ 1 140 550 MCL 1.32/0.61 53.55 85.05	**Neon** 20.179 7 **10 Ne** [Be]$2p^6$ 1 442 446 FCC 24.56 27.07
			Aluminium 26.981 538 **13 Al** [Mg]$3p^1$ 2 698 405 FCC 933.47 2 793	**Silicon** 28.085 5 **14 Si** [Mg]$3p^2$ 2 329 543 DIA 1 683 3 533	**Phosphorus** 30.973 761 **15 P** [Mg]$3p^3$ 1 820 331 ORC 1.320/3.162 317.3 550	**Sulfur** 32.066 **16 S** [Mg]$3p^4$ 2 086 1 046 ORC 2.340/1.229 388.47 717.82	**Chlorine** 35.452 7 **17 Cl** [Mg]$3p^5$ 2 030 624 ORC 1.324/0.718 172 239.1	**Argon** 39.948 **18 Ar** [Mg]$3p^6$ 1 656 532 FCC 83.81 87.30
Nickel 58.693 4 **28 Ni** [Ca]$3d^8$ 8 907 352 FCC 1 728 3 263	**Copper** 63.546 **29 Cu** [Ar]$3d^{10}4s^1$ 8 933 361 FCC 1 357.8 2 833	**Zinc** 65.39 **30 Zn** [Ca]$3d^{10}$ 7 135 266 HEX 1.856 692.68 1 183	**Gallium** 69.723 **31 Ga** [Zn]$4p^1$ 5 905 452 ORC 1.001/1.695 302.9 2 473	**Germanium** 72.61 **32 Ge** [Zn]$4p^2$ 5 323 566 DIA 1211 3103	**Arsenic** 74.921 60 **33 As** [Zn]$4p^3$ 5 776 413 RHL 54°7′ 883 (t-pt)	**Selenium** 78.96 **34 Se** [Zn]$4p^4$ 4 808 (γ) 436 HEX 1.135 493 958	**Bromine** 79.904 **35 Br** [Zn]$4p^5$ 3 120 668 ORC 1.308/0.672 265.90 332.0	**Krypton** 83.80 **36 Kr** [Zn]$4p^6$ 3 000 581 FCC 115.8 119.9
Palladium 106.42 **46 Pd** [Kr]$4d^{10}$ 11 995 389 FCC 1 828 3 233	**Silver** 107.868 2 **47 Ag** [Pd]$5s^1$ 10 500 409 FCC 1 235 2 433	**Cadmium** 112.411 **48 Cd** [Pd]$5s^2$ 8 647 298 HEX 1.886 594.2 1 043	**Indium** 114.818 **49 In** [Cd]$5p^1$ 7 290 325 TET 1.521 429.75 2 343	**Tin** 118.710 **50 Sn** [Cd]$5p^2$ 7 285 (β) 583 TET 0.546 505.08 2 893	**Antimony** 121.760 **51 Sb** [Cd]$5p^3$ 6 692 451 RHL 57°7′ 903.8 1 860	**Tellurium** 127.60 **52 Te** [Cd]$5p^4$ 6 247 446 HEX 1.33 723 1 263	**Iodine** 126.904 47 **53 I** [Cd]$5p^5$ 4 953 727 ORC 1.347/0.659 386.7 457	**Xenon** 131.29 **54 Xe** [Cd]$5p^6$ 3 560 635 FCC 161.3 165.0
Platinum 195.078 **78 Pt** [Xe]$4f^{14}5d^9 6s^1$ 21 450 392 FCC 2 041 4 093	**Gold** 196.966 55 **79 Au** [Xe]$4f^{14}5d^{10}6s^1$ 19 281 408 FCC 1 337.3 3 123	**Mercury** 200.59 **80 Hg** [Yb]$5d^{10}$ 13 546 300 RHL 70°32′ 234.32 629.9	**Thallium** 204.383 3 **81 Tl** [Hg]$6p^1$ 11 871 346 HEX 1.598 577 1743	**Lead** 207.2 **82 Pb** [Hg]$6p^2$ 11 343 495 FCC 600.7 2 023	**Bismuth** 208.980 38 **83 Bi** [Hg]$6p^3$ 9 803 475 RHL 57°14′ 544.59 1 833	**Polonium** [209] **84 Po** [Hg]$6p^4$ 9 400 337 CUB 527 1 233	**Astatine** [210] **85 At** [Hg]$6p^5$ 573 623	**Radon** [222] **86 Rn** [Hg]$6p^6$ 440 202 211
Ununnilium [271] **110 Uun**	**Unununium** [272] **111 Uuu**	**Ununbium** [285] **112 Uub**		**Ununquadium** [289] **114 Uuq**				

Europium 151.964 **63 Eu** [Ba]$4f^7$ 5 248 458 BCC 1 095 1 873	**Gadolinium** 157.25 **64 Gd** [Ba]$4f^7 5d^1$ 7 870 363 HEX 1.591 1 587 3 533	**Terbium** 158.925 34 **65 Tb** [Ba]$4f^9$ 8 267 361 HEX 1.580 1 633 3 493	**Dysprosium** 162.50 **66 Dy** [Ba]$4f^{10}$ 8 531 359 HEX 1.573 1 683 2 833	**Holmium** 164.930 32 **67 Ho** [Ba]$4f^{11}$ 8 797 358 HEX 1.570 1 743 2 973	**Erbium** 167.26 **68 Er** [Ba]$4f^{12}$ 9 044 356 HEX 1.570 1 803 3 133	**Thulium** 168.934 21 **69 Tm** [Ba]$4f^{13}$ 9 325 354 HEX 1.570 1 823 2 223	**Ytterbium** 173.04 **70 Yb** [Ba]$4f^{14}$ 6 966 (β) 549 FCC 1 097 1 473	**Lutetium** 174.967 **71 Lu** [Yb]$5d^1$ 9 842 351 HEX 1.583 1 933 3 663
Americium [243] **95 Am** [Ra]$5f^7$ 13 670 347 HEX 3.24 1 449 2 873	**Curium** [247] **96 Cm** [Rn]$5f^7 6d^1 7s^2$ 13 510 350 HEX 3.24 1 618 3 383	**Berkelium** [247] **97 Bk** [Ra]$5f^9$ 14 780 342 HEX 3.24 1 323	**Californium** [251] **98 Cf** [Ra]$5f^{10}$ 15 100 338 HEX 3.24 1 173	**Einsteinium** [252] **99 Es** [Ra]$5f^{11}$ HEX 1 133	**Fermium** [257] **100 Fm** [Ra]$5f^{12}$ 1 803	**Mendelevium** [258] **101 Md** [Ra]$5f^{13}$ 1 103	**Nobelium** [259] **102 No** [Ra]$5f^{14}$ 1 103	**Lawrencium** [262] **103 Lr** [Ra]$5f^{14}7p^1$ 1 903

6.3 Crystalline structure

Bravais lattices

Volume of primitive cell	$V=(\boldsymbol{a}\times\boldsymbol{b})\cdot\boldsymbol{c}$	(6.1)	$\boldsymbol{a},\boldsymbol{b},\boldsymbol{c}$ V	primitive base vectors volume of primitive cell
Reciprocal primitive base vectors[a]	$\boldsymbol{a}^*=2\pi\boldsymbol{b}\times\boldsymbol{c}/[(\boldsymbol{a}\times\boldsymbol{b})\cdot\boldsymbol{c}]$ $\boldsymbol{b}^*=2\pi\boldsymbol{c}\times\boldsymbol{a}/[(\boldsymbol{a}\times\boldsymbol{b})\cdot\boldsymbol{c}]$ $\boldsymbol{c}^*=2\pi\boldsymbol{a}\times\boldsymbol{b}/[(\boldsymbol{a}\times\boldsymbol{b})\cdot\boldsymbol{c}]$ $\boldsymbol{a}\cdot\boldsymbol{a}^*=\boldsymbol{b}\cdot\boldsymbol{b}^*=\boldsymbol{c}\cdot\boldsymbol{c}^*=2\pi$ $\boldsymbol{a}\cdot\boldsymbol{b}^*=\boldsymbol{a}\cdot\boldsymbol{c}^*=0$ (etc.)	(6.2) (6.3) (6.4) (6.5) (6.6)	$\boldsymbol{a}^*,\boldsymbol{b}^*,\boldsymbol{c}^*$	reciprocal primitive base vectors
Lattice vector	$\boldsymbol{R}_{uvw}=u\boldsymbol{a}+v\boldsymbol{b}+w\boldsymbol{c}$	(6.7)	$\boldsymbol{R}_{uvw}$ u,v,w	lattice vector $[uvw]$ integers
Reciprocal lattice vector	$\boldsymbol{G}_{hkl}=h\boldsymbol{a}^*+k\boldsymbol{b}^*+l\boldsymbol{c}^*$ $\exp(\mathrm{i}\boldsymbol{G}_{hkl}\cdot\boldsymbol{R}_{uvw})=1$	(6.8) (6.9)	$\boldsymbol{G}_{hkl}$ i	reciprocal lattice vector $[hkl]$ $\mathrm{i}^2=-1$
Weiss zone equation[b]	$hu+kv+lw=0$	(6.10)	(hkl)	Miller indices of plane[c]
Interplanar spacing (general)	$d_{hkl}=\dfrac{2\pi}{G_{hkl}}$	(6.11)	d_{hkl}	distance between (hkl) planes
Interplanar spacing (orthogonal basis)	$\dfrac{1}{d_{hkl}^2}=\dfrac{h^2}{a^2}+\dfrac{k^2}{b^2}+\dfrac{l^2}{c^2}$	(6.12)		

[a]Note that this is 2π times the usual definition of a "reciprocal vector" (see page 20).
[b]Condition for lattice vector $[uvw]$ to be parallel to lattice plane (hkl) in an arbitrary Bravais lattice.
[c]Miller indices are defined so that G_{hkl} is the shortest reciprocal lattice vector normal to the (hkl) planes.

Weber symbols

Converting $[uvw]$ to $[UVTW]$	$U=\frac{1}{3}(2u-v)$ $V=\frac{1}{3}(2v-u)$ $T=-\frac{1}{3}(u+v)$ $W=w$	(6.13) (6.14) (6.15) (6.16)	U,V,T,W u,v,w $[UVTW]$ $[uvw]$	Weber indices zone axis indices Weber symbol zone axis symbol
Converting $[UVTW]$ to $[uvw]$	$u=(U-T)$ $v=(V-T)$ $w=W$	(6.17) (6.18) (6.19)		
Zone law[a]	$hU+kV+iT+lW=0$	(6.20)	$(hkil)$	Miller–Bravais indices

[a]For trigonal and hexagonal systems.

Cubic lattices

lattice	primitive (P)	body-centred (I)	face-centred (F)
lattice parameter	a	a	a
volume of conventional cell	a^3	a^3	a^3
lattice points per cell	1	2	4
1st nearest neighbours[a]	6	8	12
1st n.n. distance	a	$a\sqrt{3}/2$	$a/\sqrt{2}$
2nd nearest neighbours	12	6	6
2nd n.n. distance	$a\sqrt{2}$	a	a
packing fraction[b]	$\pi/6$	$\sqrt{3}\pi/8$	$\sqrt{2}\pi/6$
reciprocal lattice[c]	P	F	I
primitive base vectors[d]	$\boldsymbol{a}_1 = a\hat{\boldsymbol{x}}$ $\boldsymbol{a}_2 = a\hat{\boldsymbol{y}}$ $\boldsymbol{a}_3 = a\hat{\boldsymbol{z}}$	$\boldsymbol{a}_1 = \frac{a}{2}(\hat{\boldsymbol{y}}+\hat{\boldsymbol{z}}-\hat{\boldsymbol{x}})$ $\boldsymbol{a}_2 = \frac{a}{2}(\hat{\boldsymbol{z}}+\hat{\boldsymbol{x}}-\hat{\boldsymbol{y}})$ $\boldsymbol{a}_3 = \frac{a}{2}(\hat{\boldsymbol{x}}+\hat{\boldsymbol{y}}-\hat{\boldsymbol{z}})$	$\boldsymbol{a}_1 = \frac{a}{2}(\hat{\boldsymbol{y}}+\hat{\boldsymbol{z}})$ $\boldsymbol{a}_2 = \frac{a}{2}(\hat{\boldsymbol{z}}+\hat{\boldsymbol{x}})$ $\boldsymbol{a}_3 = \frac{a}{2}(\hat{\boldsymbol{x}}+\hat{\boldsymbol{y}})$

[a] Or "coordination number."
[b] For close-packed spheres. The maximum possible packing fraction for spheres is $\sqrt{2}\pi/6$.
[c] The lattice parameters for the reciprocal lattices of P, I, and F are $2\pi/a$, $4\pi/a$, and $4\pi/a$ respectively.
[d] $\hat{\boldsymbol{x}}$, $\hat{\boldsymbol{y}}$, and $\hat{\boldsymbol{z}}$ are unit vectors.

Crystal systems[a]

system	*symmetry*	*unit cell*[b]	*lattices*[c]
triclinic	none	$a \neq b \neq c$; $\alpha \neq \beta \neq \gamma \neq 90°$	P
monoclinic	one diad ∥ [010]	$a \neq b \neq c$; $\alpha = \gamma = 90°$, $\beta \neq 90°$	P, C
orthorhombic	three orthogonal diads	$a \neq b \neq c$; $\alpha = \beta = \gamma = 90°$	P, C, I, F
tetragonal	one tetrad ∥ [001]	$a = b \neq c$; $\alpha = \beta = \gamma = 90°$	P, I
trigonal[d]	one triad ∥ [111]	$a = b = c$; $\alpha = \beta = \gamma < 120° \neq 90°$	P, R
hexagonal	one hexad ∥ [001]	$a = b \neq c$; $\alpha = \beta = 90°$, $\gamma = 120°$	P
cubic	four triads ∥ ⟨111⟩	$a = b = c$; $\alpha = \beta = \gamma = 90°$	P, F, I

[a] The symbol "$\neq$" implies that equality is not required by the symmetry, but neither is it forbidden.
[b] The cell axes are a, b, and c with α, β, and γ the angles between $b:c$, $c:a$, and $a:b$ respectively.
[c] The lattice types are primitive (P), body-centred (I), all face-centred (F), side-centred (C), and rhombohedral primitive (R).
[d] A primitive hexagonal unit cell, with a triad ∥ [001], is generally preferred over this rhombohedral unit cell.

6

Dislocations and cracks

Edge dislocation	$\hat{\boldsymbol{l}}\cdot\boldsymbol{b}=0$	(6.21)
Screw dislocation	$\hat{\boldsymbol{l}}\cdot\boldsymbol{b}=b$	(6.22)
Screw dislocation energy per unit length[b]	$U=\dfrac{\mu b^2}{4\pi}\ln\dfrac{R}{r_0}$	(6.23)
	$\sim\mu b^2$	(6.24)
Critical crack length[c]	$L=\dfrac{4\alpha E}{\pi(1-\sigma^2)p_0^2}$	(6.25)

$\hat{\boldsymbol{l}}$	unit vector $\parallel$ line of dislocation
$\boldsymbol{b},b$	Burgers vector[a]
U	dislocation energy per unit length
μ	shear modulus
R	outer cutoff for r
r_0	inner cutoff for r
L	critical crack length
α	surface energy per unit area
E	Young modulus
σ	Poisson ratio
p_0	applied widening stress

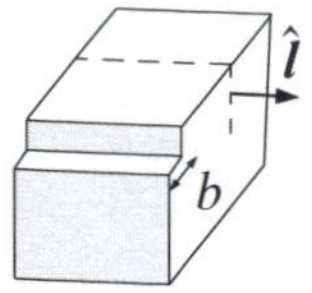

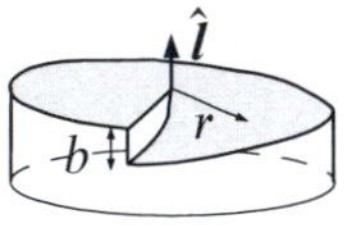

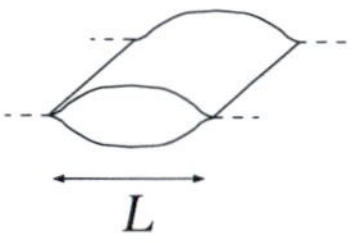

[a]The Burgers vector is a Bravais lattice vector characterising the total relative slip were the dislocation to travel throughout the crystal.
[b]Or "tension." The energy per unit length of an edge dislocation is also $\sim\mu b^2$.
[c]For a crack cavity (long $\perp L$) within an isotropic medium. Under uniform stress p_0, cracks $\geq L$ will grow and smaller cracks will shrink.

Crystal diffraction

Laue equations	$a(\cos\alpha_1-\cos\alpha_2)=h\lambda$	(6.26)
	$b(\cos\beta_1-\cos\beta_2)=k\lambda$	(6.27)
	$c(\cos\gamma_1-\cos\gamma_2)=l\lambda$	(6.28)
Bragg's law[a]	$2\boldsymbol{k}_{\text{in}}.\boldsymbol{G}+\lvert\boldsymbol{G}\rvert^2=0$	(6.29)
Atomic form factor	$f(\boldsymbol{G})=\displaystyle\int_{\text{vol}}\mathrm{e}^{-\mathrm{i}\boldsymbol{G}\cdot\boldsymbol{r}}\rho(\boldsymbol{r})\,\mathrm{d}^3r$	(6.30)
Structure factor[b]	$S(\boldsymbol{G})=\displaystyle\sum_{j=1}^{n}f_j(\boldsymbol{G})\mathrm{e}^{-\mathrm{i}\boldsymbol{G}\cdot\boldsymbol{d}_j}$	(6.31)
Scattered intensity[c]	$I(\boldsymbol{K})\propto N^2\lvert S(\boldsymbol{K})\rvert^2$	(6.32)
Debye–Waller factor[d]	$I_T=I_0\exp\left[-\dfrac{1}{3}\langle u^2\rangle\lvert\boldsymbol{G}\rvert^2\right]$	(6.33)

a,b,c	lattice parameters
$\alpha_1,\beta_1,\gamma_1$	angles between lattice base vectors and input wavevector
$\alpha_2,\beta_2,\gamma_2$	angles between lattice base vectors and output wavevector
h,k,l	integers (Laue indices)
λ	wavelength
$\boldsymbol{k}_{\text{in}}$	input wavevector
$\boldsymbol{G}$	reciprocal lattice vector
$f(\boldsymbol{G})$	atomic form factor
$\boldsymbol{r}$	position vector
$\rho(\boldsymbol{r})$	atomic electron density
$S(\boldsymbol{G})$	structure factor
n	number of atoms in basis
$\boldsymbol{d}_j$	position of jth atom within basis
$\boldsymbol{K}$	change in wavevector $(=\boldsymbol{k}_{\text{out}}-\boldsymbol{k}_{\text{in}})$
$I(\boldsymbol{K})$	scattered intensity
N	number of lattice points illuminated
I_T	intensity at temperature T
I_0	intensity from a lattice with no motion
$\langle u^2\rangle$	mean-squared thermal displacement of atoms

[a]Alternatively, see Equation (8.32).
[b]The summation is over the atoms in the basis, i.e., the atomic motif repeating with the Bravais lattice.
[c]The Bragg condition makes $\boldsymbol{K}$ a reciprocal lattice vector, with $\lvert k_{\text{in}}\rvert=\lvert k_{\text{out}}\rvert$.
[d]Effect of thermal vibrations.

6.4 Lattice dynamics

Phonon dispersion relations[a]

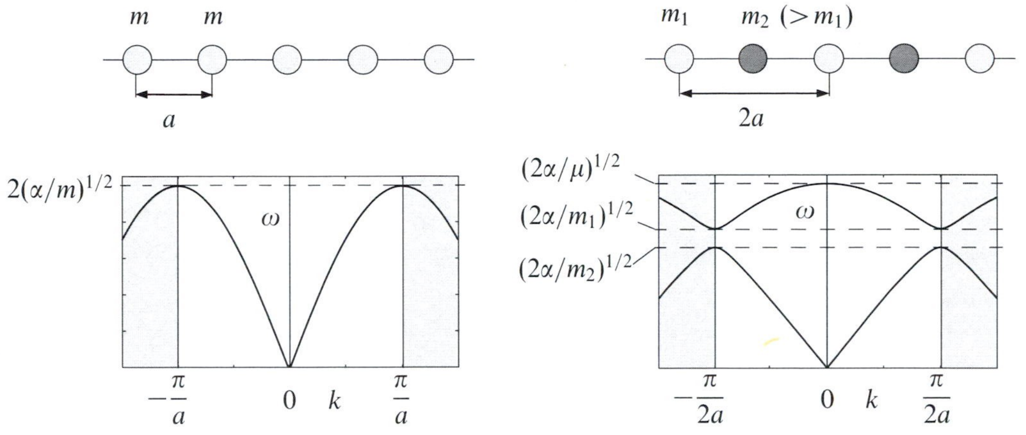

monatomic chain   diatomic chain

Monatomic linear chain	$\omega^2 = 4\frac{\alpha}{m}\sin^2\left(\frac{ka}{2}\right)$ (6.34) $v_p = \frac{\omega}{k} = a\left(\frac{\alpha}{m}\right)^{1/2}\operatorname{sinc}\left(\frac{a}{\lambda}\right)$ (6.35) $v_g = \frac{\partial\omega}{\partial k} = a\left(\frac{\alpha}{m}\right)^{1/2}\cos\left(\frac{ka}{2}\right)$ (6.36)	ω phonon angular frequency α spring constant[b] m atomic mass v_p phase speed ($\operatorname{sinc}x \equiv \frac{\sin\pi x}{\pi x}$) v_g group speed λ phonon wavelength
Diatomic linear chain[c]	$\omega^2 = \frac{\alpha}{\mu} \pm \alpha\left[\frac{1}{\mu^2} - \frac{4}{m_1 m_2}\sin^2(ka)\right]^{1/2}$ (6.37)	k wavenumber ($=2\pi/\lambda$) a atomic separation m_i atomic masses ($m_2 > m_1$) μ reduced mass [$=m_1m_2/(m_1+m_2)$]
Identical masses, alternating spring constants	$\omega^2 = \frac{\alpha_1+\alpha_2}{m} \pm \frac{1}{m}(\alpha_1^2+\alpha_2^2+2\alpha_1\alpha_2\cos ka)^{1/2}$ (6.38) $= \begin{cases} 0, \quad 2(\alpha_1+\alpha_2)/m & \text{if } k=0 \\ 2\alpha_1/m, \quad 2\alpha_2/m & \text{if } k=\pi/a \end{cases}$ (6.39)	α_i alternating spring constants 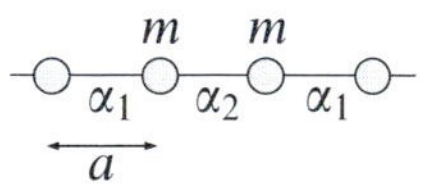

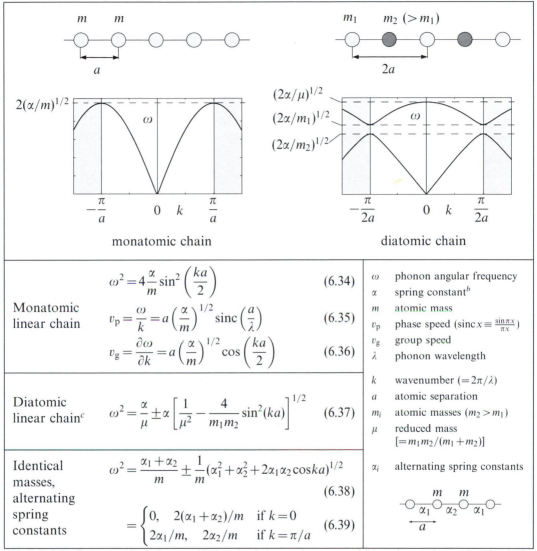

[a] Along infinite linear atomic chains, considering simple harmonic nearest-neighbour interactions only. The shaded region of the dispersion relation is outside the first Brillouin zone of the reciprocal lattice.

[b] In the sense α = restoring force/relative displacement.

[c] Note that the repeat distance for this chain is $2a$, so that the first Brillouin zone extends to $|k| < \pi/(2a)$. The optic and acoustic branches are the + and − solutions respectively.

Debye theory

Mean energy per phonon mode[a]	$\langle E\rangle=\frac{1}{2}\hbar\omega+\frac{\hbar\omega}{\exp[\hbar\omega/(k_{\mathrm{B}}T)]-1}$	(6.40)
Debye frequency	$\omega_{\mathrm{D}}=v_{\mathrm{s}}(6\pi^2N/V)^{1/3}$	(6.41)
	where $\frac{3}{v_{\mathrm{s}}^3}=\frac{1}{v_{\mathrm{l}}^3}+\frac{2}{v_{\mathrm{t}}^3}$	(6.42)
Debye temperature	$\theta_{\mathrm{D}}=\hbar\omega_{\mathrm{D}}/k_{\mathrm{B}}$	(6.43)
Phonon density of states	$g(\omega)\,\mathrm{d}\omega=\frac{3V\omega^2}{2\pi^2v_{\mathrm{s}}^3}\,\mathrm{d}\omega$	(6.44)
	(for $0<\omega<\omega_{\mathrm{D}}$, $g=0$ otherwise)	
Debye heat capacity	$C_V=9Nk_{\mathrm{B}}\frac{T^3}{\theta_{\mathrm{D}}^3}\int_0^{\theta_{\mathrm{D}}/T}\frac{x^4\mathrm{e}^x}{(\mathrm{e}^x-1)^2}\,\mathrm{d}x$	(6.45)
Dulong and Petit's law	$\simeq 3Nk_{\mathrm{B}}\quad(T\gg\theta_{\mathrm{D}})$	(6.46)
Debye T^3 law	$\simeq\frac{12\pi^4}{5}Nk_{\mathrm{B}}\frac{T^3}{\theta_{\mathrm{D}}^3}\quad(T\ll\theta_{\mathrm{D}})$	(6.47)
Internal thermal energy[b]	$U(T)=\frac{9N}{\omega_{\mathrm{D}}^3}\int_0^{\omega_{\mathrm{D}}}\frac{\hbar\omega^3}{\exp[\hbar\omega/(k_{\mathrm{B}}T)]-1}\,\mathrm{d}\omega\equiv 3Nk_{\mathrm{B}}T\,\mathrm{D}(\theta_{\mathrm{D}}/T)$	(6.48)
	where $\mathrm{D}(x)=\frac{3}{x^3}\int_0^x\frac{y^3}{\mathrm{e}^y-1}\,\mathrm{d}y$	(6.49)

$\langle E\rangle$	mean energy in a mode at ω
$\hbar$	(Planck constant)/(2π)
ω	phonon angular frequency
k_{B}	Boltzmann constant
T	temperature
ω_{D}	Debye (angular) frequency
v_{s}	effective sound speed
v_{l}	longitudinal phase speed
v_{t}	transverse phase speed
N	number of atoms in crystal
V	crystal volume
θ_{D}	Debye temperature
$g(\omega)$	density of states at ω
C_V	heat capacity, V constant
U	thermal phonon energy within crystal
$\mathrm{D}(x)$	Debye function

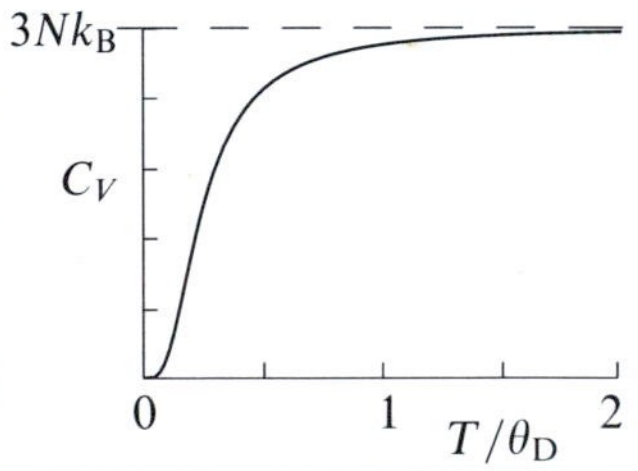

[a]Or any simple harmonic oscillator in thermal equilibrium at temperature T.
[b]Neglecting zero-point energy.

Lattice forces (simple)

Van der Waals interaction[a]	$\phi(r)=-\frac{3}{4}\frac{\alpha_p^2\hbar\omega}{(4\pi\epsilon_0)^2r^6}$	(6.50)	$\phi(r)$ two-particle potential energy; r particle separation; α_p particle polarisability; $\hbar$ (Planck constant)/(2π); ϵ_0 permittivity of free space; ω angular frequency of polarised orbital
Lennard–Jones 6-12 potential (molecular crystals)	$\phi(r)=-\frac{A}{r^6}+\frac{B}{r^{12}}$	(6.51)	A,B constants
	$=4\epsilon\left[\left(\frac{\sigma}{r}\right)^{12}-\left(\frac{\sigma}{r}\right)^6\right]$	(6.52)	ϵ,σ Lennard–Jones parameters
	$\sigma=(B/A)^{1/6};\quad \epsilon=A^2/(4B)$		
	$\phi_{\min}$ at $r=\frac{2^{1/6}}{\sigma}$	(6.53)	
De Boer parameter	$\Lambda=\frac{h}{\sigma(m\epsilon)^{1/2}}$	(6.54)	Λ de Boer parameter; h Planck constant; m particle mass
Coulomb interaction (ionic crystals)	$U_C=-\alpha_M\frac{e^2}{4\pi\epsilon_0r_0}$	(6.55)	U_C lattice Coulomb energy per ion pair; α_M Madelung constant; $-e$ electronic charge; r_0 nearest neighbour separation

[a]London's formula for fluctuating dipole interactions, neglecting the propagation time between particles.

6

Lattice thermal expansion and conduction

Grüneisen parameter[a]	$\gamma=-\frac{\partial\ln\omega}{\partial\ln V}$	(6.56)	γ Grüneisen parameter; ω normal mode frequency; V volume
Linear expansivity[b]	$\alpha=\frac{1}{3K_T}\frac{\partial p}{\partial T}\bigg\rvert_V=\frac{\gamma C_V}{3K_TV}$	(6.57)	α linear expansivity; K_T isothermal bulk modulus; p pressure; T temperature; C_V lattice heat capacity, constant V
Thermal conductivity of a phonon gas	$\lambda=\frac{1}{3}\frac{C_V}{V}v_sl$	(6.58)	λ thermal conductivity; v_s effective sound speed; l phonon mean free path
Umklapp mean free path[c]	$l_u\propto\exp(\theta_u/T)$	(6.59)	l_u umklapp mean free path; θ_u umklapp temperature ($\sim\theta_D/2$)

[a]Strictly, the Grüneisen parameter is the mean of γ over all normal modes, weighted by the mode's contribution to C_V.
[b]Or "coefficient of thermal expansion," for an isotropically expanding crystal.
[c]Mean free path determined solely by "umklapp processes" – the scattering of phonons outside the first Brillouin zone.

6.5 Electrons in solids

Free electron transport properties

Current density	$\boldsymbol{J} = -ne\boldsymbol{v}_\mathrm{d}$	(6.60)	$\boldsymbol{J}$ current density n free electron number density $-e$ electronic charge $\boldsymbol{v}_\mathrm{d}$ mean electron drift velocity
Mean electron drift velocity	$\boldsymbol{v}_\mathrm{d} = -\frac{e\tau}{m_\mathrm{e}}\boldsymbol{E}$	(6.61)	τ mean time between collisions (relaxation time) m_e electronic mass
d.c. electrical conductivity	$\sigma_0 = \frac{ne^2\tau}{m_\mathrm{e}}$	(6.62)	$\boldsymbol{E}$ applied electric field σ_0 d.c. conductivity ($\boldsymbol{J} = \sigma\boldsymbol{E}$)
a.c. electrical conductivity[a]	$\sigma(\omega) = \frac{\sigma_0}{1 - \mathrm{i}\omega\tau}$	(6.63)	ω a.c. angular frequency $\sigma(\omega)$ a.c. conductivity
Thermal conductivity	$\lambda = \frac{1}{3}\frac{C_V}{V}\langle c^2\rangle\tau$	(6.64)	C_V total electron heat capacity, V constant V volume $\langle c^2\rangle$ mean square electron speed
	$= \frac{\pi^2 n k_\mathrm{B}^2 \tau T}{3m_\mathrm{e}} \quad (T \ll T_\mathrm{F})$	(6.65)	k_B Boltzmann constant T temperature T_F Fermi temperature
Wiedemann–Franz law[b]	$\frac{\lambda}{\sigma T} = L = \frac{\pi^2 k_\mathrm{B}^2}{3e^2}$	(6.66)	L Lorenz constant ($\simeq 2.45 \times 10^{-8}\,\mathrm{W\,\Omega\,K^{-2}}$) λ thermal conductivity
Hall coefficient[c]	$R_\mathrm{H} = -\frac{1}{ne} = \frac{E_y}{J_x B_z}$	(6.67)	R_H Hall coefficient E_y Hall electric field J_x applied current density B_z magnetic flux density
Hall voltage (rectangular strip)	$V_\mathrm{H} = R_\mathrm{H}\frac{B_z I_x}{w}$	(6.68)	V_H Hall voltage I_x applied current ($= J_x \times$ cross-sectional area) w strip thickness in z

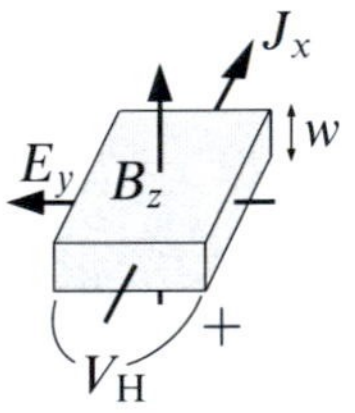

[a]For an electric field varying as $\mathrm{e}^{-\mathrm{i}\omega t}$.
[b]Holds for an arbitrary band structure.
[c]The charge on an electron is $-e$, where e is the elementary charge (approximately $+1.6 \times 10^{-19}\,\mathrm{C}$). The Hall coefficient is therefore a negative number when the dominant charge carriers are electrons.

Fermi gas

Electron density of states[a]	$g(E)=\frac{V}{2\pi^2}\left(\frac{2m_e}{\hbar^2}\right)^{3/2}E^{1/2}$	(6.69)	E electron energy (>0) $g(E)$ density of states V "gas" volume m_e electronic mass $\hbar$ (Planck constant)/(2π)
	$g(E_F)=\frac{3}{2}\frac{nV}{E_F}$	(6.70)	
Fermi wavenumber	$k_F=(3\pi^2 n)^{1/3}$	(6.71)	k_F Fermi wavenumber n number of electrons per unit volume
Fermi velocity	$v_F=\hbar k_F/m_e$	(6.72)	v_F Fermi velocity
Fermi energy ($T=0$)	$E_F=\frac{\hbar^2 k_F^2}{2m_e}=\frac{\hbar^2}{2m_e}(3\pi^2 n)^{2/3}$	(6.73)	E_F Fermi energy
Fermi temperature	$T_F=\frac{E_F}{k_B}$	(6.74)	T_F Fermi temperature k_B Boltzmann constant
Electron heat capacity[b] ($T\ll T_F$)	$C_{Ve}=\frac{\pi^2}{3}g(E_F)k_B^2 T$	(6.75)	C_{Ve} heat capacity per electron T temperature
	$=\frac{\pi^2 k_B^2}{2E_F}T$	(6.76)	
Total kinetic energy ($T=0$)	$U_0=\frac{3}{5}nVE_F$	(6.77)	U_0 total kinetic energy
Pauli paramagnetism	$\boldsymbol{M}=\chi_{HP}\boldsymbol{H}$	(6.78)	χ_{HP} Pauli magnetic susceptibility $\boldsymbol{H}$ magnetic field strength $\boldsymbol{M}$ magnetisation μ_0 permeability of free space μ_B Bohr magneton
	$=\frac{3n}{2E_F}\mu_0\mu_B^2\boldsymbol{H}$	(6.79)	
Landau diamagnetism	$\chi_{HL}=-\frac{1}{3}\chi_{HP}$	(6.80)	χ_{HL} Landau magnetic susceptibility

[a]The density of states is often quoted per unit volume in real space (i.e., $g(E)/V$ here).
[b]Equation (6.75) holds for any density of states.

Thermoelectricity

Thermopower[a]	$\mathscr{E}=\frac{\boldsymbol{J}}{\sigma}+S_T\nabla T$	(6.81)	$\mathscr{E}$ electrochemical field[b] $\boldsymbol{J}$ current density σ electrical conductivity
Peltier effect	$\boldsymbol{H}=\Pi\boldsymbol{J}-\lambda\nabla T$	(6.82)	S_T thermopower T temperature $\boldsymbol{H}$ heat flux per unit area
Kelvin relation	$\Pi=TS_T$	(6.83)	Π Peltier coefficient λ thermal conductivity

[a]Or "absolute thermoelectric power."
[b]The electrochemical field is the gradient of $(\mu/e)-\phi$, where μ is the chemical potential, $-e$ the electronic charge, and ϕ the electrical potential.

Band theory and semiconductors

Bloch's theorem	$\Psi(\boldsymbol{r}+\boldsymbol{R})=\exp(\mathrm{i}\boldsymbol{k}\cdot\boldsymbol{R})\Psi(\boldsymbol{r})$ (6.84)		Ψ electron eigenstate $\boldsymbol{k}$ Bloch wavevector $\boldsymbol{R}$ lattice vector $\boldsymbol{r}$ position vector
Electron velocity	$\boldsymbol{v}_b(\boldsymbol{k})=\frac{1}{\hbar}\nabla_{\boldsymbol{k}}E_b(\boldsymbol{k})$ (6.85)		$\boldsymbol{v}_b$ electron velocity (for wavevector $\boldsymbol{k}$) $\hbar$ (Planck constant)/2π b band index $E_b(\boldsymbol{k})$ energy band
Effective mass tensor	$m_{ij}=\hbar^2\left[\frac{\partial^2 E_b(\boldsymbol{k})}{\partial k_i \partial k_j}\right]^{-1}$ (6.86)		m_{ij} effective mass tensor k_i components of $\boldsymbol{k}$
Scalar effective mass[a]	$m^*=\hbar^2\left[\frac{\partial^2 E_b(k)}{\partial k^2}\right]^{-1}$ (6.87)		m^* scalar effective mass k $=\lvert\boldsymbol{k}\rvert$
Mobility	$\mu=\frac{\lvert\boldsymbol{v}_\mathrm{d}\rvert}{\lvert\boldsymbol{E}\rvert}=\frac{eD}{k_\mathrm{B}T}$ (6.88)		μ particle mobility v_d mean drift velocity $\boldsymbol{E}$ applied electric field $-e$ electronic charge D diffusion coefficient T temperature
Net current density	$\boldsymbol{J}=(n_\mathrm{e}\mu_\mathrm{e}+n_\mathrm{h}\mu_\mathrm{h})e\boldsymbol{E}$ (6.89)		$\boldsymbol{J}$ current density $n_\mathrm{e,h}$ electron, hole, number densities $\mu_\mathrm{e,h}$ electron, hole, mobilities
Semiconductor equation	$n_\mathrm{e}n_\mathrm{h}=\frac{(k_\mathrm{B}T)^3}{2(\pi\hbar^2)^3}(m_\mathrm{e}^* m_\mathrm{h}^*)^{3/2}\mathrm{e}^{-E_\mathrm{g}/(k_\mathrm{B}T)}$ (6.90)		k_B Boltzmann constant E_g band gap $m_\mathrm{e,h}^*$ electron, hole, effective masses
p-n junction	$I=I_0\left[\exp\left(\frac{eV}{k_\mathrm{B}T}\right)-1\right]$ (6.91) $I_0=en_\mathrm{i}^2A\left(\frac{D_\mathrm{e}}{L_\mathrm{e}N_\mathrm{a}}+\frac{D_\mathrm{h}}{L_\mathrm{h}N_\mathrm{d}}\right)$ (6.92) $L_\mathrm{e}=(D_\mathrm{e}\tau_\mathrm{e})^{1/2}$ (6.93) $L_\mathrm{h}=(D_\mathrm{h}\tau_\mathrm{h})^{1/2}$ (6.94)		I current I_0 saturation current V bias voltage (+ for forward) n_i intrinsic carrier concentration A area of junction $D_\mathrm{e,h}$ electron, hole, diffusion coefficients $L_\mathrm{e,h}$ electron, hole, diffusion lengths $\tau_\mathrm{e,h}$ electron, hole, recombination times $N_\mathrm{a,d}$ acceptor, donor, concentrations

[a]Valid for regions of k-space in which $E_b(\boldsymbol{k})$ can be taken as independent of the direction of k.

Chapter 7 Electromagnetism

7.1 Introduction

The electromagnetic force is central to nearly every physical process around us and is a major component of classical physics. In fact, the development of electromagnetic theory in the nineteenth century gave us much mathematical machinery that we now apply quite generally in other fields, including potential theory, vector calculus, and the ideas of divergence and curl.

It is therefore not surprising that this section deals with a large array of physical quantities and their relationships. As usual, SI units are assumed throughout. In the past electromagnetism has suffered from the use of a variety of systems of units, including the cgs system in both its electrostatic (esu) and electromagnetic (emu) forms. The fog has now all but cleared, but some specialised areas of research still cling to these historical measures. Readers are advised to consult the section on unit conversion if they come across such exotica in the literature.

Equations cast in the rationalised units of SI can be readily converted to the once common Gaussian (unrationalised) units by using the following symbol transformations:

7

Equation conversion: SI to Gaussian units

$\epsilon_0 \mapsto 1/(4\pi)$	$\mu_0 \mapsto 4\pi/c^2$	$\boldsymbol{B} \mapsto \boldsymbol{B}/c$
$\chi_E \mapsto 4\pi\chi_E$	$\chi_H \mapsto 4\pi\chi_H$	$\boldsymbol{H} \mapsto c\boldsymbol{H}/(4\pi)$
$\boldsymbol{A} \mapsto \boldsymbol{A}/c$	$\boldsymbol{M} \mapsto c\boldsymbol{M}$	$\boldsymbol{D} \mapsto \boldsymbol{D}/(4\pi)$

The quantities ρ, $\boldsymbol{J}$, $\boldsymbol{E}$, ϕ, σ, $\boldsymbol{P}$, ϵ_r, and μ_r are all unchanged.

7.2 Static fields

Electrostatics

Electrostatic potential	$\boldsymbol{E}=-\nabla\phi$	(7.1)	$\boldsymbol{E}$ electric field ϕ electrostatic potential
Potential difference[a]	$\phi_a-\phi_b=\int_a^b \boldsymbol{E}\cdot \mathrm{d}\boldsymbol{l}=-\int_b^a \boldsymbol{E}\cdot \mathrm{d}\boldsymbol{l}$	(7.2)	ϕ_a potential at a ϕ_b potential at b $\mathrm{d}\boldsymbol{l}$ line element
Poisson's Equation (free space)	$\nabla^2\phi=-\frac{\rho}{\epsilon_0}$	(7.3)	ρ charge density ϵ_0 permittivity of free space
Point charge at $\boldsymbol{r}'$	$\phi(\boldsymbol{r})=\frac{q}{4\pi\epsilon_0\lvert\boldsymbol{r}-\boldsymbol{r}'\rvert}$	(7.4)	q point charge
	$\boldsymbol{E}(\boldsymbol{r})=\frac{q(\boldsymbol{r}-\boldsymbol{r}')}{4\pi\epsilon_0\lvert\boldsymbol{r}-\boldsymbol{r}'\rvert^3}$	(7.5)	
Field from a charge distribution (free space)	$\boldsymbol{E}(\boldsymbol{r})=\frac{1}{4\pi\epsilon_0}\int_{\text{volume}}\frac{\rho(\boldsymbol{r}')(\boldsymbol{r}-\boldsymbol{r}')}{\lvert\boldsymbol{r}-\boldsymbol{r}'\rvert^3}\,\mathrm{d}\tau'$	(7.6)	$\mathrm{d}\tau'$ volume element $\boldsymbol{r}'$ position vector of $\mathrm{d}\tau'$

[a] Between points a and b along a path $\boldsymbol{l}$.

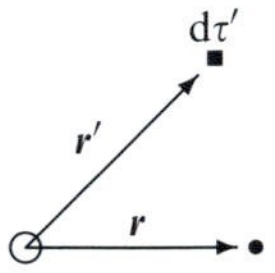

Magnetostatics[a]

Magnetic scalar potential	$\boldsymbol{B}=-\mu_0\nabla\phi_{\mathrm{m}}$	(7.7)	ϕ_{m} magnetic scalar potential $\boldsymbol{B}$ magnetic flux density
ϕ_{m} in terms of the solid angle of a generating current loop	$\phi_{\mathrm{m}}=\frac{I\Omega}{4\pi}$	(7.8)	Ω loop solid angle I current
Biot–Savart law (the field from a line current)	$\boldsymbol{B}(\boldsymbol{r})=\frac{\mu_0 I}{4\pi}\int_{\text{line}}\frac{\mathrm{d}\boldsymbol{l}\times(\boldsymbol{r}-\boldsymbol{r}')}{\lvert\boldsymbol{r}-\boldsymbol{r}'\rvert^3}$	(7.9)	$\mathrm{d}\boldsymbol{l}$ line element in the direction of the current $\boldsymbol{r}'$ position vector of $\mathrm{d}\boldsymbol{l}$
Ampère's law (differential form)	$\nabla\times\boldsymbol{B}=\mu_0\boldsymbol{J}$	(7.10)	$\boldsymbol{J}$ current density μ_0 permeability of free space
Ampère's law (integral form)	$\oint \boldsymbol{B}\cdot\mathrm{d}\boldsymbol{l}=\mu_0 I_{\mathrm{tot}}$	(7.11)	I_{tot} total current through loop

[a] In free space.

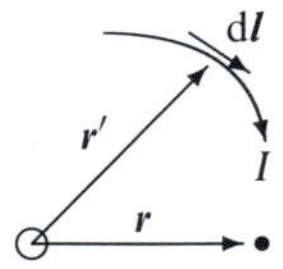

Capacitance[a]

Of sphere, radius a	$C = 4\pi\epsilon_0\epsilon_r a$	(7.12)
Of circular disk, radius a	$C = 8\epsilon_0\epsilon_r a$	(7.13)
Of two spheres, radius a, in contact	$C = 8\pi\epsilon_0\epsilon_r a \ln 2$	(7.14)
Of circular solid cylinder, radius a, length l	$C \simeq [8 + 4.1(l/a)^{0.76}]\epsilon_0\epsilon_r a$	(7.15)
Of nearly spherical surface, area S	$C \simeq 3.139 \times 10^{-11}\epsilon_r S^{1/2}$	(7.16)
Of cube, side a	$C \simeq 7.283 \times 10^{-11}\epsilon_r a$	(7.17)
Between concentric spheres, radii $a < b$	$C = 4\pi\epsilon_0\epsilon_r ab(b-a)^{-1}$	(7.18)
Between coaxial cylinders, radii $a < b$	$C = \dfrac{2\pi\epsilon_0\epsilon_r}{\ln(b/a)}$ per unit length	(7.19)
Between parallel cylinders, separation $2d$, radii a	$C = \dfrac{\pi\epsilon_0\epsilon_r}{\operatorname{arcosh}(d/a)}$ per unit length	(7.20)
	$\simeq \dfrac{\pi\epsilon_0\epsilon_r}{\ln(2d/a)}$ $(d \gg a)$	(7.21)
Between parallel, coaxial circular disks, separation d, radii a	$C \simeq \dfrac{\epsilon_0\epsilon_r\pi a^2}{d} + \epsilon_0\epsilon_r a[\ln(16\pi a/d) - 1]$	(7.22)

[a]For conductors, in an embedding medium of relative permittivity ϵ_r.

7

Inductance[a]

Of N-turn solenoid (straight or toroidal), length l, area A ($\ll l^2$)	$L = \mu_0 N^2 A/l$	(7.23)
Of coaxial cylindrical tubes, radii a, b ($a < b$)	$L = \dfrac{\mu_0}{2\pi}\ln\dfrac{b}{a}$ per unit length	(7.24)
Of parallel wires, radii a, separation $2d$	$L \simeq \dfrac{\mu_0}{\pi}\ln\dfrac{2d}{a}$ per unit length, $(2d \gg a)$	(7.25)
Of wire of radius a bent in a loop of radius $b \gg a$	$L \simeq \mu_0 b\left(\ln\dfrac{8b}{a} - 2\right)$	(7.26)

[a]For currents confined to the surfaces of perfect conductors in free space.

Electric fields[a]

Uniformly charged sphere, radius a, charge q	$\boldsymbol{E}(\boldsymbol{r}) = \begin{cases} \dfrac{q}{4\pi\epsilon_0 a^3}\boldsymbol{r} & (r<a) \\ \dfrac{q}{4\pi\epsilon_0 r^3}\boldsymbol{r} & (r\geq a) \end{cases}$	(7.27)
Uniformly charged disk, radius a, charge q (on axis, z)	$\boldsymbol{E}(z) = \dfrac{q}{2\pi\epsilon_0 a^2}\boldsymbol{z}\left(\dfrac{1}{\lvert z\rvert} - \dfrac{1}{\sqrt{z^2+a^2}}\right)$	(7.28)
Line charge, charge density λ per unit length	$\boldsymbol{E}(\boldsymbol{r}) = \dfrac{\lambda}{2\pi\epsilon_0 r^2}\boldsymbol{r}$	(7.29)
Electric dipole, moment $\boldsymbol{p}$ (spherical polar coordinates, θ angle between $\boldsymbol{p}$ and $\boldsymbol{r}$)	$E_r = \dfrac{p\cos\theta}{2\pi\epsilon_0 r^3}$	(7.30)
	$E_\theta = \dfrac{p\sin\theta}{4\pi\epsilon_0 r^3}$	(7.31)
Charge sheet, surface density σ	$E = \dfrac{\sigma}{2\epsilon_0}$	(7.32)

[a]For $\epsilon_r = 1$ in the surrounding medium.

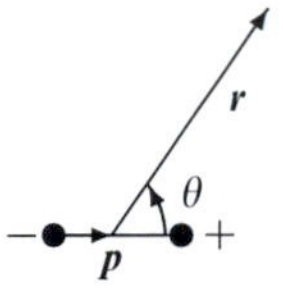

Magnetic fields[a]

Uniform infinite solenoid, current I, n turns per unit length	$B = \begin{cases} \mu_0 nI & \text{inside (axial)} \\ 0 & \text{outside} \end{cases}$	(7.33)
Uniform cylinder of current I, radius a	$B(r) = \begin{cases} \mu_0 Ir/(2\pi a^2) & r<a \\ \mu_0 I/(2\pi r) & r\geq a \end{cases}$	(7.34)
Magnetic dipole, moment $\boldsymbol{m}$ (θ angle between $\boldsymbol{m}$ and $\boldsymbol{r}$)	$B_r = \mu_0 \dfrac{m\cos\theta}{2\pi r^3}$	(7.35)
	$B_\theta = \dfrac{\mu_0 m\sin\theta}{4\pi r^3}$	(7.36)
Circular current loop of N turns, radius a, along axis, z	$B(z) = \dfrac{\mu_0 NI}{2}\dfrac{a^2}{(a^2+z^2)^{3/2}}$	(7.37)
The axis, z, of a straight solenoid, n turns per unit length, current I	$B_{\text{axis}} = \dfrac{\mu_0 nI}{2}(\cos\alpha_1 - \cos\alpha_2)$	(7.38)

[a]For $\mu_r = 1$ in the surrounding medium.

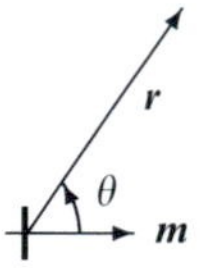

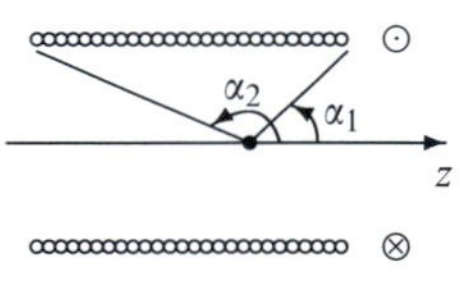

Image charges

Real charge, $+q$, at a distance:	*image point*	*image charge*
b from a conducting plane	$-b$	$-q$
b from a conducting sphere, radius a	a^2/b	$-qa/b$
b from a plane dielectric boundary:		
seen from free space	$-b$	$-q(\epsilon_r-1)/(\epsilon_r+1)$
seen from the dielectric	b	$+2q/(\epsilon_r+1)$

7.3 Electromagnetic fields (general)

Field relationships

Conservation of charge	$\nabla\cdot\boldsymbol{J}=-\dfrac{\partial\rho}{\partial t}$	(7.39)	$\boldsymbol{J}$ ρ t	current density charge density time
Magnetic vector potential	$\boldsymbol{B}=\nabla\times\boldsymbol{A}$	(7.40)	$\boldsymbol{A}$	vector potential
Electric field from potentials	$\boldsymbol{E}=-\dfrac{\partial\boldsymbol{A}}{\partial t}-\nabla\phi$	(7.41)	ϕ	electrical potential
Coulomb gauge condition	$\nabla\cdot\boldsymbol{A}=0$	(7.42)		
Lorenz gauge condition	$\nabla\cdot\boldsymbol{A}+\dfrac{1}{c^2}\dfrac{\partial\phi}{\partial t}=0$	(7.43)	c	speed of light
Potential field equations[a]	$\dfrac{1}{c^2}\dfrac{\partial^2\phi}{\partial t^2}-\nabla^2\phi=\dfrac{\rho}{\epsilon_0}$	(7.44)	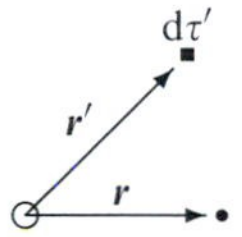	
	$\dfrac{1}{c^2}\dfrac{\partial^2\boldsymbol{A}}{\partial t^2}-\nabla^2\boldsymbol{A}=\mu_0\boldsymbol{J}$	(7.45)		
Expression for ϕ in terms of ρ[a]	$\phi(\boldsymbol{r},t)=\dfrac{1}{4\pi\epsilon_0}\displaystyle\int_{\text{volume}}\dfrac{\rho(\boldsymbol{r}',t-\lvert\boldsymbol{r}-\boldsymbol{r}'\rvert/c)}{\lvert\boldsymbol{r}-\boldsymbol{r}'\rvert}\,\mathrm{d}\tau'$	(7.46)	$\mathrm{d}\tau'$ $\boldsymbol{r}'$	volume element position vector of $\mathrm{d}\tau'$
Expression for $\boldsymbol{A}$ in terms of $\boldsymbol{J}$[a]	$\boldsymbol{A}(\boldsymbol{r},t)=\dfrac{\mu_0}{4\pi}\displaystyle\int_{\text{volume}}\dfrac{\boldsymbol{J}(\boldsymbol{r}',t-\lvert\boldsymbol{r}-\boldsymbol{r}'\rvert/c)}{\lvert\boldsymbol{r}-\boldsymbol{r}'\rvert}\,\mathrm{d}\tau'$	(7.47)	μ_0	permeability of free space

[a]Assumes the Lorenz gauge.

Liénard–Wiechert potentials[a]

Electrical potential of a moving point charge	$\phi=\dfrac{q}{4\pi\epsilon_0(\lvert\boldsymbol{r}\rvert-\boldsymbol{v}\cdot\boldsymbol{r}/c)}$	(7.48)	q $\boldsymbol{r}$ $\boldsymbol{v}$	charge vector from charge to point of observation particle velocity
Magnetic vector potential of a moving point charge	$\boldsymbol{A}=\dfrac{\mu_0 q\boldsymbol{v}}{4\pi(\lvert\boldsymbol{r}\rvert-\boldsymbol{v}\cdot\boldsymbol{r}/c)}$	(7.49)	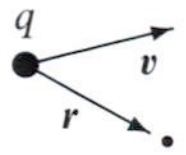	

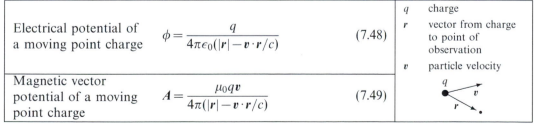

[a]In free space. The right-hand sides of these equations are evaluated at retarded times, i.e., at $t'=t-\lvert\boldsymbol{r}'\rvert/c$, where $\boldsymbol{r}'$ is the vector from the charge to the observation point at time t'.

Maxwell's equations

Differential form:		Integral form:	
$\nabla\cdot\boldsymbol{E}=\frac{\rho}{\epsilon_0}$	(7.50)	$\oint_{\text{closed surface}} \boldsymbol{E}\cdot d\boldsymbol{s}=\frac{1}{\epsilon_0}\int_{\text{volume}}\rho\, d\tau$	(7.51)
$\nabla\cdot\boldsymbol{B}=0$	(7.52)	$\oint_{\text{closed surface}} \boldsymbol{B}\cdot d\boldsymbol{s}=0$	(7.53)
$\nabla\times\boldsymbol{E}=-\frac{\partial \boldsymbol{B}}{\partial t}$	(7.54)	$\oint_{\text{loop}} \boldsymbol{E}\cdot d\boldsymbol{l}=-\frac{d\Phi}{dt}$	(7.55)
$\nabla\times\boldsymbol{B}=\mu_0\boldsymbol{J}+\mu_0\epsilon_0\frac{\partial \boldsymbol{E}}{\partial t}$	(7.56)	$\oint_{\text{loop}} \boldsymbol{B}\cdot d\boldsymbol{l}=\mu_0 I+\mu_0\epsilon_0\int_{\text{surface}}\frac{\partial \boldsymbol{E}}{\partial t}\cdot d\boldsymbol{s}$	(7.57)

	Equation (7.51) is "Gauss's law"	$d\boldsymbol{s}$	surface element
	Equation (7.55) is "Faraday's law"	$d\tau$	volume element
$\boldsymbol{E}$	electric field	$d\boldsymbol{l}$	line element
$\boldsymbol{B}$	magnetic flux density	Φ	linked magnetic flux ($=\int \boldsymbol{B}\cdot d\boldsymbol{s}$)
$\boldsymbol{J}$	current density	I	linked current ($=\int \boldsymbol{J}\cdot d\boldsymbol{s}$)
ρ	charge density	t	time

Maxwell's equations (using *D* and *H*)

Differential form:		Integral form:	
$\nabla\cdot\boldsymbol{D}=\rho_{\text{free}}$	(7.58)	$\oint_{\text{closed surface}} \boldsymbol{D}\cdot d\boldsymbol{s}=\int_{\text{volume}}\rho_{\text{free}}\, d\tau$	(7.59)
$\nabla\cdot\boldsymbol{B}=0$	(7.60)	$\oint_{\text{closed surface}} \boldsymbol{B}\cdot d\boldsymbol{s}=0$	(7.61)
$\nabla\times\boldsymbol{E}=-\frac{\partial \boldsymbol{B}}{\partial t}$	(7.62)	$\oint_{\text{loop}} \boldsymbol{E}\cdot d\boldsymbol{l}=-\frac{d\Phi}{dt}$	(7.63)
$\nabla\times\boldsymbol{H}=\boldsymbol{J}_{\text{free}}+\frac{\partial \boldsymbol{D}}{\partial t}$	(7.64)	$\oint_{\text{loop}} \boldsymbol{H}\cdot d\boldsymbol{l}=I_{\text{free}}+\int_{\text{surface}}\frac{\partial \boldsymbol{D}}{\partial t}\cdot d\boldsymbol{s}$	(7.65)

$\boldsymbol{D}$	displacement field	$\boldsymbol{E}$	electric field
ρ_{free}	free charge density (in the sense of $\rho=\rho_{\text{induced}}+\rho_{\text{free}}$)	$d\boldsymbol{s}$	surface element
$\boldsymbol{B}$	magnetic flux density	$d\tau$	volume element
$\boldsymbol{H}$	magnetic field strength	$d\boldsymbol{l}$	line element
$\boldsymbol{J}_{\text{free}}$	free current density (in the sense of $\boldsymbol{J}=\boldsymbol{J}_{\text{induced}}+\boldsymbol{J}_{\text{free}}$)	Φ	linked magnetic flux ($=\int \boldsymbol{B}\cdot d\boldsymbol{s}$)
		I_{free}	linked free current ($=\int \boldsymbol{J}_{\text{free}}\cdot d\boldsymbol{s}$)
		t	time

Relativistic electrodynamics

Lorentz transformation of electric and magnetic fields	$\boldsymbol{E}'_{\parallel}=\boldsymbol{E}_{\parallel}$ (7.66) $\boldsymbol{E}'_{\perp}=\gamma(\boldsymbol{E}+\boldsymbol{v}\times\boldsymbol{B})_{\perp}$ (7.67) $\boldsymbol{B}'_{\parallel}=\boldsymbol{B}_{\parallel}$ (7.68) $\boldsymbol{B}'_{\perp}=\gamma(\boldsymbol{B}-\boldsymbol{v}\times\boldsymbol{E}/c^2)_{\perp}$ (7.69)	$\boldsymbol{E}$ $\boldsymbol{B}$ $'$ γ $\parallel$ $\perp$	electric field magnetic flux density measured in frame moving at relative velocity $\boldsymbol{v}$ Lorentz factor $=[1-(v/c)^2]^{-1/2}$ parallel to $\boldsymbol{v}$ perpendicular to $\boldsymbol{v}$
Lorentz transformation of current and charge densities	$\rho'=\gamma(\rho-vJ_{\parallel}/c^2)$ (7.70) $J'_{\perp}=J_{\perp}$ (7.71) $J'_{\parallel}=\gamma(J_{\parallel}-v\rho)$ (7.72)	$\boldsymbol{J}$ ρ	current density charge density
Lorentz transformation of potential fields	$\phi'=\gamma(\phi-vA_{\parallel})$ (7.73) $A'_{\perp}=A_{\perp}$ (7.74) $A'_{\parallel}=\gamma(A_{\parallel}-v\phi/c^2)$ (7.75)	ϕ $\boldsymbol{A}$	electric potential magnetic vector potential
Four-vector fields[a]	$\underset{\sim}{J}=(\rho c,\boldsymbol{J})$ (7.76) $\underset{\sim}{A}=\left(\frac{\phi}{c},\boldsymbol{A}\right)$ (7.77) $\square^2=\left(\frac{1}{c^2}\frac{\partial^2}{\partial t^2},-\nabla^2\right)$ (7.78) $\square^2\underset{\sim}{A}=\mu_0\underset{\sim}{J}$ (7.79)	$\underset{\sim}{J}$ $\underset{\sim}{A}$ $\square^2$	current density four-vector potential four-vector D'Alembertian operator

[a]Other sign conventions are common here. See page 65 for a general definition of four-vectors.

7

7.4 Fields associated with media

Polarisation

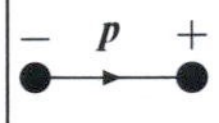

Definition of electric dipole moment	$\boldsymbol{p}=q\boldsymbol{a}$	(7.80)	$\pm q$ end charges $\boldsymbol{a}$ charge separation vector (from − to +)
Generalised electric dipole moment	$\boldsymbol{p}=\int_{\text{volume}} \boldsymbol{r}'\rho\,\mathrm{d}\tau'$	(7.81)	$\boldsymbol{p}$ dipole moment ρ charge density $\mathrm{d}\tau'$ volume element $\boldsymbol{r}'$ vector to $\mathrm{d}\tau'$
Electric dipole potential	$\phi(\boldsymbol{r})=\frac{\boldsymbol{p}\cdot\boldsymbol{r}}{4\pi\epsilon_0 r^3}$	(7.82)	ϕ dipole potential $\boldsymbol{r}$ vector from dipole ϵ_0 permittivity of free space
Dipole moment per unit volume (polarisation)[a]	$\boldsymbol{P}=n\boldsymbol{p}$	(7.83)	$\boldsymbol{P}$ polarisation n number of dipoles per unit volume
Induced volume charge density	$\nabla\cdot\boldsymbol{P}=-\rho_{\text{ind}}$	(7.84)	ρ_{ind} volume charge density
Induced surface charge density	$\sigma_{\text{ind}}=\boldsymbol{P}\cdot\hat{\boldsymbol{s}}$	(7.85)	σ_{ind} surface charge density $\hat{\boldsymbol{s}}$ unit normal to surface
Definition of electric displacement	$\boldsymbol{D}=\epsilon_0\boldsymbol{E}+\boldsymbol{P}$	(7.86)	$\boldsymbol{D}$ electric displacement $\boldsymbol{E}$ electric field
Definition of electric susceptibility	$\boldsymbol{P}=\epsilon_0\chi_E\boldsymbol{E}$	(7.87)	χ_E electrical susceptibility (may be a tensor)
Definition of relative permittivity[b]	$\epsilon_r=1+\chi_E$ $\boldsymbol{D}=\epsilon_0\epsilon_r\boldsymbol{E}$ $=\epsilon\boldsymbol{E}$	(7.88) (7.89) (7.90)	ϵ_r relative permittivity ϵ permittivity
Atomic polarisability[c]	$\boldsymbol{p}=\alpha\boldsymbol{E}_{\text{loc}}$	(7.91)	α polarisability $\boldsymbol{E}_{\text{loc}}$ local electric field
Depolarising fields	$\boldsymbol{E}_{\text{d}}=-\frac{N_{\text{d}}\boldsymbol{P}}{\epsilon_0}$	(7.92)	$\boldsymbol{E}_{\text{d}}$ depolarising field N_{d} depolarising factor =1/3 (sphere) =1 (thin slab ⊥ to $\boldsymbol{P}$) =0 (thin slab ∥ to $\boldsymbol{P}$) =1/2 (long circular cylinder, axis ⊥ to $\boldsymbol{P}$)
Clausius–Mossotti equation[d]	$\frac{n\alpha}{3\epsilon_0}=\frac{\epsilon_r-1}{\epsilon_r+2}$	(7.93)	

[a] Assuming dipoles are parallel. The equivalent of Equation (7.112) holds for a hot gas of electric dipoles.
[b] Relative permittivity as defined here is for a linear isotropic medium.
[c] The polarisability of a conducting sphere radius a is $\alpha=4\pi\epsilon_0 a^3$. The definition $\boldsymbol{p}=\alpha\epsilon_0\boldsymbol{E}_{\text{loc}}$ is also used.
[d] With the substitution $\eta^2=\epsilon_r$ [cf. Equation (7.195) with $\mu_r=1$] this is also known as the "Lorentz–Lorenz formula."

Magnetisation

Definition of magnetic dipole moment	$\mathrm{d}\boldsymbol{m} = I\,\mathrm{d}\boldsymbol{s}$	(7.94)	$\mathrm{d}\boldsymbol{m}$ dipole moment I loop current $\mathrm{d}\boldsymbol{s}$ loop area (right-hand sense with respect to loop current)
Generalised magnetic dipole moment	$\boldsymbol{m} = \frac{1}{2}\int_{\text{volume}} \boldsymbol{r}' \times \boldsymbol{J}\,\mathrm{d}\tau'$	(7.95)	$\boldsymbol{m}$ dipole moment $\boldsymbol{J}$ current density $\mathrm{d}\tau'$ volume element $\boldsymbol{r}'$ vector to $\mathrm{d}\tau'$
Magnetic dipole (scalar) potential	$\phi_{\mathrm{m}}(\boldsymbol{r}) = \frac{\mu_0 \boldsymbol{m}\cdot\boldsymbol{r}}{4\pi r^3}$	(7.96)	ϕ_{m} magnetic scalar potential $\boldsymbol{r}$ vector from dipole μ_0 permeability of free space
Dipole moment per unit volume (magnetisation)[a]	$\boldsymbol{M} = n\boldsymbol{m}$	(7.97)	$\boldsymbol{M}$ magnetisation n number of dipoles per unit volume
Induced volume current density	$\boldsymbol{J}_{\mathrm{ind}} = \nabla \times \boldsymbol{M}$	(7.98)	$\boldsymbol{J}_{\mathrm{ind}}$ volume current density (i.e., A m^{-2})
Induced surface current density	$\boldsymbol{j}_{\mathrm{ind}} = \boldsymbol{M} \times \hat{\boldsymbol{s}}$	(7.99)	$\boldsymbol{j}_{\mathrm{ind}}$ surface current density (i.e., A m^{-1}) $\hat{\boldsymbol{s}}$ unit normal to surface
Definition of magnetic field strength, $\boldsymbol{H}$	$\boldsymbol{B} = \mu_0(\boldsymbol{H} + \boldsymbol{M})$	(7.100)	$\boldsymbol{B}$ magnetic flux density $\boldsymbol{H}$ magnetic field strength
Definition of magnetic susceptibility	$\boldsymbol{M} = \chi_H \boldsymbol{H}$ $= \frac{\chi_B \boldsymbol{B}}{\mu_0}$ $\chi_B = \frac{\chi_H}{1+\chi_H}$	(7.101) (7.102) (7.103)	χ_H magnetic susceptibility. χ_B is also used (both may be tensors)
Definition of relative permeability[b]	$\boldsymbol{B} = \mu_0 \mu_{\mathrm{r}} \boldsymbol{H}$ $= \mu \boldsymbol{H}$ $\mu_{\mathrm{r}} = 1 + \chi_H$ $= \frac{1}{1-\chi_B}$	(7.104) (7.105) (7.106) (7.107)	μ_{r} relative permeability μ permeability

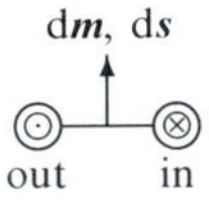

[a]Assuming all the dipoles are parallel. See Equation (7.112) for a classical paramagnetic gas and page 101 for the quantum generalisation.
[b]Relative permeability as defined here is for a linear isotropic medium.

7

Paramagnetism and diamagnetism

Diamagnetic moment of an atom	$\boldsymbol{m} = -\frac{e^2}{6m_e} Z \langle r^2 \rangle \boldsymbol{B}$	(7.108)	$\boldsymbol{m}$ magnetic moment; $\langle r^2 \rangle$ mean squared orbital radius (of all electrons); Z atomic number; $\boldsymbol{B}$ magnetic flux density; m_e electron mass; $-e$ electronic charge
Intrinsic electron magnetic moment[a]	$\boldsymbol{m} \simeq -\frac{e}{2m_e} g \boldsymbol{J}$	(7.109)	$\boldsymbol{J}$ total angular momentum; g Landé g-factor (=2 for spin, =1 for orbital momentum)
Langevin function	$\mathscr{L}(x) = \coth x - \frac{1}{x}$	(7.110)	$\mathscr{L}(x)$ Langevin function
	$\simeq x/3 \qquad (x \lesssim 1)$	(7.111)	
Classical gas paramagnetism ($\lvert \boldsymbol{J} \rvert \gg \hbar$)	$\langle M \rangle = n m_0 \mathscr{L}\left(\frac{m_0 B}{kT}\right)$	(7.112)	$\langle M \rangle$ apparent magnetisation; m_0 magnitude of magnetic dipole moment; n dipole number density
Curie's law	$\chi_H = \frac{\mu_0 n m_0^2}{3kT}$	(7.113)	T temperature; k Boltzmann constant; χ_H magnetic susceptibility
Curie–Weiss law	$\chi_H = \frac{\mu_0 n m_0^2}{3k(T - T_c)}$	(7.114)	μ_0 permeability of free space; T_c Curie temperature

[a] See also page 100.

Boundary conditions for $\boldsymbol{E}$, $\boldsymbol{D}$, $\boldsymbol{B}$, and $\boldsymbol{H}$[a]

Parallel component of the electric field	$E_{\parallel}$ continuous	(7.115)	$\parallel$ component parallel to interface
Perpendicular component of the magnetic flux density	$B_{\perp}$ continuous	(7.116)	$\perp$ component perpendicular to interface
Electric displacement[b]	$\hat{\boldsymbol{s}} \cdot (\boldsymbol{D}_2 - \boldsymbol{D}_1) = \sigma$	(7.117)	$\boldsymbol{D}_{1,2}$ electrical displacements in media 1 & 2; $\hat{\boldsymbol{s}}$ unit normal to surface, directed $1 \rightarrow 2$; σ surface density of free charge
Magnetic field strength[c]	$\hat{\boldsymbol{s}} \times (\boldsymbol{H}_2 - \boldsymbol{H}_1) = \boldsymbol{j}_s$	(7.118)	$\boldsymbol{H}_{1,2}$ magnetic field strengths in media 1 & 2; $\boldsymbol{j}_s$ surface current per unit width

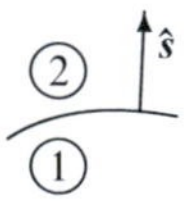

[a] At the plane surface between two uniform media.
[b] If $\sigma = 0$, then $D_{\perp}$ is continuous.
[c] If $\boldsymbol{j}_s = \boldsymbol{0}$ then $H_{\parallel}$ is continuous.

7.5 Force, torque, and energy

Electromagnetic force and torque

Force between two static charges: Coulomb's law	$\boldsymbol{F}_2=\frac{q_1q_2}{4\pi\epsilon_0 r_{12}^2}\hat{\boldsymbol{r}}_{12}$	(7.119)	$\boldsymbol{F}_2$ force on q_2 $q_{1,2}$ charges $\boldsymbol{r}_{12}$ vector from 1 to 2 $\hat{}$ unit vector ϵ_0 permittivity of free space
Force between two current-carrying elements	$\mathrm{d}\boldsymbol{F}_2=\frac{\mu_0 I_1 I_2}{4\pi r_{12}^2}[\mathrm{d}\boldsymbol{l}_2\times(\mathrm{d}\boldsymbol{l}_1\times\hat{\boldsymbol{r}}_{12})]$	(7.120)	$\mathrm{d}\boldsymbol{l}_{1,2}$ line elements $I_{1,2}$ currents flowing along $\mathrm{d}\boldsymbol{l}_1$ and $\mathrm{d}\boldsymbol{l}_2$ $\mathrm{d}\boldsymbol{F}_2$ force on $\mathrm{d}\boldsymbol{l}_2$ μ_0 permeability of free space
Force on a current-carrying element in a magnetic field	$\mathrm{d}\boldsymbol{F}=I\,\mathrm{d}\boldsymbol{l}\times\boldsymbol{B}$	(7.121)	$\mathrm{d}\boldsymbol{l}$ line element $\boldsymbol{F}$ force I current flowing along $\mathrm{d}\boldsymbol{l}$ $\boldsymbol{B}$ magnetic flux density
Force on a charge (Lorentz force)	$\boldsymbol{F}=q(\boldsymbol{E}+\boldsymbol{v}\times\boldsymbol{B})$	(7.122)	$\boldsymbol{E}$ electric field $\boldsymbol{v}$ charge velocity
Force on an electric dipole[a]	$\boldsymbol{F}=(\boldsymbol{p}\cdot\nabla)\boldsymbol{E}$	(7.123)	$\boldsymbol{p}$ electric dipole moment
Force on a magnetic dipole[b]	$\boldsymbol{F}=(\boldsymbol{m}\cdot\nabla)\boldsymbol{B}$	(7.124)	$\boldsymbol{m}$ magnetic dipole moment
Torque on an electric dipole	$\boldsymbol{G}=\boldsymbol{p}\times\boldsymbol{E}$	(7.125)	$\boldsymbol{G}$ torque
Torque on a magnetic dipole	$\boldsymbol{G}=\boldsymbol{m}\times\boldsymbol{B}$	(7.126)	
Torque on a current loop	$\boldsymbol{G}=I_\mathrm{L}\oint_{\mathrm{loop}}\boldsymbol{r}\times(\mathrm{d}\boldsymbol{l}_\mathrm{L}\times\boldsymbol{B})$	(7.127)	$\mathrm{d}\boldsymbol{l}_\mathrm{L}$ line-element (of loop) $\boldsymbol{r}$ position vector of $\mathrm{d}\boldsymbol{l}_\mathrm{L}$ I_L current around loop

[a] $\boldsymbol{F}$ simplifies to $\nabla(\boldsymbol{p}\cdot\boldsymbol{E})$ if p is intrinsic, $\nabla(pE/2)$ if $\boldsymbol{p}$ is induced by $\boldsymbol{E}$ and the medium is isotropic.
[b] $\boldsymbol{F}$ simplifies to $\nabla(\boldsymbol{m}\cdot\boldsymbol{B})$ if m is intrinsic, $\nabla(mB/2)$ if $\boldsymbol{m}$ is induced by $\boldsymbol{B}$ and the medium is isotropic.

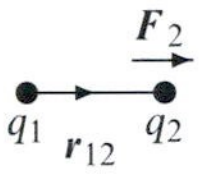

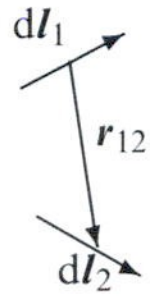

Electromagnetic energy

Electromagnetic field energy density (in free space)	$u=\frac{1}{2}\epsilon_0 E^2+\frac{1}{2}\frac{B^2}{\mu_0}$	(7.128)	u energy density E electric field B magnetic flux density
Energy density in media	$u=\frac{1}{2}(\boldsymbol{D}\cdot\boldsymbol{E}+\boldsymbol{B}\cdot\boldsymbol{H})$	(7.129)	ϵ_0 permittivity of free space μ_0 permeability of free space $\boldsymbol{D}$ electric displacement $\boldsymbol{H}$ magnetic field strength
Energy flow (Poynting) vector	$\boldsymbol{N}=\boldsymbol{E}\times\boldsymbol{H}$	(7.130)	c speed of light $\boldsymbol{N}$ energy flow rate per unit area $\perp$ to the flow direction
Mean flux density at a distance r from a short oscillating dipole	$\langle\boldsymbol{N}\rangle=\frac{\omega^4 p_0^2\sin^2\theta}{32\pi^2\epsilon_0 c^3 r^3}\boldsymbol{r}$	(7.131)	p_0 amplitude of dipole moment $\boldsymbol{r}$ vector from dipole ($\gg$wavelength) θ angle between $\boldsymbol{p}$ and $\boldsymbol{r}$ ω oscillation frequency
Total mean power from oscillating dipole[a]	$W=\frac{\omega^4 p_0^2/2}{6\pi\epsilon_0 c^3}$	(7.132)	W total mean radiated power
Self-energy of a charge distribution	$U_{\text{tot}}=\frac{1}{2}\int_{\text{volume}}\phi(\boldsymbol{r})\rho(\boldsymbol{r})\,\mathrm{d}\tau$	(7.133)	U_{tot} total energy $\mathrm{d}\tau$ volume element $\boldsymbol{r}$ position vector of $\mathrm{d}\tau$ ϕ electrical potential ρ charge density
Energy of an assembly of capacitors[b]	$U_{\text{tot}}=\frac{1}{2}\sum_i\sum_j C_{ij}V_iV_j$	(7.134)	V_i potential of ith capacitor C_{ij} mutual capacitance between capacitors i and j
Energy of an assembly of inductors[c]	$U_{\text{tot}}=\frac{1}{2}\sum_i\sum_j L_{ij}I_iI_j$	(7.135)	L_{ij} mutual inductance between inductors i and j
Intrinsic dipole in an electric field	$U_{\text{dip}}=-\boldsymbol{p}\cdot\boldsymbol{E}$	(7.136)	U_{dip} energy of dipole $\boldsymbol{p}$ electric dipole moment
Intrinsic dipole in a magnetic field	$U_{\text{dip}}=-\boldsymbol{m}\cdot\boldsymbol{B}$	(7.137)	$\boldsymbol{m}$ magnetic dipole moment
Hamiltonian of a charged particle in an EM field[d]	$H=\frac{\lvert\boldsymbol{p}_m-q\boldsymbol{A}\rvert^2}{2m}+q\phi$	(7.138)	H Hamiltonian $\boldsymbol{p}_m$ particle momentum q particle charge m particle mass $\boldsymbol{A}$ magnetic vector potential

[a]Sometimes called "Larmor's formula."
[b]C_{ii} is the self-capacitance of the ith body. Note that $C_{ij}=C_{ji}$.
[c]L_{ii} is the self-inductance of the ith body. Note that $L_{ij}=L_{ji}$.
[d]Newtonian limit, i.e., velocity $\ll c$.

7.6 LCR circuits

LCR definitions

Current	$I=\frac{dQ}{dt}$	(7.139)	I current Q charge
Ohm's law	$V=IR$	(7.140)	R resistance V potential difference over R I current through R
Ohm's law (field form)	$\boldsymbol{J}=\sigma\boldsymbol{E}$	(7.141)	$\boldsymbol{J}$ current density $\boldsymbol{E}$ electric field σ conductivity
Resistivity	$\rho=\frac{1}{\sigma}=\frac{RA}{l}$	(7.142)	ρ resistivity A area of face (I is normal to face) l length
Capacitance	$C=\frac{Q}{V}$	(7.143)	C capacitance V potential difference across C
Current through capacitor	$I=C\frac{dV}{dt}$	(7.144)	I current through C t time
Self-inductance	$L=\frac{\Phi}{I}$	(7.145)	Φ total linked flux I current through inductor
Voltage across inductor	$V=-L\frac{dI}{dt}$	(7.146)	V potential difference over L
Mutual inductance	$L_{12}=\frac{\Phi_1}{I_2}=L_{21}$	(7.147)	Φ_1 total flux from loop 2 linked by loop 1 L_{12} mutual inductance I_2 current through loop 2
Coefficient of coupling	$\lvert L_{12}\rvert=k\sqrt{L_1L_2}$	(7.148)	k coupling coefficient between L_1 and L_2 (≤ 1)
Linked magnetic flux through a coil	$\Phi=N\phi$	(7.149)	Φ linked flux N number of turns around ϕ ϕ flux through area of turns

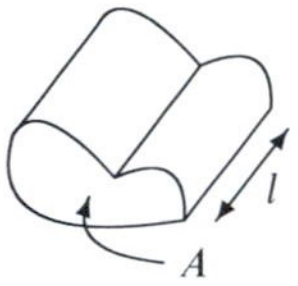

Resonant LCR circuits

Phase resonant frequency[a]	$\omega_0^2 = \begin{cases} 1/LC & \text{(series)} \\ 1/LC - R^2/L^2 & \text{(parallel)} \end{cases}$ (7.150)	ω_0 resonant angular frequency; L inductance; C capacitance; R resistance
Tuning[b]	$\frac{\delta\omega}{\omega_0} = \frac{1}{Q} = \frac{R}{\omega_0 L}$ (7.151)	$\delta\omega$ half-power bandwidth; Q quality factor
Quality factor	$Q = 2\pi \frac{\text{stored energy}}{\text{energy lost per cycle}}$ (7.152)	

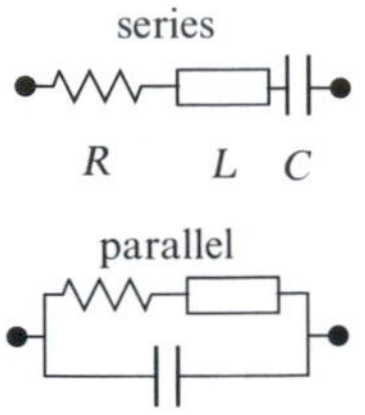

[a]At which the impedance is purely real.
[b]Assuming the capacitor is purely reactive. If L and R are parallel, then $1/Q = \omega_0 L/R$.

Energy in capacitors, inductors, and resistors

Energy stored in a capacitor	$U = \frac{1}{2}CV^2 = \frac{1}{2}QV = \frac{1}{2}\frac{Q^2}{C}$	(7.153)	U stored energy; C capacitance; Q charge; V potential difference
Energy stored in an inductor	$U = \frac{1}{2}LI^2 = \frac{1}{2}\Phi I = \frac{1}{2}\frac{\Phi^2}{L}$	(7.154)	L inductance; Φ linked magnetic flux; I current
Power dissipated in a resistor[a] (Joule's law)	$W = IV = I^2R = \frac{V^2}{R}$	(7.155)	W power dissipated; R resistance
Relaxation time	$\tau = \frac{\epsilon_0\epsilon_r}{\sigma}$	(7.156)	τ relaxation time; ϵ_r relative permittivity; σ conductivity

[a]This is d.c., or instantaneous a.c., power.

Electrical impedance

Impedances in series	$\boldsymbol{Z}_{\text{tot}} = \sum_n \boldsymbol{Z}_n$	(7.157)
Impedances in parallel	$\boldsymbol{Z}_{\text{tot}} = \left(\sum_n \boldsymbol{Z}_n^{-1}\right)^{-1}$	(7.158)
Impedance of capacitance	$\boldsymbol{Z}_{\text{C}} = -\frac{\mathbf{i}}{\omega C}$	(7.159)
Impedance of inductance	$\boldsymbol{Z}_{\text{L}} = \mathbf{i}\omega L$	(7.160)

Impedance: $\boldsymbol{Z}$ — Capacitance: C
Inductance: L — Resistance: $R = \text{Re}[\boldsymbol{Z}]$
Conductance: $G = 1/R$ — Reactance: $X = \text{Im}[\boldsymbol{Z}]$
Admittance: $\boldsymbol{Y} = 1/\boldsymbol{Z}$ — Susceptance: $S = 1/X$

Kirchhoff's laws

Current law	$\sum_{\text{node}} I_i = 0$	(7.161)	I_i	currents impinging on node
Voltage law	$\sum_{\text{loop}} V_i = 0$	(7.162)	V_i	potential differences around loop

Transformers[a]

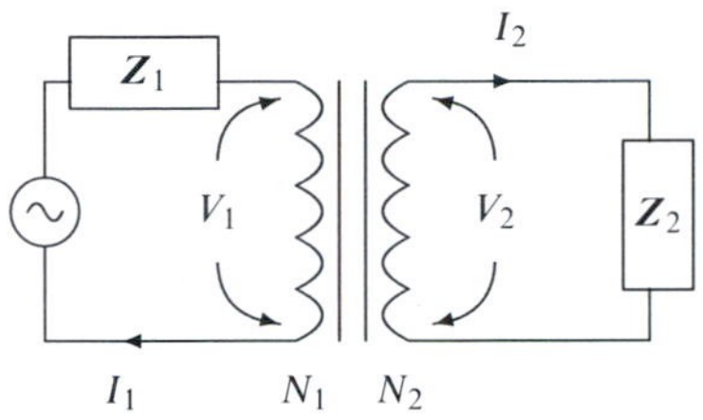

n	turns ratio
N_1	number of primary turns
N_2	number of secondary turns
V_1	primary voltage
V_2	secondary voltage
I_1	primary current
I_2	secondary current
$\boldsymbol{Z}_{\text{out}}$	output impedance
$\boldsymbol{Z}_{\text{in}}$	input impedance
$\boldsymbol{Z}_1$	source impedance
$\boldsymbol{Z}_2$	load impedance

Turns ratio	$n = N_2/N_1$	(7.163)
Transformation of voltage and current	$V_2 = nV_1$	(7.164)
	$I_2 = I_1/n$	(7.165)
Output impedance (seen by $\boldsymbol{Z}_2$)	$\boldsymbol{Z}_{\text{out}} = n^2 \boldsymbol{Z}_1$	(7.166)
Input impedance (seen by $\boldsymbol{Z}_1$)	$\boldsymbol{Z}_{\text{in}} = \boldsymbol{Z}_2/n^2$	(7.167)

[a]Ideal, with a coupling constant of 1 between loss-free windings.

Star–delta transformation

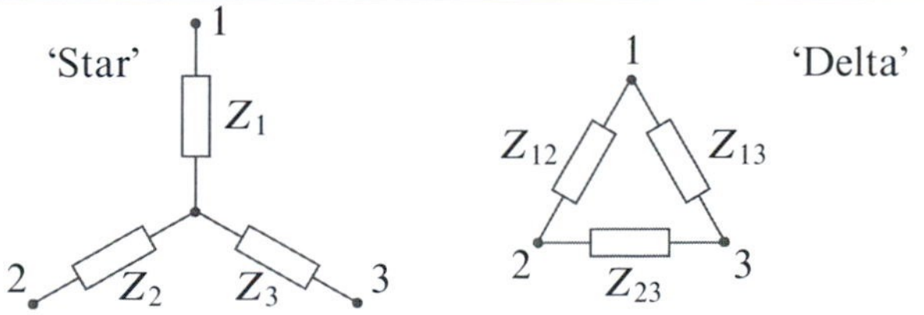

i,j,k	node indices (1,2, or 3)
Z_i	impedance on node i
Z_{ij}	impedance connecting nodes i and j

Star impedances	$Z_i = \dfrac{Z_{ij}Z_{ik}}{Z_{ij}+Z_{ik}+Z_{jk}}$	(7.168)
Delta impedances	$Z_{ij} = Z_i Z_j \left(\dfrac{1}{Z_i}+\dfrac{1}{Z_j}+\dfrac{1}{Z_k}\right)$	(7.169)

7

7.7 Transmission lines and waveguides

Transmission line relations

Loss-free transmission line equations	$\frac{\partial V}{\partial x}=-L\frac{\partial I}{\partial t}$	(7.170)	V potential difference across line I current in line L inductance per unit length C capacitance per unit length
	$\frac{\partial I}{\partial x}=-C\frac{\partial V}{\partial t}$	(7.171)	
Wave equation for a lossless transmission line	$\frac{1}{LC}\frac{\partial^2 V}{\partial x^2}=\frac{\partial^2 V}{\partial t^2}$	(7.172)	x distance along line t time
	$\frac{1}{LC}\frac{\partial^2 I}{\partial x^2}=\frac{\partial^2 I}{\partial t^2}$	(7.173)	
Characteristic impedance of lossless line	$Z_c=\sqrt{\frac{L}{C}}$	(7.174)	Z_c characteristic impedance
Characteristic impedance of lossy line	$\boldsymbol{Z}_c=\sqrt{\frac{R+\mathrm{i}\omega L}{G+\mathrm{i}\omega C}}$	(7.175)	R resistance per unit length of conductor G conductance per unit length of insulator ω angular frequency
Wave speed along a lossless line	$v_p=v_g=\frac{1}{\sqrt{LC}}$	(7.176)	v_p phase speed v_g group speed
Input impedance of a terminated lossless line	$\boldsymbol{Z}_{in}=Z_c\frac{\boldsymbol{Z}_t\cos kl-\mathrm{i}Z_c\sin kl}{Z_c\cos kl-\mathrm{i}\boldsymbol{Z}_t\sin kl}$	(7.177)	$\boldsymbol{Z}_{in}$ (complex) input impedance $\boldsymbol{Z}_t$ (complex) terminating impedance k wavenumber ($=2\pi/\lambda$)
	$=Z_c^2/\boldsymbol{Z}_t \quad \text{if } l=\lambda/4$	(7.178)	
Reflection coefficient from a terminated line	$\boldsymbol{r}=\frac{\boldsymbol{Z}_t-\boldsymbol{Z}_c}{\boldsymbol{Z}_t+\boldsymbol{Z}_c}$	(7.179)	l distance from termination $\boldsymbol{r}$ (complex) voltage reflection coefficient
Line voltage standing wave ratio	$\mathrm{VSWR}=\frac{1+\lvert\boldsymbol{r}\rvert}{1-\lvert\boldsymbol{r}\rvert}$	(7.180)	

Transmission line impedances[a]

Coaxial line	$Z_c=\sqrt{\frac{\mu}{4\pi^2\epsilon}}\ln\frac{b}{a}\simeq\frac{60}{\sqrt{\epsilon_r}}\ln\frac{b}{a}$	(7.181)	Z_c characteristic impedance (Ω) a radius of inner conductor b radius of outer conductor ϵ permittivity ($=\epsilon_0\epsilon_r$)
Open wire feeder	$Z_c=\sqrt{\frac{\mu}{\pi^2\epsilon}}\ln\frac{l}{r}\simeq\frac{120}{\sqrt{\epsilon_r}}\ln\frac{l}{r}$	(7.182)	μ permeability ($=\mu_0\mu_r$) r radius of wires l distance between wires ($\gg r$)
Paired strip	$Z_c=\sqrt{\frac{\mu}{\epsilon}}\frac{d}{w}\simeq\frac{377}{\sqrt{\epsilon_r}}\frac{d}{w}$	(7.183)	d strip separation w strip width ($\gg d$)
Microstrip line	$Z_c\simeq\frac{377}{\sqrt{\epsilon_r}[(w/h)+2]}$	(7.184)	h height above earth plane ($\ll w$)

[a] For lossless lines.

Waveguides[a]

Waveguide equation	$k_{\mathrm{g}}^2 = \frac{\omega^2}{c^2} - \frac{m^2\pi^2}{a^2} - \frac{n^2\pi^2}{b^2}$	(7.185)	k_{g} ω a b m,n c	wavenumber in guide angular frequency guide height guide width mode indices with respect to a and b (integers) speed of light
Guide cutoff frequency	$\nu_{\mathrm{c}} = c\sqrt{\left(\frac{m}{2a}\right)^2 + \left(\frac{n}{2b}\right)^2}$	(7.186)	ν_{c} ω_{c}	cutoff frequency $2\pi\nu_{\mathrm{c}}$
Phase velocity above cutoff	$v_{\mathrm{p}} = \frac{c}{\sqrt{1-(\nu_{\mathrm{c}}/\nu)^2}}$	(7.187)	v_{p} ν	phase velocity frequency
Group velocity above cutoff	$v_{\mathrm{g}} = c^2/v_{\mathrm{p}} = c\sqrt{1-(\nu_{\mathrm{c}}/\nu)^2}$	(7.188)	v_{g}	group velocity
Wave impedances[b]	$Z_{\mathrm{TM}} = Z_0\sqrt{1-(\nu_{\mathrm{c}}/\nu)^2}$	(7.189)	Z_{TM} Z_{TE}	wave impedance for transverse magnetic modes wave impedance for transverse electric modes
	$Z_{\mathrm{TE}} = Z_0/\sqrt{1-(\nu_{\mathrm{c}}/\nu)^2}$	(7.190)	Z_0	impedance of free space $(=\sqrt{\mu_0/\epsilon_0})$

Field solutions for TE_{mn} modes[c]

$$B_x = \frac{\mathbf{i}k_{\mathrm{g}}c^2}{\omega_{\mathrm{c}}^2}\frac{\partial B_z}{\partial x} \qquad E_x = \frac{\mathbf{i}\omega c^2}{\omega_{\mathrm{c}}^2}\frac{\partial B_z}{\partial y}$$

$$B_y = \frac{\mathbf{i}k_{\mathrm{g}}c^2}{\omega_{\mathrm{c}}^2}\frac{\partial B_z}{\partial y} \qquad E_y = \frac{-\mathbf{i}\omega c^2}{\omega_{\mathrm{c}}^2}\frac{\partial B_z}{\partial x} \qquad (7.191)$$

$$B_z = B_0 \cos\frac{m\pi x}{a}\cos\frac{n\pi y}{b} \qquad E_z = 0$$

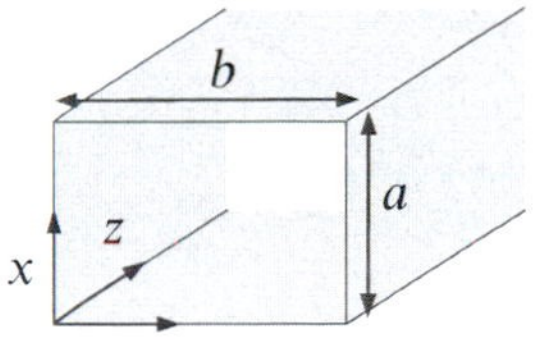

Field solutions for TM_{mn} modes[c]

$$E_x = \frac{\mathbf{i}k_{\mathrm{g}}c^2}{\omega_{\mathrm{c}}^2}\frac{\partial E_z}{\partial x} \qquad B_x = \frac{-\mathbf{i}\omega}{\omega_c^2}\frac{\partial E_z}{\partial y}$$

$$E_y = \frac{\mathbf{i}k_{\mathrm{g}}c^2}{\omega_{\mathrm{c}}^2}\frac{\partial E_z}{\partial y} \qquad B_y = \frac{\mathbf{i}\omega}{\omega_c^2}\frac{\partial E_z}{\partial x} \qquad (7.192)$$

$$E_z = E_0 \sin\frac{m\pi x}{a}\sin\frac{n\pi y}{b} \qquad B_z = 0$$

[a] Equations are for lossless waveguides with rectangular cross sections and no dielectric.
[b] The ratio of the electric field to the magnetic field strength in the xy plane.
[c] Both TE and TM modes propagate in the z direction with a further factor of $\exp[\mathbf{i}(k_{\mathrm{g}}z-\omega t)]$ on all components. B_0 and E_0 are the amplitudes of the z components of magnetic flux density and electric field respectively.

7.8 Waves in and out of media

Waves in lossless media

Electric field	$\nabla^2 \boldsymbol{E} = \mu\epsilon \frac{\partial^2 \boldsymbol{E}}{\partial t^2}$	(7.193)	$\boldsymbol{E}$ electric field; μ permeability $(=\mu_0\mu_r)$; ϵ permittivity $(=\epsilon_0\epsilon_r)$
Magnetic field	$\nabla^2 \boldsymbol{B} = \mu\epsilon \frac{\partial^2 \boldsymbol{B}}{\partial t^2}$	(7.194)	$\boldsymbol{B}$ magnetic flux density; t time
Refractive index	$\eta = \sqrt{\epsilon_r \mu_r}$	(7.195)	
Wave speed	$v = \frac{1}{\sqrt{\mu\epsilon}} = \frac{c}{\eta}$	(7.196)	v wave phase speed; η refractive index; c speed of light
Impedance of free space	$Z_0 = \sqrt{\frac{\mu_0}{\epsilon_0}} \simeq 376.7\,\Omega$	(7.197)	Z_0 impedance of free space
Wave impedance	$Z = \frac{E}{H} = Z_0 \sqrt{\frac{\mu_r}{\epsilon_r}}$	(7.198)	Z wave impedance; H magnetic field strength

Radiation pressure[a]

Radiation momentum density	$\boldsymbol{G} = \frac{\boldsymbol{N}}{c^2}$	(7.199)	$\boldsymbol{G}$ momentum density; $\boldsymbol{N}$ Poynting vector; c speed of light
Isotropic radiation	$p_n = \frac{1}{3}u(1+R)$	(7.200)	p_n normal pressure; u incident radiation energy density; R (power) reflectance coefficient
Specular reflection	$p_n = u(1+R)\cos^2\theta_i$	(7.201)	p_t tangential pressure; θ_i angle of incidence
	$p_t = u(1-R)\sin\theta_i\cos\theta_i$	(7.202)	
From an extended source[b]	$p_n = \frac{1+R}{c} \iint I_\nu(\theta,\phi)\cos^2\theta \, d\Omega \, d\nu$	(7.203)	I_ν specific intensity; ν frequency; Ω solid angle; θ angle between $d\Omega$ and normal to plane
From a point source,[c] luminosity L	$p_n = \frac{L(1+R)}{4\pi r^2 c}$	(7.204)	L source luminosity (i.e., radiant power); r distance from source

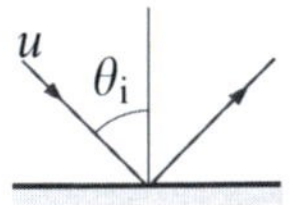

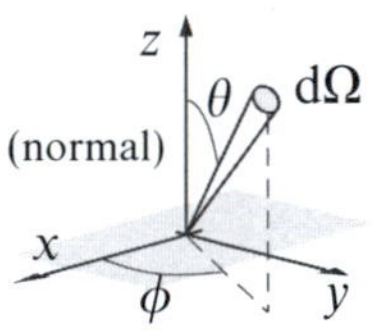

[a]On an opaque surface.
[b]In spherical polar coordinates. See page 120 for the meaning of specific intensity.
[c]Normal to the plane.

Antennas

Spherical polar geometry:

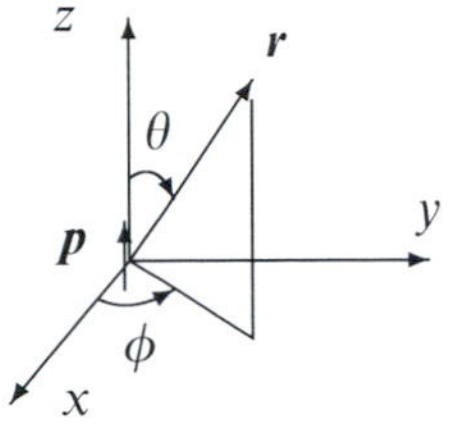

Field from a short ($l \ll \lambda$) electric dipole in free space[a]	$$E_r = \frac{1}{2\pi\epsilon_0}\left(\frac{[\dot{p}]}{r^2c}+\frac{[p]}{r^3}\right)\cos\theta \quad (7.205)$$ $$E_\theta = \frac{1}{4\pi\epsilon_0}\left(\frac{[\ddot{p}]}{rc^2}+\frac{[\dot{p}]}{r^2c}+\frac{[p]}{r^3}\right)\sin\theta \quad (7.206)$$ $$B_\phi = \frac{\mu_0}{4\pi}\left(\frac{[\ddot{p}]}{rc}+\frac{[\dot{p}]}{r^2}\right)\sin\theta \quad (7.207)$$	r θ $[p]$ c	distance from dipole angle between $\boldsymbol{r}$ and $\boldsymbol{p}$ retarded dipole moment $[p]=p(t-r/c)$ speed of light
Radiation resistance of a short electric dipole in free space	$$R = \frac{\omega^2 l^2}{6\pi\epsilon_0 c^3} = \frac{2\pi Z_0}{3}\left(\frac{l}{\lambda}\right)^2 \quad (7.208)$$ $$\simeq 789\left(\frac{l}{\lambda}\right)^2 \text{ ohm} \quad (7.209)$$	l ω λ Z_0	dipole length ($\ll \lambda$) angular frequency wavelength impedance of free space
Beam solid angle	$$\Omega_A = \int_{4\pi} P_n(\theta,\phi)\,d\Omega \quad (7.210)$$	Ω_A P_n $d\Omega$	beam solid angle normalised antenna power pattern $P_n(0,0)=1$ differential solid angle
Forward power gain	$$G(0) = \frac{4\pi}{\Omega_A} \quad (7.211)$$	G	antenna gain
Antenna effective area	$$A_e = \frac{\lambda^2}{\Omega_A} \quad (7.212)$$	A_e	effective area
Power gain of a short dipole	$$G(\theta) = \frac{3}{2}\sin^2\theta \quad (7.213)$$		
Beam efficiency	$$\text{efficiency} = \frac{\Omega_M}{\Omega_A} \quad (7.214)$$	Ω_M	main lobe solid angle
Antenna temperature[b]	$$T_A = \frac{1}{\Omega_A}\int_{4\pi} T_b(\theta,\phi)P_n(\theta,\phi)\,d\Omega \quad (7.215)$$	T_A T_b	antenna temperature sky brightness temperature

[a] All field components propagate with a further phase factor equal to $\exp i(kr-\omega t)$, where $k=2\pi/\lambda$.
[b] The brightness temperature of a source of specific intensity I_ν is $T_b=\lambda^2 I_\nu/(2k_B)$.

7

Reflection, refraction, and transmission[a]

parallel incidence

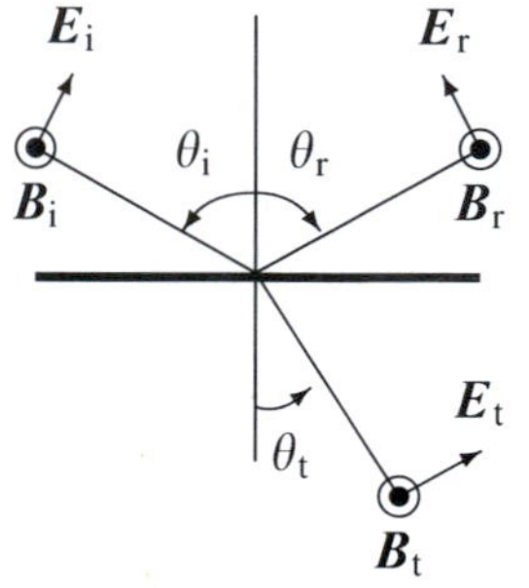

perpendicular incidence

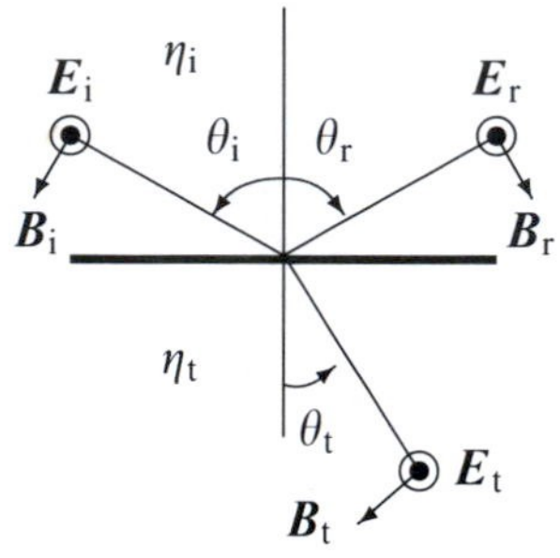

$\boldsymbol{E}$	electric field
$\boldsymbol{B}$	magnetic flux density
$\eta_{\rm i}$	refractive index on incident side
$\eta_{\rm t}$	refractive index on transmitted side
$\theta_{\rm i}$	angle of incidence
$\theta_{\rm r}$	angle of reflection
$\theta_{\rm t}$	angle of refraction

Law of reflection $\qquad \theta_{\rm i}=\theta_{\rm r} \qquad$ (7.216)

Snell's law[b] $\qquad \eta_{\rm i}\sin\theta_{\rm i}=\eta_{\rm t}\sin\theta_{\rm t} \qquad$ (7.217)

Brewster's law $\qquad \tan\theta_{\rm B}=\eta_{\rm t}/\eta_{\rm i} \qquad$ (7.218)

$\theta_{\rm B}$ Brewster's angle of incidence for plane-polarised reflection ($r_\parallel=0$)

Fresnel equations of reflection and refraction

$$r_\parallel=\frac{\sin 2\theta_{\rm i}-\sin 2\theta_{\rm t}}{\sin 2\theta_{\rm i}+\sin 2\theta_{\rm t}} \qquad (7.219)$$

$$t_\parallel=\frac{4\cos\theta_{\rm i}\sin\theta_{\rm t}}{\sin 2\theta_{\rm i}+\sin 2\theta_{\rm t}} \qquad (7.220)$$

$$R_\parallel=r_\parallel^2 \qquad (7.221)$$

$$T_\parallel=\frac{\eta_{\rm t}\cos\theta_{\rm t}}{\eta_{\rm i}\cos\theta_{\rm i}}t_\parallel^2 \qquad (7.222)$$

$$r_\perp=-\frac{\sin(\theta_{\rm i}-\theta_{\rm t})}{\sin(\theta_{\rm i}+\theta_{\rm t})} \qquad (7.223)$$

$$t_\perp=\frac{2\cos\theta_{\rm i}\sin\theta_{\rm t}}{\sin(\theta_{\rm i}+\theta_{\rm t})} \qquad (7.224)$$

$$R_\perp=r_\perp^2 \qquad (7.225)$$

$$T_\perp=\frac{\eta_{\rm t}\cos\theta_{\rm t}}{\eta_{\rm i}\cos\theta_{\rm i}}t_\perp^2 \qquad (7.226)$$

Coefficients for normal incidence[c]

$$R=\frac{(\eta_{\rm i}-\eta_{\rm t})^2}{(\eta_{\rm i}+\eta_{\rm t})^2} \qquad (7.227)$$

$$T=\frac{4\eta_{\rm i}\eta_{\rm t}}{(\eta_{\rm i}+\eta_{\rm t})^2} \qquad (7.228)$$

$$R+T=1 \qquad (7.229)$$

$$r=\frac{\eta_{\rm i}-\eta_{\rm t}}{\eta_{\rm i}+\eta_{\rm t}} \qquad (7.230)$$

$$t=\frac{2\eta_{\rm i}}{\eta_{\rm i}+\eta_{\rm t}} \qquad (7.231)$$

$$t-r=1 \qquad (7.232)$$

$\parallel$	electric field parallel to the plane of incidence	$\perp$	electric field perpendicular to the plane of incidence
R	(power) reflectance coefficient	r	amplitude reflection coefficient
T	(power) transmittance coefficient	t	amplitude transmission coefficient

[a] For the plane boundary between lossless dielectric media. All coefficients refer to the electric field component and whether it is parallel or perpendicular to the plane of incidence. Perpendicular components are out of the paper.
[b] The incident wave suffers total internal reflection if $\frac{\eta_{\rm i}}{\eta_{\rm t}}\sin\theta_{\rm i}>1$.
[c] I.e., $\theta_{\rm i}=0$. Use the diagram labelled "perpendicular incidence" for correct phases.

Propagation in conducting media[a]

Electrical conductivity ($B=0$)	$\sigma = n_e e \mu = \dfrac{n_e e^2}{m_e}\tau_c$	(7.233)
Refractive index of an ohmic conductor[b]	$\eta = (1+\mathrm{i})\left(\dfrac{\sigma}{4\pi\nu\epsilon_0}\right)^{1/2}$	(7.234)
Skin depth in an ohmic conductor	$\delta = (\mu_0 \sigma \pi \nu)^{-1/2}$	(7.235)

σ	electrical conductivity
n_e	electron number density
τ_c	electron relaxation time
μ	electron mobility
B	magnetic flux density
m_e	electron mass
$-e$	electronic charge
η	refractive index
ϵ_0	permittivity of free space
ν	frequency
δ	skin depth
μ_0	permeability of free space

[a]Assuming a relative permeability, μ_r, of 1.
[b]Taking the wave to have an $e^{-i\omega t}$ time dependence, and the low-frequency limit ($\sigma \gg 2\pi\nu\epsilon_0$).

Electron scattering processes[a]

Rayleigh scattering cross section[b]	$\sigma_R = \dfrac{\omega^4 \alpha^2}{6\pi\epsilon_0 c^4}$	(7.236)
Thomson scattering cross section[c]	$\sigma_T = \dfrac{8\pi}{3}\left(\dfrac{e^2}{4\pi\epsilon_0 m_e c^2}\right)^2$	(7.237)
	$= \dfrac{8\pi}{3} r_e^2 \simeq 6.652\times 10^{-29}\,\mathrm{m}^2$	(7.238)
Inverse Compton scattering[d]	$P_{tot} = \dfrac{4}{3}\sigma_T c u_{rad}\gamma^2\left(\dfrac{v^2}{c^2}\right)$	(7.239)
Compton scattering[e]	$\lambda' - \lambda = \dfrac{h}{m_e c}(1-\cos\theta)$	(7.240)
	$h\nu' = \dfrac{m_e c^2}{1-\cos\theta + (1/\varepsilon)}$	(7.241)
	$\cot\phi = (1+\varepsilon)\tan\dfrac{\theta}{2}$	(7.242)
Klein–Nishina cross section (for a free electron)	$\sigma_{KN} = \dfrac{\pi r_e^2}{\varepsilon}\left\{\left[1-\dfrac{2(\varepsilon+1)}{\varepsilon^2}\right]\ln(2\varepsilon+1) + \dfrac{1}{2} + \dfrac{4}{\varepsilon} - \dfrac{1}{2(2\varepsilon+1)^2}\right\}$	(7.243)
	$\simeq \sigma_T \quad (\varepsilon \ll 1)$	(7.244)
	$\simeq \dfrac{\pi r_e^2}{\varepsilon}\left(\ln 2\varepsilon + \dfrac{1}{2}\right) \quad (\varepsilon \gg 1)$	(7.245)

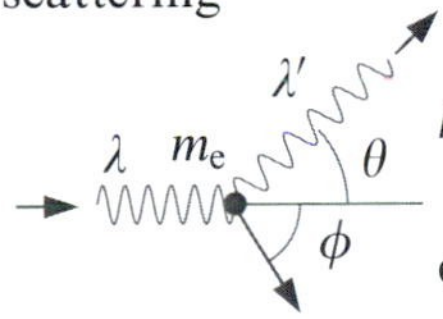

σ_R	Rayleigh cross section
ω	radiation angular frequency
α	particle polarisability
ϵ_0	permittivity of free space
σ_T	Thomson cross section
m_e	electron (rest) mass
r_e	classical electron radius
c	speed of light
P_{tot}	electron energy loss rate
u_{rad}	radiation energy density
γ	Lorentz factor $=[1-(v/c)^2]^{-1/2}$
v	electron speed
λ, λ'	incident & scattered wavelengths
ν, ν'	incident & scattered frequencies
θ	photon scattering angle
$\dfrac{h}{m_e c}$	electron Compton wavelength
ε	$= h\nu/(m_e c^2)$
σ_{KN}	Klein–Nishina cross section

[a]For Rutherford scattering see page 72.
[b]Scattering by bound electrons.
[c]Scattering from free electrons, $\varepsilon \ll 1$.
[d]Electron energy loss rate due to photon scattering in the Thomson limit ($\gamma h\nu \ll m_e c^2$).
[e]From an electron at rest.

Cherenkov radiation

Cherenkov cone angle	$\cos\theta = \frac{c}{\eta v}$	(7.246)	θ ray cone semi-angle c (vacuum) speed of light $\eta(\omega)$ refractive index v particle velocity
Radiated power[a]	$P_{\rm tot} = \frac{e^2\mu_0}{4\pi} v \int_0^{\omega_c} \left[1 - \frac{c^2}{v^2\eta^2(\omega)}\right] \omega \, d\omega$ where $\eta(\omega) \geq \frac{c}{v}$ for $0 < \omega < \omega_c$	(7.247)	$P_{\rm tot}$ total radiated power $-e$ electronic charge μ_0 free space permeability ω angular frequency ω_c cutoff frequency

[a]From a point charge, e, travelling at speed v through a medium of refractive index $\eta(\omega)$.

7.9 Plasma physics

Warm plasmas

Landau length	$l_L = \frac{e^2}{4\pi\epsilon_0 k_B T_e}$ $\simeq 1.67 \times 10^{-5} T_e^{-1}$ m	(7.248) (7.249)	l_L Landau length $-e$ electronic charge ϵ_0 permittivity of free space k_B Boltzmann constant T_e electron temperature (K)
Electron Debye length	$\lambda_{\rm De} = \left(\frac{\epsilon_0 k_B T_e}{n_e e^2}\right)^{1/2}$ $\simeq 69(T_e/n_e)^{1/2}$ m	(7.250) (7.251)	$\lambda_{\rm De}$ electron Debye length n_e electron number density (m^{-3})
Debye screening[a]	$\phi(r) = \frac{q\exp(-2^{1/2}r/\lambda_{\rm De})}{4\pi\epsilon_0 r}$	(7.252)	ϕ effective potential q point charge r distance from q
Debye number	$N_{\rm De} = \frac{4}{3}\pi n_e \lambda_{\rm De}^3$	(7.253)	$N_{\rm De}$ electron Debye number
Relaxation times ($B=0$)[b]	$\tau_e = 3.44 \times 10^5 \frac{T_e^{3/2}}{n_e \ln\Lambda}$ s $\tau_i = 2.09 \times 10^7 \frac{T_i^{3/2}}{n_e \ln\Lambda}\left(\frac{m_i}{m_p}\right)^{1/2}$ s	(7.254) (7.255)	τ_e electron relaxation time τ_i ion relaxation time T_i ion temperature (K) $\ln\Lambda$ Coulomb logarithm (typically 10 to 20) B magnetic flux density
Characteristic electron thermal speed[c]	$v_{\rm te} = \left(\frac{2k_B T_e}{m_e}\right)^{1/2}$ $\simeq 5.51 \times 10^3 T_e^{1/2}$ m s^{-1}	(7.256) (7.257)	$v_{\rm te}$ electron thermal speed m_e electron mass

[a]Effective (Yukawa) potential from a point charge q immersed in a plasma.
[b]Collision times for electrons and *singly* ionised ions with Maxwellian speed distributions, $T_i \lesssim T_e$. The Spitzer conductivity can be calculated from Equation (7.233).
[c]Defined so that the Maxwellian velocity distribution $\propto \exp(-v^2/v_{\rm te}^2)$. There are other definitions (see *Maxwell–Boltzmann distribution* on page 112).

Electromagnetic propagation in cold plasmas[a]

Plasma frequency	$(2\pi\nu_{\mathrm{p}})^2 = \dfrac{n_{\mathrm{e}}e^2}{\epsilon_0 m_{\mathrm{e}}} = \omega_{\mathrm{p}}^2$ (7.258) $\nu_{\mathrm{p}} \simeq 8.98 n_{\mathrm{e}}^{1/2}$ Hz (7.259)		ν_{p} plasma frequency ω_{p} plasma angular frequency n_{e} electron number density (m^{-3}) m_{e} electron mass
Plasma refractive index ($B=0$)	$\eta = \left[1-(\nu_{\mathrm{p}}/\nu)^2\right]^{1/2}$	(7.260)	$-e$ electronic charge ϵ_0 permittivity of free space η refractive index ν frequency
Plasma dispersion relation ($B=0$)	$c^2k^2 = \omega^2 - \omega_{\mathrm{p}}^2$	(7.261)	k wavenumber ($=2\pi/\lambda$) ω angular frequency ($=2\pi/\nu$) c speed of light
Plasma phase velocity ($B=0$)	$v_\phi = c/\eta$	(7.262)	v_ϕ phase velocity
Plasma group velocity ($B=0$)	$v_{\mathrm{g}} = c\eta$ (7.263) $v_\phi v_{\mathrm{g}} = c^2$ (7.264)		v_{g} group velocity
Cyclotron (Larmor, or gyro-) frequency	$2\pi\nu_{\mathrm{C}} = \dfrac{qB}{m} = \omega_{\mathrm{C}}$ (7.265) $\nu_{\mathrm{Ce}} \simeq 28\times10^9 B$ Hz (7.266) $\nu_{\mathrm{Cp}} \simeq 15.2\times10^6 B$ Hz (7.267)		ν_{C} cyclotron frequency ω_{C} cyclotron angular frequency ν_{Ce} electron ν_{C} ν_{Cp} proton ν_{C} q particle charge B magnetic flux density (T)
Larmor (cyclotron, or gyro-) radius	$r_{\mathrm{L}} = \dfrac{v_\perp}{\omega_{\mathrm{C}}} = v_\perp \dfrac{m}{qB}$ (7.268) $r_{\mathrm{Le}} = 5.69\times10^{-12}\left(\dfrac{v_\perp}{B}\right)$ m (7.269) $r_{\mathrm{Lp}} = 10.4\times10^{-9}\left(\dfrac{v_\perp}{B}\right)$ m (7.270)		m particle mass (γm if relativistic) r_{L} Larmor radius r_{Le} electron r_{L} r_{Lp} proton r_{L} $v_\perp$ speed $\perp$ to $\boldsymbol{B}$ ($\mathrm{m\,s^{-1}}$)
Mixed propagation modes[b]	$\eta^2 = 1 - \dfrac{X(1-X)}{(1-X)-\frac{1}{2}Y^2\sin^2\theta_B \pm S}$, (7.271) where $X=(\omega_{\mathrm{p}}/\omega)^2$, $Y=\omega_{\mathrm{Ce}}/\omega$, and $S^2 = \dfrac{1}{4}Y^4\sin^4\theta_B + Y^2(1-X)^2\cos^2\theta_B$		θ_B angle between wavefront normal ($\hat{\boldsymbol{k}}$) and $\boldsymbol{B}$
Faraday rotation[c]	$\Delta\psi = \underbrace{\dfrac{\mu_0 e^3}{8\pi^2 m_{\mathrm{e}}^2 c}}_{2.63\times10^{-13}} \lambda^2 \int_{\mathrm{line}} n_{\mathrm{e}}\boldsymbol{B}\cdot\mathrm{d}\boldsymbol{l}$ (7.272) $= R\lambda^2$ (7.273)		$\Delta\psi$ rotation angle λ wavelength ($=2\pi/k$) $\mathrm{d}\boldsymbol{l}$ line element in direction of wave propagation R rotation measure

[a] I.e., plasmas in which electromagnetic force terms dominate over thermal pressure terms. Also taking $\mu_{\mathrm{r}}=1$.
[b] In a collisionless electron plasma. The ordinary and extraordinary modes are the + and − roots of S^2 when $\theta_B=\pi/2$. When $\theta_B=0$, these roots are the right and left circularly polarised modes respectively, using the optical convention for handedness.
[c] In a tenuous plasma, SI units throughout. $\Delta\psi$ is taken positive if $\boldsymbol{B}$ is directed towards the observer.

Magnetohydrodynamics[a]

Sound speed	$v_s = \left(\frac{\gamma p}{\rho}\right)^{1/2} = \left(\frac{2\gamma k_B T}{m_p}\right)^{1/2}$	(7.274)	v_s sound (wave) speed γ ratio of heat capacities p hydrostatic pressure ρ plasma mass density k_B Boltzmann constant T temperature (K) m_p proton mass
	$\simeq 166 T^{1/2}\,\mathrm{m\,s^{-1}}$	(7.275)	
Alfvén speed	$v_A = \frac{B}{(\mu_0 \rho)^{1/2}}$	(7.276)	v_A Alfvén speed B magnetic flux density (T) μ_0 permeability of free space n_e electron number density ($\mathrm{m^{-3}}$)
	$\simeq 2.18 \times 10^{16} B n_e^{-1/2} \quad \mathrm{m\,s^{-1}}$	(7.277)	
Plasma beta	$\beta = \frac{2\mu_0 p}{B^2} = \frac{4\mu_0 n_e k_B T}{B^2} = \frac{2v_s^2}{\gamma v_A^2}$	(7.278)	β plasma beta (ratio of hydrostatic to magnetic pressure)
Direct electrical conductivity	$\sigma_d = \frac{n_e^2 e^2 \sigma}{n_e^2 e^2 + \sigma^2 B^2}$	(7.279)	$-e$ electronic charge σ_d direct conductivity σ conductivity ($B=0$)
Hall electrical conductivity	$\sigma_H = \frac{\sigma B}{n_e e}\sigma_d$	(7.280)	σ_H Hall conductivity
Generalised Ohm's law	$\boldsymbol{J} = \sigma_d(\boldsymbol{E}+\boldsymbol{v}\times\boldsymbol{B}) + \sigma_H \hat{\boldsymbol{B}}\times(\boldsymbol{E}+\boldsymbol{v}\times\boldsymbol{B})$	(7.281)	$\boldsymbol{J}$ current density $\boldsymbol{E}$ electric field $\boldsymbol{v}$ plasma velocity field $\hat{\boldsymbol{B}}$ $=\boldsymbol{B}/\vert\boldsymbol{B}\vert$
Resistive MHD equations (single-fluid model)[b]	$\frac{\partial \boldsymbol{B}}{\partial t} = \nabla\times(\boldsymbol{v}\times\boldsymbol{B}) + \eta\nabla^2\boldsymbol{B}$	(7.282)	μ_0 permeability of free space η magnetic diffusivity [$=1/(\mu_0\sigma)$] ν kinematic viscosity $\boldsymbol{g}$ gravitational field strength
	$\frac{\partial \boldsymbol{v}}{\partial t} + (\boldsymbol{v}\cdot\nabla)\boldsymbol{v} = -\frac{\nabla p}{\rho} + \frac{1}{\mu_0\rho}(\nabla\times\boldsymbol{B})\times\boldsymbol{B} + \nu\nabla^2\boldsymbol{v} + \frac{1}{3}\nu\nabla(\nabla\cdot\boldsymbol{v}) + \boldsymbol{g}$	(7.283)	
Shear Alfvénic dispersion relation[c]	$\omega = k v_A \cos\theta_B$	(7.284)	ω angular frequency ($=2\pi\nu$) $\boldsymbol{k}$ wavevector ($k=2\pi/\lambda$) θ_B angle between $\boldsymbol{k}$ and $\boldsymbol{B}$
Magnetosonic dispersion relation[d]	$\omega^2 k^2 (v_s^2 + v_A^2) - \omega^4 = v_s^2 v_A^2 k^4 \cos^2\theta_B$	(7.285)	

[a] For a warm, fully ionised, electrically neutral p^+/e^- plasma, $\mu_r = 1$. Relativistic and displacement current effects are assumed to be negligible and all oscillations are taken as being well below all resonance frequencies.

[b] Neglecting bulk (second) viscosity.

[c] Nonresistive, inviscid flow.

[d] Nonresistive, inviscid flow. The greater and lesser solutions for ω^2 are the fast and slow magnetosonic waves respectively.

Synchrotron radiation

Power radiated by a single electron[a]	$P_{\text{tot}} = 2\sigma_{\text{T}} c u_{\text{mag}} \gamma^2 \left(\frac{v}{c}\right)^2 \sin^2\theta$	(7.286)
	$\simeq 1.59 \times 10^{-14} B^2 \gamma^2 \left(\frac{v}{c}\right)^2 \sin^2\theta$ W	(7.287)
... averaged over pitch angles	$P_{\text{tot}} = \frac{4}{3}\sigma_{\text{T}} c u_{\text{mag}} \gamma^2 \left(\frac{v}{c}\right)^2$	(7.288)
	$\simeq 1.06 \times 10^{-14} B^2 \gamma^2 \left(\frac{v}{c}\right)^2$ W	(7.289)
Single electron emission spectrum[b]	$P(\nu) = \frac{3^{1/2} e^3 B \sin\theta}{4\pi\epsilon_0 c m_{\text{e}}} F(\nu/\nu_{\text{ch}})$	(7.290)
	$\simeq 2.34 \times 10^{-25} B \sin\theta F(\nu/\nu_{\text{ch}})$ W Hz^{-1}	(7.291)
Characteristic frequency	$\nu_{\text{ch}} = \frac{3}{2}\gamma^2 \frac{eB}{2\pi m_{\text{e}}} \sin\theta$	(7.292)
	$\simeq 4.2 \times 10^{10} \gamma^2 B \sin\theta$ Hz	(7.293)
Spectral function	$F(x) = x \int_x^\infty K_{5/3}(y)\,\mathrm{d}y$	(7.294)
	$\simeq \begin{cases} 2.15 x^{1/3} & (x \ll 1) \\ 1.25 x^{1/2} \mathrm{e}^{-x} & (x \gg 1) \end{cases}$	(7.295)

P_{tot}	total radiated power
σ_{T}	Thomson cross section
u_{mag}	magnetic energy density $= B^2/(2\mu_0)$
$\boldsymbol{v}$	electron velocity ($\sim c$)
γ	Lorentz factor $= [1-(v/c)^2]^{-1/2}$
θ	pitch angle (angle between $\boldsymbol{v}$ and $\boldsymbol{B}$)
$\boldsymbol{B}$	magnetic flux density
c	speed of light
$P(\nu)$	emission spectrum
ν	frequency
ν_{ch}	characteristic frequency
$-e$	electronic charge
ϵ_0	free space permittivity
m_{e}	electronic (rest) mass
F	spectral function
$K_{5/3}$	modified Bessel fn. of the 2nd kind, order 5/3

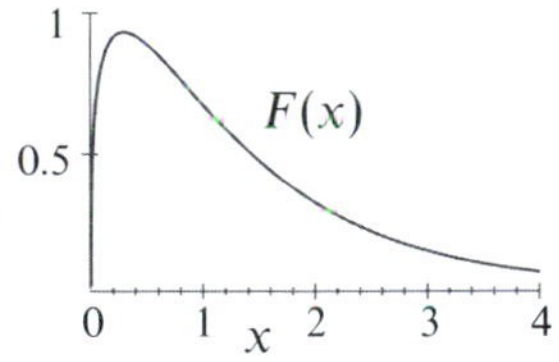

[a]This expression also holds for cyclotron radiation ($v \ll c$).
[b]I.e., total radiated power per unit frequency interval.

Bremsstrahlung[a]

Single electron and ion[b]

$$\frac{\mathrm{d}W}{\mathrm{d}\omega}=\frac{Z^2e^6}{24\pi^4\epsilon_0^3c^3m_e^2}\frac{\omega^2}{\gamma^2v^4}\left[\frac{1}{\gamma^2}K_0^2\left(\frac{\omega b}{\gamma v}\right)+K_1^2\left(\frac{\omega b}{\gamma v}\right)\right] \quad (7.296)$$

$$\simeq\frac{Z^2e^6}{24\pi^4\epsilon_0^3c^3m_e^2b^2v^2} \quad (\omega b \ll \gamma v) \quad (7.297)$$

Thermal bremsstrahlung radiation ($v \ll c$; Maxwellian distribution)

$$\frac{\mathrm{d}P}{\mathrm{d}V\,\mathrm{d}\nu}=6.8\times10^{-51}Z^2T^{-1/2}n_in_eg(\nu,T)\exp\left(\frac{-h\nu}{kT}\right) \quad \mathrm{W\,m^{-3}\,Hz^{-1}} \quad (7.298)$$

$$\text{where} \quad g(\nu,T)\simeq\begin{cases}0.28[\ln(4.4\times10^{16}T^3\nu^{-2}Z^{-2})-0.76] & (h\nu\ll kT\lesssim10^5kZ^2)\\ 0.55\ln(2.1\times10^{10}T\nu^{-1}) & (h\nu\ll10^5kZ^2\lesssim kT)\\ (2.1\times10^{10}T\nu^{-1})^{-1/2} & (h\nu\gg kT)\end{cases} \quad (7.299)$$

$$\frac{\mathrm{d}P}{\mathrm{d}V}\simeq1.7\times10^{-40}Z^2T^{1/2}n_in_e \quad \mathrm{W\,m^{-3}} \quad (7.300)$$

ω	angular frequency ($=2\pi\nu$)	v	electron velocity	W	energy radiated
Ze	ionic charge	K_i	modified Bessel functions of order i (see page 47)	T	electron temperature (K)
$-e$	electronic charge	γ	Lorentz factor $=[1-(v/c)^2]^{-1/2}$	n_i	ion number density ($\mathrm{m^{-3}}$)
ϵ_0	permittivity of free space	P	power radiated	n_e	electron number density ($\mathrm{m^{-3}}$)
c	speed of light	V	volume	k	Boltzmann constant
m_e	electronic mass	ν	frequency (Hz)	h	Planck constant
b	collision parameter[c]			g	Gaunt factor

[a]Classical treatment. The ions are at rest, and all frequencies are above the plasma frequency.
[b]The spectrum is approximately flat at low frequencies and drops exponentially at frequencies $\gtrsim \gamma v/b$.
[c]Distance of closest approach.

Chapter 8 Optics

8.1 Introduction

Any attempt to unify the notations and terminology of optics is doomed to failure. This is partly due to the long and illustrious history of the subject (a pedigree shared only with mechanics), which has allowed a variety of approaches to develop, and partly due to the disparate fields of physics to which its basic principles have been applied. Optical ideas find their way into most wave-based branches of physics, from quantum mechanics to radio propagation.

Nowhere is the lack of convention more apparent than in the study of polarisation, and so a cautionary note follows. The conventions used here can be taken largely from context, but the reader should be aware that alternative sign and handedness conventions do exist and are widely used. In particular we will take a circularly polarised wave as being right-handed if, for an observer looking *towards* the source, the electric field vector in a plane perpendicular to the line of sight rotates clockwise. This convention is often used in optics textbooks and has the conceptual advantage that the electric field orientation describes a right-hand corkscrew in space, with the direction of energy flow defining the screw direction. It is however opposite to the system widely used in radio engineering, where the handedness of a helical antenna generating or receiving the wave defines the handedness and is also in the opposite sense to the wave's own angular momentum vector.

8.2 Interference

Newton's rings[a]

nth dark ring	$r_n^2 = nR\lambda_0$	(8.1)	r_n	radius of nth ring
			n	integer (≥ 0)
			R	lens radius of curvature
nth bright ring	$r_n^2 = \left(n+\frac{1}{2}\right)R\lambda_0$	(8.2)	λ_0	wavelength in external medium

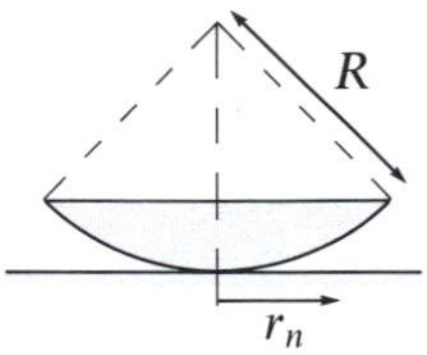

[a]Viewed in reflection.

Dielectric layers[a]

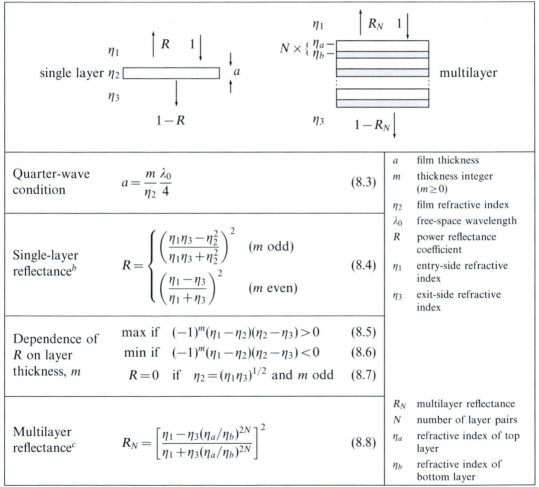

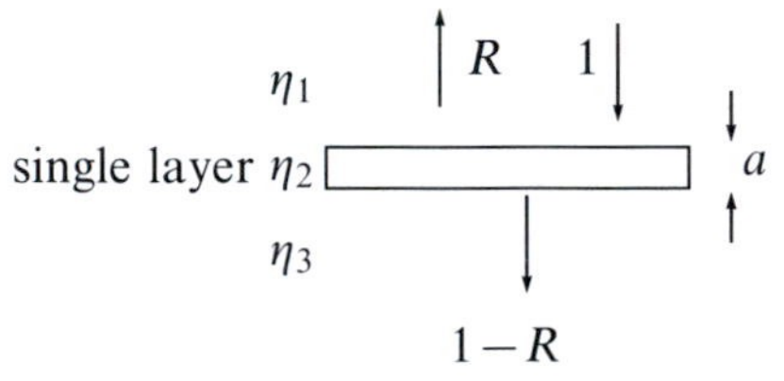

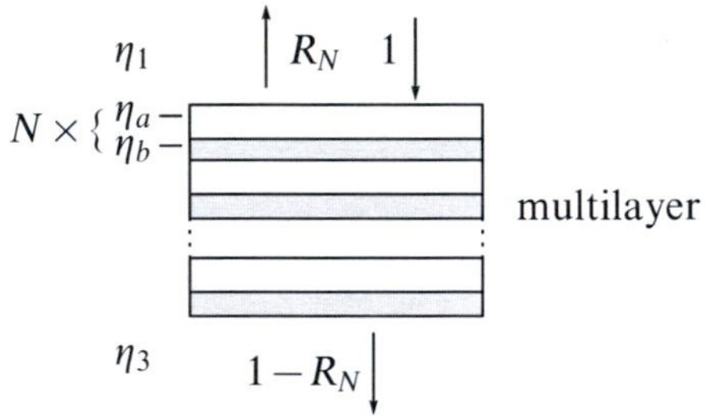

Quarter-wave condition	$a = \frac{m}{\eta_2}\frac{\lambda_0}{4}$	(8.3)	a	film thickness
			m	thickness integer ($m \geq 0$)
			η_2	film refractive index
			λ_0	free-space wavelength
Single-layer reflectance[b]	$R = \begin{cases} \left(\frac{\eta_1\eta_3 - \eta_2^2}{\eta_1\eta_3 + \eta_2^2}\right)^2 & (m \text{ odd}) \\ \left(\frac{\eta_1 - \eta_3}{\eta_1 + \eta_3}\right)^2 & (m \text{ even}) \end{cases}$	(8.4)	R	power reflectance coefficient
			η_1	entry-side refractive index
			η_3	exit-side refractive index
Dependence of R on layer thickness, m	max if $(-1)^m(\eta_1-\eta_2)(\eta_2-\eta_3) > 0$	(8.5)		
	min if $(-1)^m(\eta_1-\eta_2)(\eta_2-\eta_3) < 0$	(8.6)		
	$R=0$ if $\eta_2 = (\eta_1\eta_3)^{1/2}$ and m odd	(8.7)		
Multilayer reflectance[c]	$R_N = \left[\frac{\eta_1 - \eta_3(\eta_a/\eta_b)^{2N}}{\eta_1 + \eta_3(\eta_a/\eta_b)^{2N}}\right]^2$	(8.8)	R_N	multilayer reflectance
			N	number of layer pairs
			η_a	refractive index of top layer
			η_b	refractive index of bottom layer

[a]For normal incidence, assuming the quarter-wave condition. The media are also assumed lossless, with $\mu_r = 1$.
[b]See page 154 for the definition of R.
[c]For a stack of N layer pairs, giving an overall refractive index sequence $\eta_1\eta_a, \eta_b\eta_a \ldots \eta_a\eta_b\eta_3$ (see right-hand diagram). Each layer in the stack meets the quarter-wave condition with $m=1$.

Fabry-Perot etalon[a]

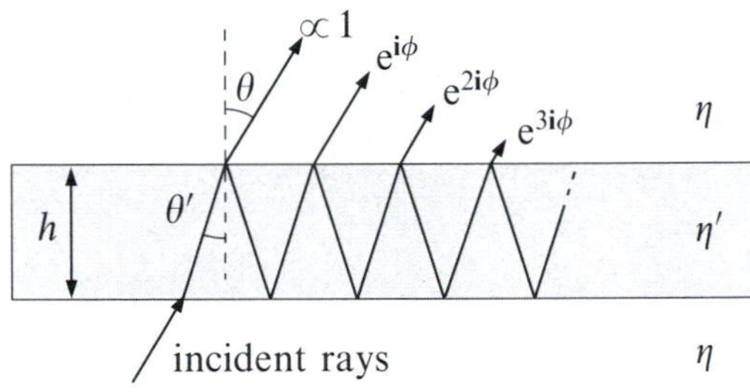

Incremental phase difference[b]	$\phi = 2k_0 h\eta' \cos\theta'$ (8.9) $= 2k_0 h\eta' \left[1 - \left(\frac{\eta \sin\theta}{\eta'}\right)^2\right]^{1/2}$ (8.10) $= 2\pi n$ for a maximum (8.11)	ϕ incremental phase difference k_0 free-space wavenumber ($= 2\pi/\lambda_0$) h cavity width θ fringe inclination (usually $\ll 1$) θ' internal angle of refraction η' cavity refractive index η external refractive index n fringe order (integer)
Coefficient of finesse	$F = \frac{4R}{(1-R)^2}$ (8.12)	F coefficient of finesse R interface power reflectance
Finesse	$\mathcal{F} = \frac{\pi}{2} F^{1/2}$ (8.13) $= \frac{\lambda_0}{\eta' h} Q$ (8.14)	$\mathcal{F}$ finesse λ_0 free-space wavelength Q cavity quality factor
Transmitted intensity	$I(\theta) = \frac{I_0(1-R)^2}{1 + R^2 - 2R\cos\phi}$ (8.15) $= \frac{I_0}{1 + F\sin^2(\phi/2)}$ (8.16) $= I_0 A(\theta)$ (8.17)	I transmitted intensity I_0 incident intensity A Airy function
Fringe intensity profile	$\Delta\phi = 2\arcsin(F^{-1/2})$ (8.18) $\simeq 2F^{-1/2}$ (8.19)	$\Delta\phi$ phase difference at half intensity point
Chromatic resolving power	$\frac{\lambda_0}{\delta\lambda} \simeq \frac{R^{1/2}\pi n}{1-R} = n\mathcal{F}$ (8.20) $\simeq \frac{2\mathcal{F} h\eta'}{\lambda_0}$ $(\theta \ll 1)$ (8.21)	$\delta\lambda$ minimum resolvable wavelength difference
Free spectral range[c]	$\delta\lambda_f = \mathcal{F}\delta\lambda$ (8.22) $\delta\nu_f = \frac{c}{2\eta' h}$ (8.23)	$\delta\lambda_f$ wavelength free spectral range $\delta\nu_f$ frequency free spectral range

[a]Neglecting any effects due to surface coatings on the etalon. See also *Lasers* on page 174.
[b]Between adjacent rays. Highest order fringes are near the centre of the pattern.
[c]At near-normal incidence ($\theta \simeq 0$), the orders of two spectral components separated by $< \delta\lambda_f$ will not overlap.

8.3 Fraunhofer diffraction

Gratings[a]

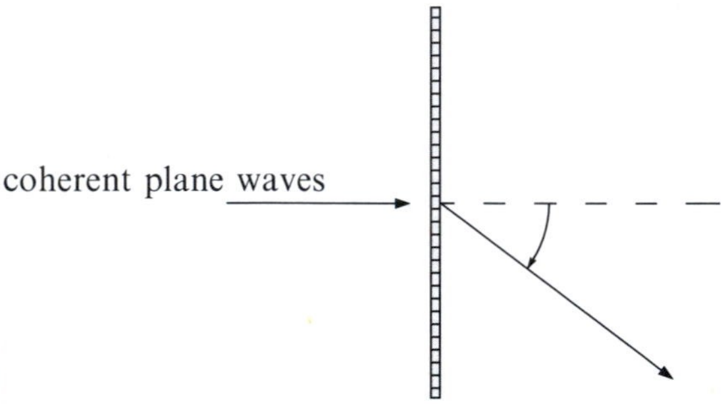

Young's double slits[b]	$I(s)=I_0\cos^2\frac{kDs}{2}$	(8.24)	$I(s)$ diffracted intensity I_0 peak intensity θ diffraction angle s $=\sin\theta$ D slit separation	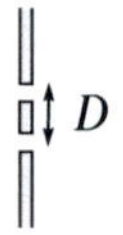
N equally spaced narrow slits	$I(s)=I_0\left[\frac{\sin(Nkds/2)}{N\sin(kds/2)}\right]^2$	(8.25)	λ wavelength N number of slits k wavenumber $(=2\pi/\lambda)$ d slit spacing	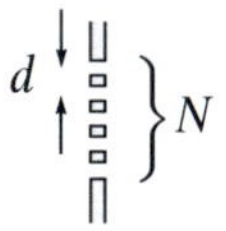
Infinite grating	$I(s)=I_0\sum_{n=-\infty}^{\infty}\delta\left(s-\frac{n\lambda}{d}\right)$	(8.26)	n diffraction order δ Dirac delta function	
Normal incidence	$\sin\theta_n=\frac{n\lambda}{d}$	(8.27)	θ_n angle of diffracted maximum	
Oblique incidence	$\sin\theta_n+\sin\theta_i=\frac{n\lambda}{d}$	(8.28)	θ_i angle of incident illumination	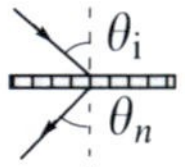
Reflection grating	$\sin\theta_n-\sin\theta_i=\frac{n\lambda}{d}$	(8.29)		
Chromatic resolving power	$\frac{\lambda}{\delta\lambda}=Nn$	(8.30)	$\delta\lambda$ diffraction peak width	
Grating dispersion	$\frac{\partial\theta}{\partial\lambda}=\frac{n}{d\cos\theta}$	(8.31)		
Bragg's law[c]	$2a\sin\theta_n=n\lambda$	(8.32)	a atomic plane spacing	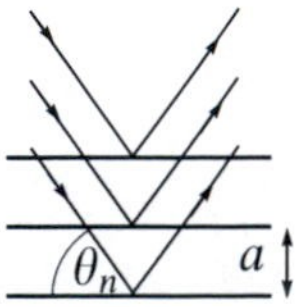

[a]Unless stated otherwise, the illumination is normal to the grating.
[b]Two narrow slits separated by D.
[c]The condition is for Bragg reflection, with $\theta_n=\theta_i$.

Aperture diffraction

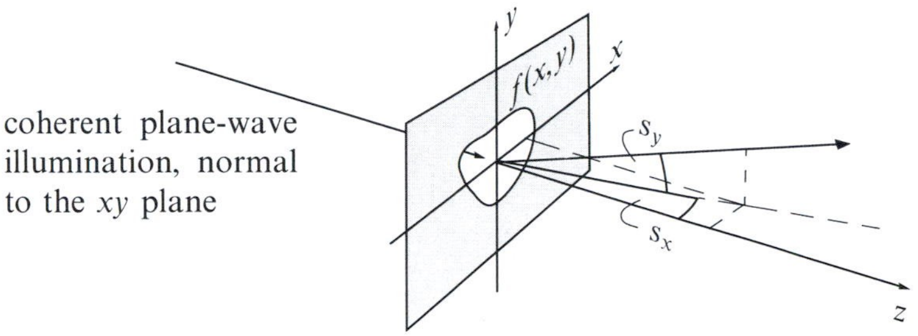

General 1-D aperture[a]	$$\psi(s) \propto \int_{-\infty}^{\infty} f(x) e^{-iksx}\, dx \qquad (8.33)$$ $$I(s) \propto \psi\psi^*(s) \qquad (8.34)$$		ψ diffracted wavefunction I diffracted intensity θ diffraction angle s $=\sin\theta$
General 2-D aperture in (x,y) plane (small angles)	$$\psi(s_x,s_y) \propto \iint_{\infty} f(x,y) e^{-ik(s_x x + s_y y)}\, dx\, dy \qquad (8.35)$$		f aperture amplitude transmission function x,y distance across aperture k wavenumber $(=2\pi/\lambda)$ s_x deflection $\parallel$ xz plane s_y deflection $\perp$ xz plane
Broad 1-D slit[b]	$$I(s) = I_0 \frac{\sin^2(kas/2)}{(kas/2)^2} \qquad (8.36)$$ $$\equiv I_0 \operatorname{sinc}^2(as/\lambda) \qquad (8.37)$$		I_0 peak intensity a slit width (in x) λ wavelength
Sidelobe intensity	$$\frac{I_n}{I_0} = \left(\frac{2}{\pi}\right)^2 \frac{1}{(2n+1)^2} \quad (n>0) \qquad (8.38)$$		I_n nth sidelobe intensity
Rectangular aperture (small angles)	$$I(s_x,s_y) = I_0 \operatorname{sinc}^2 \frac{as_x}{\lambda} \operatorname{sinc}^2 \frac{bs_y}{\lambda} \qquad (8.39)$$		a aperture width in x b aperture width in y
Circular aperture[c]	$$I(s) = I_0 \left[\frac{2J_1(kDs/2)}{kDs/2}\right]^2 \qquad (8.40)$$		J_1 first-order Bessel function D aperture diameter
First minimum[d]	$$s = 1.22\frac{\lambda}{D} \qquad (8.41)$$		λ wavelength
First subsid. maximum	$$s = 1.64\frac{\lambda}{D} \qquad (8.42)$$		
Weak 1-D phase object	$$f(x) = \exp[i\phi(x)] \simeq 1 + i\phi(x) \qquad (8.43)$$		$\phi(x)$ phase distribution i $i^2 = -1$
Fraunhofer limit[e]	$$L \gg \frac{(\Delta x)^2}{\lambda} \qquad (8.44)$$		L distance of aperture from observation point Δx aperture size

[a]The Fraunhofer integral.

[b]Note that $\operatorname{sinc} x = (\sin \pi x)/(\pi x)$.

[c]The central maximum is known as the "Airy disk."

[d]The "Rayleigh resolution criterion" states that two point sources of equal intensity can just be resolved with diffraction-limited optics if separated in angle by $1.22\lambda/D$.

[e]Plane-wave illumination.

8.4 Fresnel diffraction

Kirchhoff's diffraction formula[a]

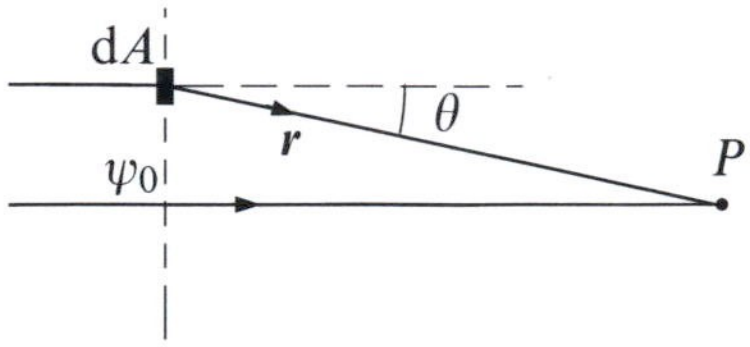

(source at infinity)

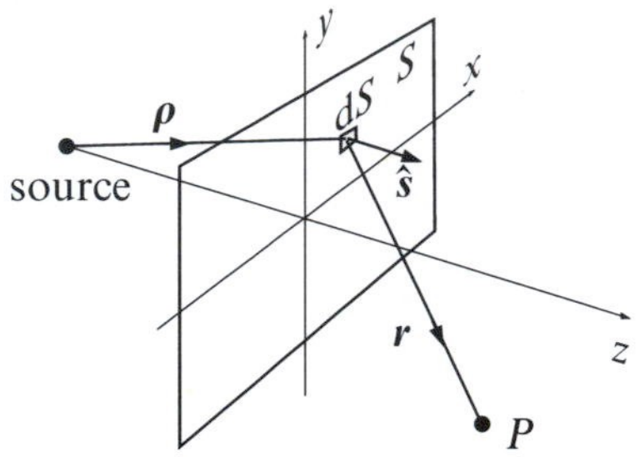

Source at infinity	$\psi_P = -\frac{\mathrm{i}}{\lambda}\psi_0 \int_{\text{plane}} K(\theta)\frac{\mathrm{e}^{\mathrm{i}kr}}{r}\,\mathrm{d}A$	(8.45)	ψ_P complex amplitude at P λ wavelength k wavenumber ($=2\pi/\lambda$) ψ_0 incident amplitude θ obliquity angle r distance of $\mathrm{d}A$ from P ($\gg\lambda$) $\mathrm{d}A$ area element on incident wavefront K obliquity factor $\mathrm{d}S$ element of closed surface $\hat{}$ unit vector $\boldsymbol{s}$ vector normal to $\mathrm{d}S$ $\boldsymbol{r}$ vector from P to $\mathrm{d}S$ $\boldsymbol{\rho}$ vector from source to $\mathrm{d}S$ E_0 amplitude (see footnote)
where:			
Obliquity factor (cardioid)	$K(\theta) = \frac{1}{2}(1+\cos\theta)$	(8.46)	
Source at finite distance[b]	$\psi_P = -\frac{\mathrm{i}E_0}{\lambda}\oint_{\text{closed surface}} \frac{\mathrm{e}^{\mathrm{i}k(\rho+r)}}{2\rho r}[\cos(\hat{\boldsymbol{s}}\cdot\hat{\boldsymbol{r}})-\cos(\hat{\boldsymbol{s}}\cdot\hat{\boldsymbol{\rho}})]\,\mathrm{d}S$	(8.47)	

[a]Also known as the "Fresnel–Kirchhoff formula." Diffraction by an obstacle coincident with the integration surface can be approximated by omitting that part of the surface from the integral.
[b]The source amplitude at $\boldsymbol{\rho}$ is $\psi(\rho) = E_0\mathrm{e}^{\mathrm{i}k\rho}/\rho$. The integral is taken over a surface enclosing the point P.

Fresnel zones

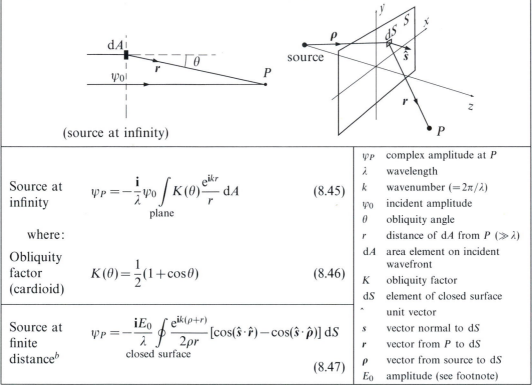

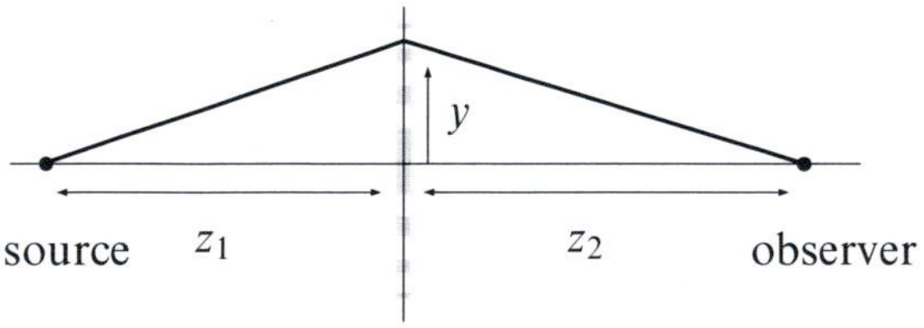

Effective aperture distance[a]	$\frac{1}{z} = \frac{1}{z_1} + \frac{1}{z_2}$	(8.48)	z effective distance z_1 source–aperture distance z_2 aperture–observer distance n half-period zone number λ wavelength y_n nth half-period zone radius z_m distance of mth zero from aperture R aperture radius
Half-period zone radius	$y_n = (n\lambda z)^{1/2}$	(8.49)	
Axial zeros (circular aperture)	$z_m = \frac{R^2}{2m\lambda}$	(8.50)	

[a]I.e., the aperture–observer distance to be employed when the source is not at infinity.

Cornu spiral

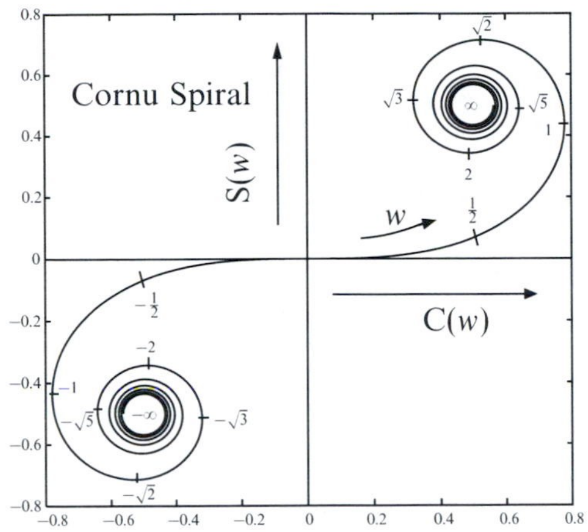

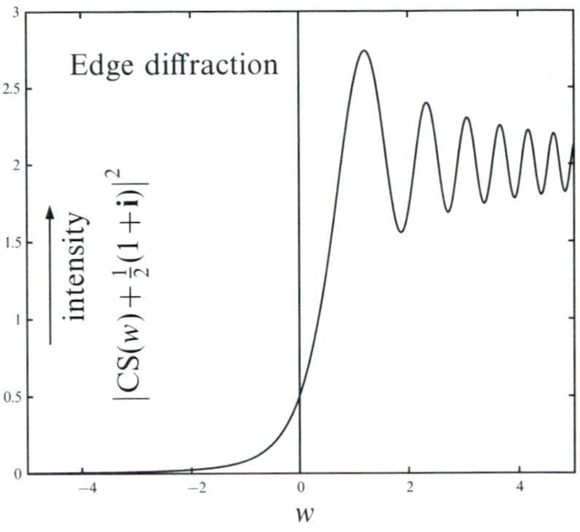

Fresnel integrals[a]	$\mathrm{C}(w)=\int_0^w \cos\frac{\pi t^2}{2}\,\mathrm{d}t$ (8.51) $\mathrm{S}(w)=\int_0^w \sin\frac{\pi t^2}{2}\,\mathrm{d}t$ (8.52)	C Fresnel cosine integral S Fresnel sine integral
Cornu spiral	$\mathrm{CS}(w)=\mathrm{C}(w)+\mathbf{i}\mathrm{S}(w)$ (8.53) $\mathrm{CS}(\pm\infty)=\pm\frac{1}{2}(1+\mathbf{i})$ (8.54)	CS Cornu spiral v,w length along spiral
Edge diffraction	$\psi_P=\frac{\psi_0}{2^{1/2}}[\mathrm{CS}(w)+\frac{1}{2}(1+\mathbf{i})]$ (8.55) where $w=y\left(\frac{2}{\lambda z}\right)^{1/2}$ (8.56)	ψ_P complex amplitude at P ψ_0 unobstructed amplitude λ wavelength z distance of P from aperture plane [see (8.48)] y position of edge
Diffraction from a long slit[b]	$\psi_P=\frac{\psi_0}{2^{1/2}}[\mathrm{CS}(w_2)-\mathrm{CS}(w_1)]$ (8.57) where $w_i=y_i\left(\frac{2}{\lambda z}\right)^{1/2}$ (8.58)	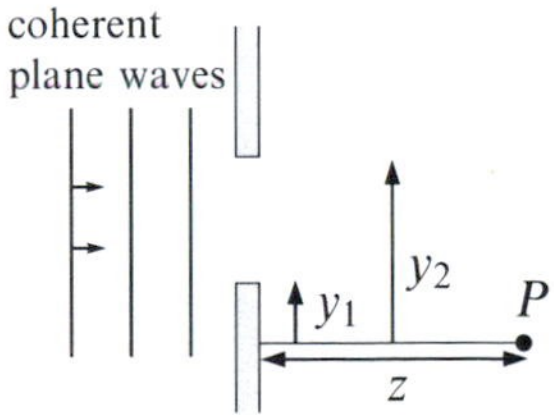
Diffraction from a rectangular aperture	$\psi_P=\frac{\psi_0}{2}[\mathrm{CS}(v_2)-\mathrm{CS}(v_1)]\times$ (8.59) $[\mathrm{CS}(w_2)-\mathrm{CS}(w_1)]$ (8.60) where $v_i=x_i\left(\frac{2}{\lambda z}\right)^{1/2}$ (8.61) and $w_i=y_i\left(\frac{2}{\lambda z}\right)^{1/2}$ (8.62)	x_i positions of slit sides y_i positions of slit top/bottom

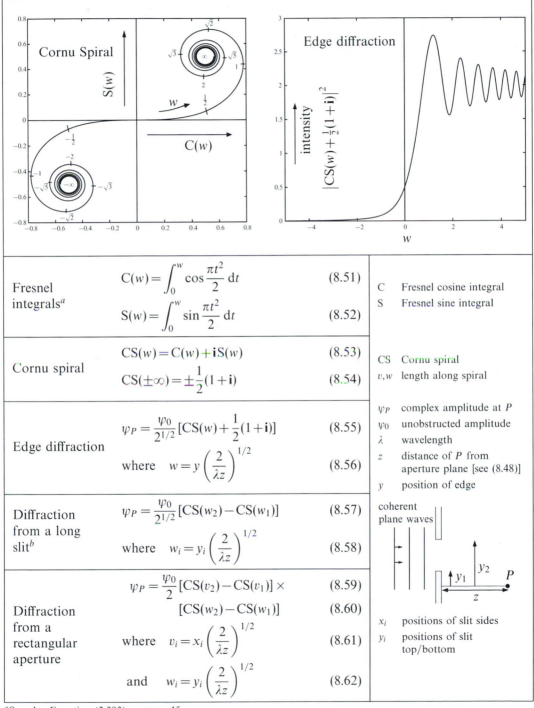

[a]See also Equation (2.393) on page 45.
[b]Slit long in x.

8

8.5 Geometrical optics

Lenses and mirrors[a]

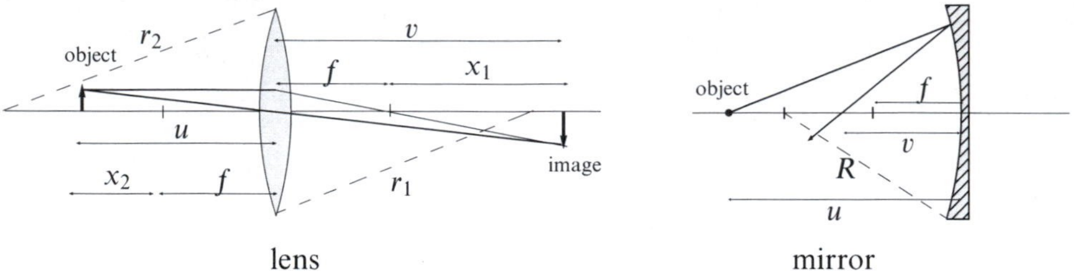

sign convention	+	−
r	centred to right	centred to left
u	real object	virtual object
v	real image	virtual image
f	converging lens/ concave mirror	diverging lens/ convex mirror
M_T	erect image	inverted image

Fermat's principle[b]	$L=\int \eta\, dl$ is stationary	(8.63)	L optical path length; η refractive index; dl ray path element
Gauss's lens formula	$\frac{1}{u}+\frac{1}{v}=\frac{1}{f}$	(8.64)	u object distance; v image distance; f focal length
Newton's lens formula	$x_1x_2=f^2$	(8.65)	x_1 $=v-f$; x_2 $=u-f$
Lensmaker's formula	$\frac{1}{u}+\frac{1}{v}=(\eta-1)\left(\frac{1}{r_1}-\frac{1}{r_2}\right)$	(8.66)	r_i radii of curvature of lens surfaces
Mirror formula[c]	$\frac{1}{u}+\frac{1}{v}=-\frac{2}{R}=\frac{1}{f}$	(8.67)	R mirror radius of curvature
Dioptre number	$D=\frac{1}{f}\ \mathrm{m}^{-1}$	(8.68)	D dioptre number (f in metres)
Focal ratio[d]	$n=\frac{f}{d}$	(8.69)	n focal ratio; d lens or mirror diameter
Transverse linear magnification	$M_T=-\frac{v}{u}$	(8.70)	M_T transverse magnification
Longitudinal linear magnification	$M_L=-M_T^2$	(8.71)	M_L longitudinal magnification

[a]Formulas assume "Gaussian optics," i.e., all lenses are thin and all angles small. Light enters from the left.
[b]A stationary optical path length has, to first order, a length identical to that of adjacent paths.
[c]The mirror is concave if $R<0$, convex if $R>0$.
[d]Or "f-number," written $f/2$ if $n=2$ etc.

Prisms (dispersing)

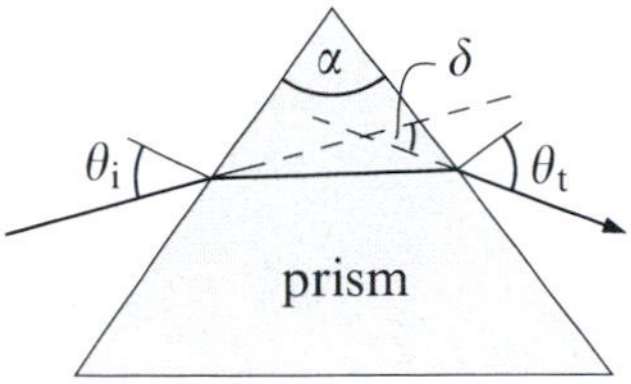

Transmission angle	$\sin\theta_t = (\eta^2 - \sin^2\theta_i)^{1/2}\sin\alpha - \sin\theta_i\cos\alpha$ (8.72)	θ_i angle of incidence θ_t angle of transmission α apex angle η refractive index
Deviation	$\delta = \theta_i + \theta_t - \alpha$ (8.73)	δ angle of deviation
Minimum deviation condition	$\sin\theta_i = \sin\theta_t = \eta\sin\frac{\alpha}{2}$ (8.74)	
Refractive index	$\eta = \frac{\sin[(\delta_m + \alpha)/2]}{\sin(\alpha/2)}$ (8.75)	δ_m minimum deviation
Angular dispersion[a]	$D = \frac{d\delta}{d\lambda} = \frac{2\sin(\alpha/2)}{\cos[(\delta_m + \alpha)/2]}\frac{d\eta}{d\lambda}$ (8.76)	D dispersion λ wavelength

[a] At minimum deviation.

Optical fibres

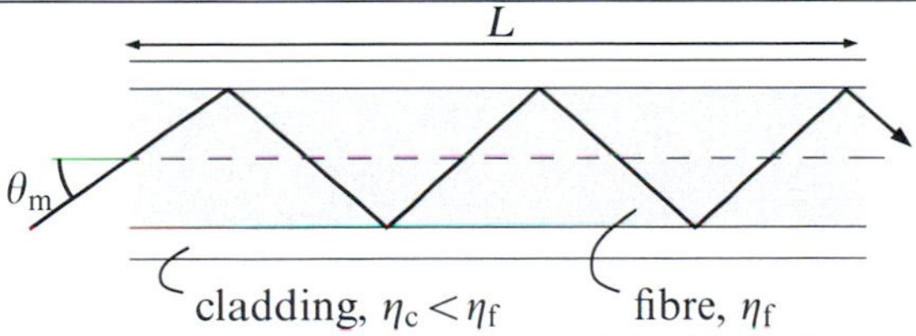

Acceptance angle	$\sin\theta_m = \frac{1}{\eta_0}(\eta_f^2 - \eta_c^2)^{1/2}$ (8.77)	θ_m maximum angle of incidence η_0 exterior refractive index η_f fibre refractive index η_c cladding refractive index
Numerical aperture	$N = \eta_0 \sin\theta_m$ (8.78)	N numerical aperture
Multimode dispersion[a]	$\frac{\Delta t}{L} = \frac{\eta_f}{c}\left(\frac{\eta_f}{\eta_c} - 1\right)$ (8.79)	Δt temporal dispersion L fibre length c speed of light

[a] Of a pulse with a given wavelength, caused by the range of incident angles up to θ_m. Sometimes called "intermodal dispersion" or "modal dispersion."

8.6 Polarisation

Elliptical polarisation[a]

Elliptical polarisation	$\boldsymbol{E} = (E_{0x}, E_{0y}e^{i\delta})e^{i(kz-\omega t)}$	(8.80)
Polarisation angle[b]	$\tan 2\alpha = \dfrac{2E_{0x}E_{0y}}{E_{0x}^2 - E_{0y}^2}\cos\delta$	(8.81)
Ellipticity[c]	$e = \dfrac{a-b}{a}$	(8.82)
Malus's law[d]	$I(\theta) = I_0 \cos^2\theta$	(8.83)

$\boldsymbol{E}$	electric field
k	wavevector
z	propagation axis
ωt	angular frequency × time
E_{0x}	x amplitude of $\boldsymbol{E}$
E_{0y}	y amplitude of $\boldsymbol{E}$
δ	relative phase of E_y with respect to E_x
α	polarisation angle
e	ellipticity
a	semi-major axis
b	semi-minor axis
$I(\theta)$	transmitted intensity
I_0	incident intensity
θ	polariser–analyser angle

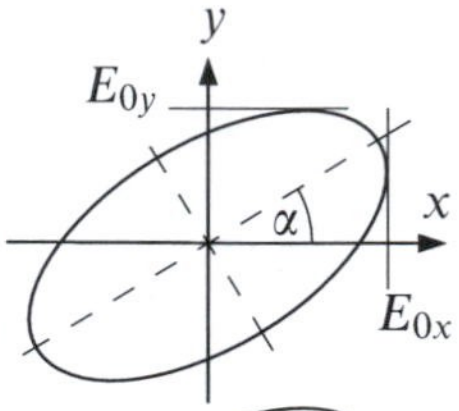

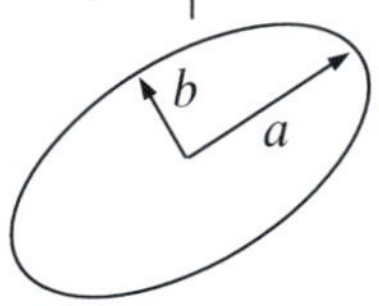

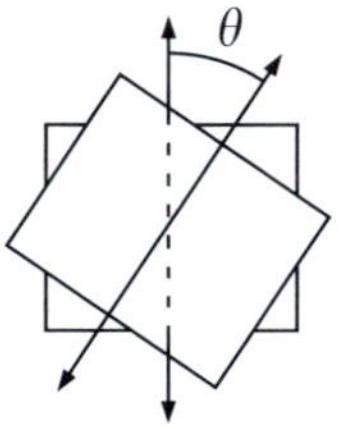

[a]See the introduction (page 161) for a discussion of sign and handedness conventions.
[b]Angle between ellipse major axis and x axis. Sometimes the polarisation angle is defined as $\pi/2-\alpha$.
[c]This is one of several definitions for ellipticity.
[d]Transmission through skewed polarisers for unpolarised incident light.

Jones vectors and matrices

Normalised electric field[a]	$\boldsymbol{E} = \begin{pmatrix} E_x \\ E_y \end{pmatrix}; \quad \lvert\boldsymbol{E}\rvert = 1$	(8.84)
Example vectors:	$E_x = \begin{pmatrix} 1 \\ 0 \end{pmatrix}$ $\quad E_{45} = \frac{1}{\sqrt{2}}\begin{pmatrix} 1 \\ 1 \end{pmatrix}$ $E_r = \frac{1}{\sqrt{2}}\begin{pmatrix} 1 \\ -\mathbf{i} \end{pmatrix}$ $\quad E_l = \frac{1}{\sqrt{2}}\begin{pmatrix} 1 \\ \mathbf{i} \end{pmatrix}$	
Jones matrix	$\boldsymbol{E}_t = \mathbf{A}\boldsymbol{E}_i$	(8.85)

$\boldsymbol{E}$	electric field
E_x	x component of $\boldsymbol{E}$
E_y	y component of $\boldsymbol{E}$
E_{45}	45° to x axis
E_r	right-hand circular
E_l	left-hand circular
$\boldsymbol{E}_t$	transmitted vector
$\boldsymbol{E}_i$	incident vector
$\mathbf{A}$	Jones matrix

Example matrices:

Linear polariser $\parallel x$	$\begin{pmatrix} 1 & 0 \\ 0 & 0 \end{pmatrix}$	Linear polariser $\parallel y$	$\begin{pmatrix} 0 & 0 \\ 0 & 1 \end{pmatrix}$
Linear polariser at 45°	$\frac{1}{2}\begin{pmatrix} 1 & 1 \\ 1 & 1 \end{pmatrix}$	Linear polariser at −45°	$\frac{1}{2}\begin{pmatrix} 1 & -1 \\ -1 & 1 \end{pmatrix}$
Right circular polariser	$\frac{1}{2}\begin{pmatrix} 1 & \mathbf{i} \\ -\mathbf{i} & 1 \end{pmatrix}$	Left circular polariser	$\frac{1}{2}\begin{pmatrix} 1 & -\mathbf{i} \\ \mathbf{i} & 1 \end{pmatrix}$
$\lambda/4$ plate (fast $\parallel x$)	$e^{i\pi/4}\begin{pmatrix} 1 & 0 \\ 0 & \mathbf{i} \end{pmatrix}$	$\lambda/4$ plate (fast $\perp x$)	$e^{i\pi/4}\begin{pmatrix} 1 & 0 \\ 0 & -\mathbf{i} \end{pmatrix}$

[a]Known as the "normalised Jones vector."

Stokes parameters[a]

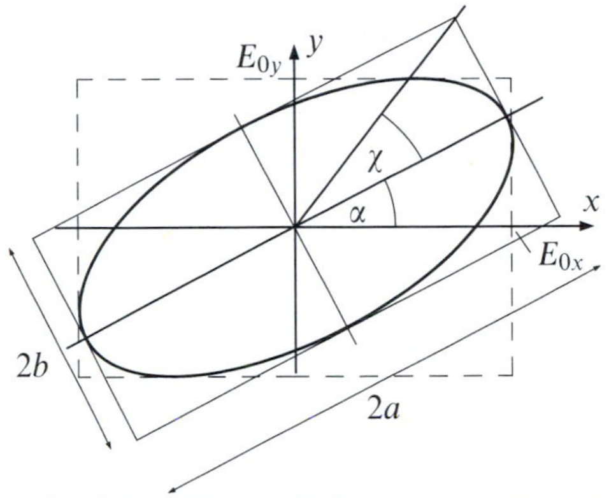

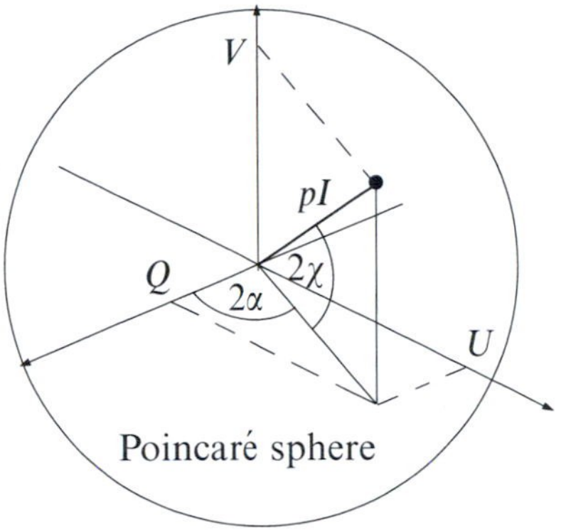

Electric fields	$E_x = E_{0x} e^{i(kz-\omega t)}$	(8.86)	k	wavevector
	$E_y = E_{0y} e^{i(kz-\omega t+\delta)}$	(8.87)	ωt	angular frequency $\times$ time
			δ	relative phase of E_y with respect to E_x
Axial ratio[b]	$\tan\chi = \pm r = \pm\frac{b}{a}$	(8.88)	χ	(see diagram)
			r	axial ratio
Stokes parameters	$I = \langle E_x^2 \rangle + \langle E_y^2 \rangle$	(8.89)		
	$Q = \langle E_x^2 \rangle - \langle E_y^2 \rangle$	(8.90)	E_x	electric field component $\parallel x$
	$= pI \cos 2\chi \cos 2\alpha$	(8.91)	E_y	electric field component $\parallel y$
	$U = 2\langle E_x E_y \rangle \cos\delta$	(8.92)	E_{0x}	field amplitude in x direction
	$= pI \cos 2\chi \sin 2\alpha$	(8.93)	E_{0y}	field amplitude in y direction
	$V = 2\langle E_x E_y \rangle \sin\delta$	(8.94)	α	polarisation angle
	$= pI \sin 2\chi$	(8.95)	p	degree of polarisation
			$\langle\cdot\rangle$	mean over time
Degree of polarisation	$p = \frac{(Q^2+U^2+V^2)^{1/2}}{I} \leq 1$	(8.96)		

	Q/I	U/I	V/I		Q/I	U/I	V/I
left circular	0	0	−1	right circular	0	0	1
linear $\parallel x$	1	0	0	linear $\parallel y$	−1	0	0
linear 45° to x	0	1	0	linear −45° to x	0	−1	0
unpolarised	0	0	0				

[a]Using the convention that right-handed circular polarisation corresponds to a clockwise rotation of the electric field in a given plane when looking towards the source. The propagation direction in the diagram is out of the plane. The parameters I, Q, U, and V are sometimes denoted s_0, s_1, s_2, and s_3, and other nomenclatures exist. There is no generally accepted definition – often the parameters are scaled to be dimensionless, with $s_0 = 1$, or to represent power flux through a plane $\perp$ the beam, i.e., $I = (\langle E_x^2 \rangle + \langle E_y^2 \rangle)/Z_0$ etc., where Z_0 is the impedance of free space.

[b]The axial ratio is positive for right-handed polarisation and negative for left-handed polarisation using our definitions.

8

8.7 Coherence (scalar theory)

Mutual coherence function	$$\Gamma_{12}(\tau)=\langle\psi_1(t)\psi_2^*(t+\tau)\rangle \qquad (8.97)$$	Γ_{ij} mutual coherence function τ temporal interval ψ_i (complex) wave disturbance at spatial point i
Complex degree of coherence	$$\gamma_{12}(\tau)=\frac{\langle\psi_1(t)\psi_2^*(t+\tau)\rangle}{[\langle\|\psi_1\|^2\rangle\langle\|\psi_2\|^2\rangle]^{1/2}} \qquad (8.98)$$ $$=\frac{\Gamma_{12}(\tau)}{[\Gamma_{11}(0)\Gamma_{22}(0)]^{1/2}} \qquad (8.99)$$	t time $\langle\cdot\rangle$ mean over time γ_{ij} complex degree of coherence $*$ complex conjugate
Combined intensity[a]	$$I_{\text{tot}}=I_1+I_2+2(I_1I_2)^{1/2}\Re[\gamma_{12}(\tau)] \qquad (8.100)$$	I_{tot} combined intensity I_i intensity of disturbance at point i $\Re$ real part of
Fringe visibility	$$V(\tau)=\frac{2(I_1I_2)^{1/2}}{I_1+I_2}\|\gamma_{12}(\tau)\| \qquad (8.101)$$	
if $\|\gamma_{12}(\tau)\|$ is a constant:	$$V=\frac{I_{\max}-I_{\min}}{I_{\max}+I_{\min}} \qquad (8.102)$$	$I_{\max}$ max. combined intensity $I_{\min}$ min. combined intensity
if $I_1=I_2$:	$$V(\tau)=\|\gamma_{12}(\tau)\| \qquad (8.103)$$	
Complex degree of temporal coherence[b]	$$\gamma(\tau)=\frac{\langle\psi_1(t)\psi_1^*(t+\tau)\rangle}{\langle\|\psi_1(t)^2\|\rangle} \qquad (8.104)$$ $$=\frac{\int I(\omega)e^{-i\omega\tau}\,d\omega}{\int I(\omega)\,d\omega} \qquad (8.105)$$	$\gamma(\tau)$ degree of temporal coherence $I(\omega)$ specific intensity ω radiation angular frequency c speed of light
Coherence time and length	$$\Delta\tau_c=\frac{\Delta l_c}{c}\sim\frac{1}{\Delta\nu} \qquad (8.106)$$	$\Delta\tau_c$ coherence time Δl_c coherence length $\Delta\nu$ spectral bandwidth
Complex degree of spatial coherence[c]	$$\gamma(\boldsymbol{D})=\frac{\langle\psi_1\psi_2^*\rangle}{[\langle\|\psi_1\|^2\rangle\langle\|\psi_2\|^2\rangle]^{1/2}} \qquad (8.107)$$ $$=\frac{\int I(\hat{\boldsymbol{s}})e^{ik\boldsymbol{D}\cdot\hat{\boldsymbol{s}}}\,d\Omega}{\int I(\hat{\boldsymbol{s}})\,d\Omega} \qquad (8.108)$$	$\gamma(\boldsymbol{D})$ degree of spatial coherence $\boldsymbol{D}$ spatial separation of points 1 and 2 $I(\hat{\boldsymbol{s}})$ specific intensity of distant extended source in direction $\hat{\boldsymbol{s}}$ $d\Omega$ differential solid angle $\hat{\boldsymbol{s}}$ unit vector in the direction of $d\Omega$
Intensity correlation[d]	$$\frac{\langle I_1I_2\rangle}{[\langle I_1\rangle^2\langle I_2\rangle^2]^{1/2}}=1+\gamma^2(\boldsymbol{D}) \qquad (8.109)$$	k wavenumber
Speckle intensity distribution[e]	$$\text{pr}(I)=\frac{1}{\langle I\rangle}e^{-I/\langle I\rangle} \qquad (8.110)$$	pr probability density
Speckle size (coherence width)	$$\Delta w_c\simeq\frac{\lambda}{\alpha} \qquad (8.111)$$	Δw_c characteristic speckle size λ wavelength α source angular size as seen from the screen

[a] From interfering the disturbances at points 1 and 2 with a relative delay τ.
[b] Or "autocorrelation function."
[c] Between two points on a wavefront, separated by $\boldsymbol{D}$. The integral is over the entire extended source.
[d] For wave disturbances that have a Gaussian probability distribution in amplitude. This is "Gaussian light" such as from a thermal source.
[e] Also for Gaussian light.

8.8 Line radiation

Spectral line broadening

Natural broadening[a]	$I(\omega)=\frac{(2\pi\tau)^{-1}}{(2\tau)^{-2}+(\omega-\omega_0)^2}$	(8.112)
Natural half-width	$\Delta\omega=\frac{1}{2\tau}$	(8.113)
Collision broadening	$I(\omega)=\frac{(\pi\tau_c)^{-1}}{(\tau_c)^{-2}+(\omega-\omega_0)^2}$	(8.114)
Collision and pressure half-width[c]	$\Delta\omega=\frac{1}{\tau_c}=p\pi d^2\left(\frac{\pi mkT}{16}\right)^{-1/2}$	(8.115)
Doppler broadening	$I(\omega)=\left(\frac{mc^2}{2kT\omega_0^2\pi}\right)^{1/2}\exp\left[-\frac{mc^2}{2kT}\frac{(\omega-\omega_0)^2}{\omega_0^2}\right]$	(8.116)
Doppler half-width	$\Delta\omega=\omega_0\left(\frac{2kT\ln 2}{mc^2}\right)^{1/2}$	(8.117)

$I(\omega)$	normalised intensity[b]
τ	lifetime of excited state
ω	angular frequency ($=2\pi\nu$)
$\Delta\omega$	half-width at half-power
ω_0	centre frequency
τ_c	mean time between collisions
p	pressure
d	effective atomic diameter
m	gas particle mass
k	Boltzmann constant
T	temperature
c	speed of light

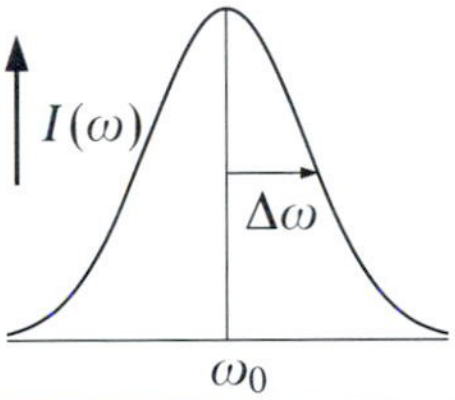

[a]The transition probability per unit time for the state is $=1/\tau$. In the classical limit of a damped oscillator, the e-folding time of the electric field is 2τ. Both the natural and collision profiles described here are Lorentzian.
[b]The intensity spectra are normalised so that $\int I(\omega)\,d\omega=1$, assuming $\Delta\omega/\omega_0 \ll 1$.
[c]The pressure-broadening relation combines Equations (5.78), (5.86) and (5.89) and assumes an otherwise perfect gas of finite-sized atoms. More accurate expressions are considerably more complicated.

Einstein coefficients[a]

Absorption	$R_{12}=B_{12}I_\nu n_1$	(8.118)
Spontaneous emission	$R_{21}=A_{21}n_2$	(8.119)
Stimulated emission	$R'_{21}=B_{21}I_\nu n_2$	(8.120)
Coefficient ratios	$\frac{A_{21}}{B_{12}}=\frac{2h\nu^3}{c^2}\frac{g_1}{g_2}$	(8.121)
	$\frac{B_{21}}{B_{12}}=\frac{g_1}{g_2}$	(8.122)

R_{ij}	transition rate, level $i\rightarrow j$ ($m^{-3}s^{-1}$)
B_{ij}	Einstein B coefficients
I_ν	specific intensity of radiation field
A_{21}	Einstein A coefficient
n_i	number density of atoms in quantum level i (m^{-3})
h	Planck constant
ν	frequency
c	speed of light
g_i	degeneracy of ith level

[a]Note that the coefficients can also be defined in terms of spectral energy density, $u_\nu=4\pi I_\nu/c$ rather than I_ν. In this case $\frac{A_{21}}{B_{12}}=\frac{8\pi h\nu^3}{c^3}\frac{g_1}{g_2}$. See also *Population densities* on page 116.

8

Lasers[a]

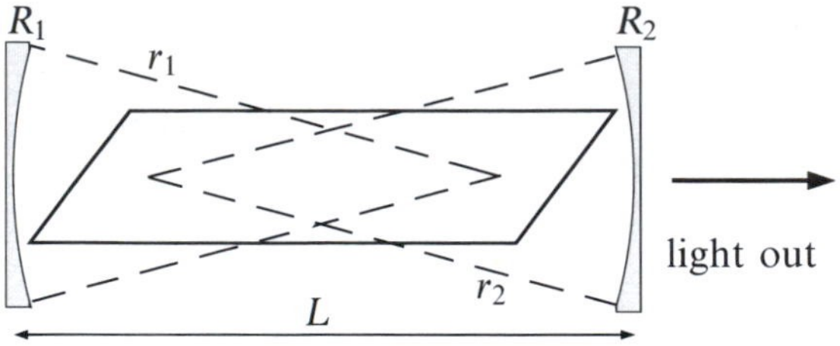

Cavity stability condition	$$0 \le \left(1-\frac{L}{r_1}\right)\left(1-\frac{L}{r_2}\right) \le 1 \quad (8.123)$$	$r_{1,2}$ radii of curvature of end-mirrors L distance between mirror centres
Longitudinal cavity modes[b]	$$\nu_n = \frac{c}{2L}n \quad (8.124)$$	ν_n mode frequency n integer c speed of light
Cavity Q	$$Q = \frac{2\pi L(R_1R_2)^{1/4}}{\lambda[1-(R_1R_2)^{1/2}]} \quad (8.125)$$ $$\simeq \frac{4\pi L}{\lambda(1-R_1R_2)} \quad (8.126)$$	Q quality factor $R_{1,2}$ mirror (power) reflectances λ wavelength
Cavity line width	$$\Delta\nu_c = \frac{\nu_n}{Q} = 1/(2\pi\tau_c) \quad (8.127)$$	$\Delta\nu_c$ cavity line width (FWHP) τ_c cavity photon lifetime
Schawlow–Townes line width	$$\frac{\Delta\nu}{\nu_n} = \frac{2\pi h(\Delta\nu_c)^2}{P}\left(\frac{g_l N_u}{g_l N_u - g_u N_l}\right) \quad (8.128)$$	$\Delta\nu$ line width (FWHP) P laser power $g_{u,l}$ degeneracy of upper/lower levels $N_{u,l}$ number density of upper/lower levels
Threshold lasing condition	$$R_1R_2\exp[2(\alpha-\beta)L] > 1 \quad (8.129)$$	α gain per unit length of medium β loss per unit length of medium

[a]Also see the *Fabry-Perot etalon* on page 163. Note that "cavity" refers to the empty cavity, with no lasing medium present.
[b]The mode spacing equals the cavity free spectral range.

Chapter 9 Astrophysics

9.1 Introduction

Many of the formulas associated with astronomy and astrophysics are either too specialised for a general work such as this or are common to other fields and can therefore be found elsewhere in this book. The following section includes many of the relationships that fall into neither of these categories, including equations to convert between various astronomical coordinate systems and some basic formulas associated with cosmology.

Exceptionally, this section also includes data on the Sun, Earth, Moon, and planets. Observational astrophysics remains a largely inexact science, and parameters of these (and other) bodies are often used as approximate base units in measurements. For example, the masses of stars and galaxies are frequently quoted as multiples of the mass of the Sun ($1M_\odot = 1.989 \times 10^{30}$ kg), extra-solar system planets in terms of the mass of Jupiter, and so on. Astronomers seem to find it particularly difficult to drop arcane units and conventions, resulting in a profusion of measures and nomenclatures throughout the subject. However, the convention of using suitable astronomical objects in this way is both useful and widely accepted.

9.2 Solar system data

Solar data

equatorial radius	$R_\odot$	=	6.960×10^8 m		=	$109.1 R_\oplus$
mass	$M_\odot$	=	1.9891×10^{30} kg		=	$3.32946 \times 10^5 M_\oplus$
polar moment of inertia	$I_\odot$	=	5.7×10^{46} kg m^2		=	$7.09 \times 10^8 I_\oplus$
bolometric luminosity	$L_\odot$	=	3.826×10^{26} W			
effective surface temperature	$T_\odot$	=	5770 K			
solar constant[a]			1.368×10^3 W m^{-2}			
absolute magnitude	M_V	=	$+4.83$;	M_{bol}	=	$+4.75$
apparent magnitude	m_V	=	-26.74;	m_{bol}	=	-26.82

[a]Bolometric flux at a distance of 1 astronomical unit (AU).

Earth data

equatorial radius	$R_\oplus$	=	6.37814×10^6 m	=	$9.166 \times 10^{-3} R_\odot$
flattening[a]	f	=	0.00335364	=	1/298.183
mass	$M_\oplus$	=	5.9742×10^{24} kg	=	$3.0035 \times 10^{-6} M_\odot$
polar moment of inertia	$I_\oplus$	=	8.037×10^{37} kg m^2	=	$1.41 \times 10^{-9} I_\odot$
orbital semi-major axis[b]	1 AU	=	1.495979×10^{11} m	=	$214.9 R_\odot$
mean orbital velocity			2.979×10^4 m s^{-1}		
equatorial surface gravity	g_e	=	9.780327 m s^{-2}		(includes rotation)
polar surface gravity	g_p	=	9.832186 m s^{-2}		
rotational angular velocity	ω_e	=	7.292115×10^{-5} rad s^{-1}		

[a]f equals $(R_\oplus - R_{polar})/R_\oplus$. The mean radius of the Earth is 6.3710×10^6 m.
[b]About the Sun.

Moon data

equatorial radius	R_m	=	1.7374×10^6 m	=	$0.27240 R_\oplus$
mass	M_m	=	7.3483×10^{22} kg	=	$1.230 \times 10^{-2} M_\oplus$
mean orbital radius[a]	a_m	=	3.84400×10^8 m	=	$60.27 R_\oplus$
mean orbital velocity			1.03×10^3 m s^{-1}		
orbital period (sidereal)			27.32166 d		
equatorial surface gravity			1.62 m s^{-2}	=	$0.166 g_e$

[a]About the Earth.

Planetary data[a]

	$M/M_\oplus$	$R/R_\oplus$	T(d)	P(yr)	a(AU)
Mercury	0.055274	0.38251	58.646	0.24085	0.38710
Venus[b]	0.81500	0.94883	243.018	0.615228	0.72335
Earth	1	1	0.99727	1.00004	1.00000
Mars	0.10745	0.53260	1.02596	1.88093	1.52371
Jupiter	317.85	11.209	0.41354	11.8613	5.20253
Saturn	95.159	9.4491	0.44401	29.6282	9.57560
Uranus[b]	14.500	4.0073	0.71833	84.7466	19.2934
Neptune	17.204	3.8826	0.67125	166.344	30.2459
Pluto[b]	0.00251	0.18736	6.3872	248.348	39.5090

M	mass
R	equatorial radius
T	rotational period
P	orbital period
a	mean distance
$M_\oplus$	5.9742×10^{24} kg
$R_\oplus$	6.37814×10^6 m
1 d	86400 s
1 yr	3.15569×10^7 s
1 AU	1.495979×10^{11} m

[a]Using the osculating orbital elements for 1998. Note that P is the instantaneous orbital period, calculated from the planet's daily motion. The radii of gas giants are taken at 1 atmosphere pressure.
[b]Retrograde rotation.

9.3 Coordinate transformations (astronomical)

Time in astronomy

Julian day number[a]	$JD = D - 32075 + 1461 * (Y + 4800 + (M-14)/12)/4$ $+ 367 * (M - 2 - (M-14)/12 * 12)/12$ $- 3 * ((Y + 4900 + (M-14)/12)/100)/4$	(9.1)	JD Julian day number D day of month number Y calendar year, e.g., 1963 M calendar month (Jan=1) $*$ integer multiply $/$ integer divide
Modified Julian day number	$MJD = JD - 2400000.5$	(9.2)	MJD modified Julian day number
Day of week	$W = (JD + 1) \mod 7$	(9.3)	W day of week (0=Sunday, 1=Monday, ...)
Local civil time	$\mathrm{LCT} = \mathrm{UTC} + \mathrm{TZC} + \mathrm{DSC}$	(9.4)	LCT local civil time UTC coordinated universal time TZC time zone correction DSC daylight saving correction
Julian centuries	$T = \dfrac{JD - 2451545.5}{36525}$	(9.5)	T Julian centuries between 12^{h} UTC 1 Jan 2000 and 0^{h} UTC $D/M/Y$
Greenwich sidereal time	$\mathrm{GMST} = 6^{\mathrm{h}}41^{\mathrm{m}}50^{\mathrm{s}}.54841$ $+ 8640184^{\mathrm{s}}.812866T$ $+ 0^{\mathrm{s}}.093104T^2$ $- 0^{\mathrm{s}}.0000062T^3$	(9.6)	GMST Greenwich mean sidereal time at 0^{h} UTC $D/M/Y$ (for later times use $1\,\mathrm{s} = 1.002738$ sidereal seconds)
Local sidereal time	$\mathrm{LST} = \mathrm{GMST} + \dfrac{\lambda^\circ}{15^\circ}$	(9.7)	LST local sidereal time λ° geographic longitude, degrees east of Greenwich

[a] For the Julian day starting at noon on the calendar day in question. The routine is designed around integer arithmetic with "truncation towards zero" (so that $-5/3 = -1$) and is valid for dates from the onset of the Gregorian calendar, 15 October 1582. JD represents the number of days since Greenwich mean noon 1 Jan 4713 BC. For reference, noon, 1 Jan 2000 $= JD2451545$ and was a Saturday ($W = 6$).

Horizon coordinates[a]

Hour angle	$H = \mathrm{LST} - \alpha$	(9.8)	LST local sidereal time H (local) hour angle
Equatorial to horizon	$\sin a = \sin\delta \sin\phi + \cos\delta \cos\phi \cos H$ $\tan A \equiv \dfrac{-\cos\delta \sin H}{\sin\delta \cos\phi - \sin\phi \cos\delta \cos H}$	(9.9) (9.10)	α right ascension δ declination a altitude A azimuth (E from N) ϕ observer's latitude
Horizon to equatorial	$\sin\delta = \sin a \sin\phi + \cos a \cos\phi \cos A$ $\tan H \equiv \dfrac{-\cos a \sin A}{\sin a \cos\phi - \sin\phi \cos a \cos A}$	(9.11) (9.12)	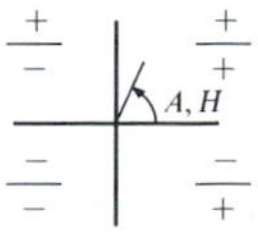

[a] Conversions between horizon or alt–azimuth coordinates, (a, A), and celestial equatorial coordinates, (δ, α). There are a number of conventions for defining azimuth. For example, it is sometimes taken as the angle west from south rather than east from north. The quadrants for A and H can be obtained from the signs of the numerators and denominators in Equations (9.10) and (9.12) (see diagram).

Ecliptic coordinates[a]

Obliquity of the ecliptic	$\varepsilon = 23^\circ 26' 21''.45 - 46''.815\,T$ $-0''.0006\,T^2$ $+0''.00181\,T^3$	(9.13)	ε mean ecliptic obliquity T Julian centuries since J2000.0[b]
Equatorial to ecliptic	$\sin\beta = \sin\delta\cos\varepsilon - \cos\delta\sin\varepsilon\sin\alpha$	(9.14)	α right ascension δ declination
	$\tan\lambda \equiv \dfrac{\sin\alpha\cos\varepsilon + \tan\delta\sin\varepsilon}{\cos\alpha}$	(9.15)	λ ecliptic longitude β ecliptic latitude
Ecliptic to equatorial	$\sin\delta = \sin\beta\cos\varepsilon + \cos\beta\sin\varepsilon\sin\lambda$	(9.16)	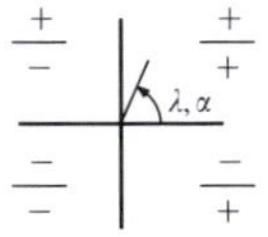
	$\tan\alpha \equiv \dfrac{\sin\lambda\cos\varepsilon - \tan\beta\sin\varepsilon}{\cos\lambda}$	(9.17)	

[a]Conversions between ecliptic, (β,λ), and celestial equatorial, (δ,α), coordinates. β is positive above the ecliptic and λ increases eastwards. The quadrants for λ and α can be obtained from the signs of the numerators and denominators in Equations (9.15) and (9.17) (see diagram).
[b]See Equation (9.5).

Galactic coordinates[a]

Galactic frame	$\alpha_g = 192^\circ 15'$	(9.18)	α_g right ascension of north galactic pole
	$\delta_g = 27^\circ 24'$	(9.19)	δ_g declination of north galactic pole
	$l_g = 33^\circ$	(9.20)	
Equatorial to galactic	$\sin b = \cos\delta\cos\delta_g\cos(\alpha - \alpha_g) + \sin\delta\sin\delta_g$	(9.21)	l_g ascending node of galactic plane on equator
	$\tan(l - l_g) \equiv \dfrac{\tan\delta\cos\delta_g - \cos(\alpha - \alpha_g)\sin\delta_g}{\sin(\alpha - \alpha_g)}$	(9.22)	
Galactic to equatorial	$\sin\delta = \cos b\cos\delta_g\sin(l - l_g) + \sin b\sin\delta_g$	(9.23)	δ declination α right ascension
	$\tan(\alpha - \alpha_g) \equiv \dfrac{\cos(l - l_g)}{\tan b\cos\delta_g - \sin\delta_g\sin(l - l_g)}$	(9.24)	b galactic latitude l galactic longitude

[a]Conversions between galactic, (b,l), and celestial equatorial, (δ,α), coordinates. The galactic frame is defined at epoch B1950.0. The quadrants of l and α can be obtained from the signs of the numerators and denominators in Equations (9.22) and (9.24).

Precession of equinoxes[a]

In right ascension	$\alpha \simeq \alpha_0 + (3^s.075 + 1^s.336\sin\alpha_0\tan\delta_0)N$	(9.25)	α right ascension of date α_0 right ascension at J2000.0 N number of years since J2000.0
In declination	$\delta \simeq \delta_0 + (20''.043\cos\alpha_0)N$	(9.26)	δ declination of date δ_0 declination at J2000.0

[a]Right ascension in hours, minutes, and seconds; declination in degrees, arcminutes, and arcseconds. These equations are valid for several hundred years each side of J2000.0.

9.4 Observational astrophysics

Astronomical magnitudes

Apparent magnitude	$m_1 - m_2 = -2.5\log_{10}\frac{F_1}{F_2}$	(9.27)	m_i F_i	apparent magnitude of object i energy flux from object i
Distance modulus[a]	$m - M = 5\log_{10} D - 5$ $= -5\log_{10} p - 5$	(9.28) (9.29)	M $m-M$ D p	absolute magnitude distance modulus distance to object (parsec) annual parallax (arcsec)
Luminosity–magnitude relation	$M_{\mathrm{bol}} = 4.75 - 2.5\log_{10}\frac{L}{L_\odot}$ $L \simeq 3.04 \times 10^{(28-0.4M_{\mathrm{bol}})}$	(9.30) (9.31)	M_{bol} L $L_\odot$	bolometric absolute magnitude luminosity (W) solar luminosity (3.826×10^{26} W)
Flux–magnitude relation	$F_{\mathrm{bol}} \simeq 2.559 \times 10^{-(8+0.4m_{\mathrm{bol}})}$	(9.32)	F_{bol} m_{bol}	bolometric flux ($\mathrm{W\,m^{-2}}$) bolometric apparent magnitude
Bolometric correction	$BC = m_{\mathrm{bol}} - m_{\mathrm{V}}$ $= M_{\mathrm{bol}} - M_{\mathrm{V}}$	(9.33) (9.34)	BC m_{V} M_{V}	bolometric correction V-band apparent magnitude V-band absolute magnitude
Colour index[b]	$B - V = m_{\mathrm{B}} - m_{\mathrm{V}}$ $U - B = m_{\mathrm{U}} - m_{\mathrm{B}}$	(9.35) (9.36)	$B-V$ $U-B$	observed $B-V$ colour index observed $U-B$ colour index
Colour excess[c]	$E = (B-V) - (B-V)_0$	(9.37)	E $(B-V)_0$	$B-V$ colour excess intrinsic $B-V$ colour index

[a]Neglecting extinction.
[b]Using the UBV magnitude system. The bands are centred around 365 nm (U), 440 nm (B), and 550 nm (V).
[c]The $U-B$ colour excess is defined similarly.

Photometric wavelengths

Mean wavelength	$\lambda_0 = \frac{\int \lambda R(\lambda)\,\mathrm{d}\lambda}{\int R(\lambda)\,\mathrm{d}\lambda}$	(9.38)	λ_0 λ R	mean wavelength wavelength system spectral response
Isophotal wavelength	$F(\lambda_{\mathrm{i}}) = \frac{\int F(\lambda)R(\lambda)\,\mathrm{d}\lambda}{\int R(\lambda)\,\mathrm{d}\lambda}$	(9.39)	$F(\lambda)$ λ_{i}	flux density of source (in terms of wavelength) isophotal wavelength
Effective wavelength	$\lambda_{\mathrm{eff}} = \frac{\int \lambda F(\lambda)R(\lambda)\,\mathrm{d}\lambda}{\int F(\lambda)R(\lambda)\,\mathrm{d}\lambda}$	(9.40)	λ_{eff}	effective wavelength

Planetary bodies

Bode's law[a]	$D_{\mathrm{AU}} = \dfrac{4+3\times 2^n}{10}$	(9.41)	D_{AU}	planetary orbital radius (AU)
			n	index: Mercury $=-\infty$, Venus $=0$, Earth $=1$, Mars $=2$, Ceres $=3$, Jupiter$=4$, ...
Roche limit	$R \gtrsim \left(\dfrac{100M}{9\pi\rho}\right)^{1/3}$	(9.42)	R	satellite orbital radius
			M	central mass
	$\gtrsim 2.46R_0$ (if densities equal)	(9.43)	ρ	satellite density
			R_0	central body radius
Synodic period[b]	$\dfrac{1}{S} = \left\|\dfrac{1}{P} - \dfrac{1}{P_\oplus}\right\|$	(9.44)	S	synodic period
			P	planetary orbital period
			$P_\oplus$	Earth's orbital period

[a]Also known as the "Titius–Bode rule." Note that the asteroid Ceres is counted as a planet in this scheme. The relationship breaks down for Neptune and Pluto.
[b]Of a planet.

Distance indicators

Hubble law	$v = H_0 d$	(9.45)	v	cosmological recession velocity
			H_0	Hubble parameter (present epoch)
			d	(proper) distance
Annual parallax	$D_{\mathrm{pc}} = p^{-1}$	(9.46)	D_{pc}	distance (parsec)
			p	annual parallax ($\pm p$ arcsec from mean)
Cepheid variables[a]	$\log_{10}\dfrac{\langle L\rangle}{L_\odot} \simeq 1.15\log_{10}P_{\mathrm{d}} + 2.47$	(9.47)	$\langle L\rangle$	mean cepheid luminosity
			$L_\odot$	Solar luminosity
	$M_{\mathrm{V}} \simeq -2.76\log_{10}P_{\mathrm{d}} - 1.40$	(9.48)	P_{d}	pulsation period (days)
			M_{V}	absolute visual magnitude
Tully–Fisher relation[b]	$M_{\mathrm{I}} \simeq -7.68\log_{10}\left(\dfrac{2v_{\mathrm{rot}}}{\sin i}\right) - 2.58$	(9.49)	M_{I}	I-band absolute magnitude
			v_{rot}	observed maximum rotation velocity ($\mathrm{km\,s^{-1}}$)
			i	galactic inclination (90° when edge-on)
Einstein rings	$\theta^2 = \dfrac{4GM}{c^2}\left(\dfrac{d_{\mathrm{s}} - d_{\mathrm{l}}}{d_{\mathrm{s}}d_{\mathrm{l}}}\right)$	(9.50)	θ	ring angular radius
			M	lens mass
			d_{s}	distance from observer to source
			d_{l}	distance from observer to lens
Sunyaev–Zel'dovich effect[c]	$\dfrac{\Delta T}{T} = -2\displaystyle\int \dfrac{n_{\mathrm{e}}kT_{\mathrm{e}}\sigma_{\mathrm{T}}}{m_{\mathrm{e}}c^2}\,\mathrm{d}l$	(9.51)	T	apparent CMBR temperature
			$\mathrm{d}l$	path element through cloud
			R	cloud radius
			n_{e}	electron number density
			k	Boltzmann constant
... for a homogeneous sphere	$\dfrac{\Delta T}{T} = -\dfrac{4Rn_{\mathrm{e}}kT_{\mathrm{e}}\sigma_{\mathrm{T}}}{m_{\mathrm{e}}c^2}$	(9.52)	T_{e}	electron temperature
			σ_{T}	Thomson cross section
			m_{e}	electron mass
			c	speed of light

[a]Period–luminosity relation for classical Cepheids. Uncertainty in M_{V} is ± 0.27 (Madore & Freedman, 1991, Publications of the Astronomical Society of the Pacific, **103**, 933).
[b]Galaxy rotation velocity–magnitude relation in the infrared I waveband, centred at $0.90\,\mu\mathrm{m}$. The coefficients depend on waveband and galaxy type (see Giovanelli *et al.*, 1997, The Astronomical Journal, **113**, 1).
[c]Scattering of the cosmic microwave background radiation (CMBR) by a cloud of electrons, seen as a temperature decrement, ΔT, in the Rayleigh–Jeans limit ($\lambda \gg 1\,\mathrm{mm}$).

9.5 Stellar evolution

Evolutionary timescales

Free-fall timescale[a]	$\tau_{\rm ff}=\left(\frac{3\pi}{32G\rho_0}\right)^{1/2}$	(9.53)	$\tau_{\rm ff}$	free-fall timescale
			G	constant of gravitation
			ρ_0	initial mass density
Kelvin–Helmholtz timescale	$\tau_{\rm KH}=\frac{-U_{\rm g}}{L}$	(9.54)	$\tau_{\rm KH}$	Kelvin–Helmholtz timescale
	$\simeq\frac{GM^2}{R_0L}$	(9.55)	$U_{\rm g}$	gravitational potential energy
			M	body's mass
			R_0	body's initial radius
			L	body's luminosity

[a]For the gravitational collapse of a uniform sphere.

Star formation

Jeans length[a]	$\lambda_{\rm J}=\left(\frac{\pi}{G\rho}\frac{{\rm d}p}{{\rm d}\rho}\right)^{1/2}$	(9.56)	$\lambda_{\rm J}$	Jeans length
			G	constant of gravitation
			ρ	cloud mass density
			p	pressure
Jeans mass	$M_{\rm J}=\frac{\pi}{6}\rho\lambda_{\rm J}^3$	(9.57)	$M_{\rm J}$	(spherical) Jeans mass
Eddington limiting luminosity[b]	$L_{\rm E}=\frac{4\pi GMm_{\rm p}c}{\sigma_{\rm T}}$	(9.58)	$L_{\rm E}$	Eddington luminosity
	$\simeq 1.26\times10^{31}\frac{M}{M_\odot}$ W	(9.59)	M	stellar mass
			$M_\odot$	solar mass
			$m_{\rm p}$	proton mass
			c	speed of light
			$\sigma_{\rm T}$	Thomson cross section

[a]Note that $({\rm d}p/{\rm d}\rho)^{1/2}$ is the sound speed in the cloud.
[b]Assuming the opacity is mostly from Thomson scattering.

Stellar theory[a]

Conservation of mass	$\frac{{\rm d}M_r}{{\rm d}r}=4\pi\rho r^2$	(9.60)	r	radial distance
			M_r	mass interior to r
			ρ	mass density
Hydrostatic equilibrium	$\frac{{\rm d}p}{{\rm d}r}=\frac{-G\rho M_r}{r^2}$	(9.61)	p	pressure
			G	constant of gravitation
Energy release	$\frac{{\rm d}L_r}{{\rm d}r}=4\pi\rho r^2\epsilon$	(9.62)	L_r	luminosity interior to r
			ϵ	power generated per unit mass
Radiative transport	$\frac{{\rm d}T}{{\rm d}r}=\frac{-3}{16\sigma}\frac{\langle\kappa\rangle\rho}{T^3}\frac{L_r}{4\pi r^2}$	(9.63)	T	temperature
			σ	Stefan–Boltzmann constant
			$\langle\kappa\rangle$	mean opacity
Convective transport	$\frac{{\rm d}T}{{\rm d}r}=\frac{\gamma-1}{\gamma}\frac{T}{p}\frac{{\rm d}p}{{\rm d}r}$	(9.64)	γ	ratio of heat capacities, c_p/c_V

[a]For stars in static equilibrium with adiabatic convection. Note that ρ is a function of r. κ and ϵ are functions of temperature and composition.

Stellar fusion processes[a]

PP I chain	PP II chain	PP III chain
$p^+ + p^+ \rightarrow {}^2_1H + e^+ + \nu_e$ ${}^2_1H + p^+ \rightarrow {}^3_2He + \gamma$ ${}^3_2He + {}^3_2He \rightarrow {}^4_2He + 2p^+$	$p^+ + p^+ \rightarrow {}^2_1H + e^+ + \nu_e$ ${}^2_1H + p^+ \rightarrow {}^3_2He + \gamma$ ${}^3_2He + {}^4_2He \rightarrow {}^7_4Be + \gamma$ ${}^7_4Be + e^- \rightarrow {}^7_3Li + \nu_e$ ${}^7_3Li + p^+ \rightarrow 2{}^4_2He$	$p^+ + p^+ \rightarrow {}^2_1H + e^+ + \nu_e$ ${}^2_1H + p^+ \rightarrow {}^3_2He + \gamma$ ${}^3_2He + {}^4_2He \rightarrow {}^7_4Be + \gamma$ ${}^7_4Be + p^+ \rightarrow {}^8_5B + \gamma$ ${}^8_5B \rightarrow {}^8_4Be + e^+ + \nu_e$ ${}^8_4Be \rightarrow 2{}^4_2He$
CNO cycle	**triple-α process**	
${}^{12}_6C + p^+ \rightarrow {}^{13}_7N + \gamma$ ${}^{13}_7N \rightarrow {}^{13}_6C + e^+ + \nu_e$ ${}^{13}_6C + p^+ \rightarrow {}^{14}_7N + \gamma$ ${}^{14}_7N + p^+ \rightarrow {}^{15}_8O + \gamma$ ${}^{15}_8O \rightarrow {}^{15}_7N + e^+ + \nu_e$ ${}^{15}_7N + p^+ \rightarrow {}^{12}_6C + {}^4_2He$	${}^4_2He + {}^4_2He \rightleftharpoons {}^8_4Be + \gamma$ ${}^8_4Be + {}^4_2He \rightleftharpoons {}^{12}_6C^*$ ${}^{12}_6C^* \rightarrow {}^{12}_6C + \gamma$	γ photon p^+ proton e^+ positron e^- electron ν_e electron neutrino

[a]All species are taken as fully ionised.

Pulsars

Braking index	$\dot{\omega} \propto -\omega^n$	(9.65)	ω rotational angular velocity P rotational period $(=2\pi/\omega)$ n braking index
	$n = 2 - \frac{P\ddot{P}}{\dot{P}^2}$	(9.66)	
Characteristic age[a]	$T = \frac{1}{n-1}\frac{P}{\dot{P}}$	(9.67)	T characteristic age L luminosity μ_0 permeability of free space c speed of light
Magnetic dipole radiation	$L = \frac{\mu_0 \|\ddot{m}\|^2 \sin^2\theta}{6\pi c^3}$	(9.68)	m pulsar magnetic dipole moment R pulsar radius B_p magnetic flux density at magnetic pole θ angle between magnetic and rotational axes
	$= \frac{2\pi R^6 B_p^2 \omega^4 \sin^2\theta}{3c^3 \mu_0}$	(9.69)	
Dispersion measure	$DM = \int_0^D n_e \, dl$	(9.70)	DM dispersion measure D path length to pulsar dl path element n_e electron number density
Dispersion[b]	$\frac{d\tau}{d\nu} = \frac{-e^2}{4\pi^2 \epsilon_0 m_e c \nu^3} DM$	(9.71)	τ pulse arrival time $\Delta\tau$ difference in pulse arrival time ν_i observing frequencies m_e electron mass
	$\Delta\tau = \frac{e^2}{8\pi^2 \epsilon_0 m_e c}\left(\frac{1}{\nu_1^2} - \frac{1}{\nu_2^2}\right) DM$	(9.72)	

[a]Assuming $n \neq 1$ and that the pulsar has already slowed significantly. Usually n is assumed to be 3 (magnetic dipole radiation), giving $T = P/(2\dot{P})$.
[b]The pulse arrives first at the higher observing frequency.

Compact objects and black holes

Schwarzschild radius	$r_s = \frac{2GM}{c^2} \simeq 3\frac{M}{M_\odot}$ km (9.73)	r_s Schwarzschild radius G constant of gravitation M mass of body c speed of light $M_\odot$ solar mass
Gravitational redshift	$\frac{\nu_\infty}{\nu_r} = \left(1 - \frac{2GM}{rc^2}\right)^{1/2}$ (9.74)	r distance from mass centre ν_∞ frequency at infinity ν_r frequency at r
Gravitational wave radiation[a]	$L_g = \frac{32}{5}\frac{G^4}{c^5}\frac{m_1^2 m_2^2 (m_1+m_2)}{a^5}$ (9.75)	m_i orbiting masses a mass separation L_g gravitational luminosity
Rate of change of orbital period	$\dot{P} = -\frac{96}{5}(4\pi^2)^{4/3}\frac{G^{5/3}}{c^5}\frac{m_1 m_2 P^{-5/3}}{(m_1+m_2)^{1/3}}$ (9.76)	P orbital period
Neutron star degeneracy pressure (nonrelativistic)	$p = \frac{(3\pi^2)^{2/3}}{5}\frac{\hbar^2}{m_n}\left(\frac{\rho}{m_n}\right)^{5/3} = \frac{2}{3}u$ (9.77)	p pressure $\hbar$ (Planck constant)/(2π) m_n neutron mass ρ density
Relativistic[b]	$p = \frac{\hbar c (3\pi^2)^{1/3}}{4}\left(\frac{\rho}{m_n}\right)^{4/3} = \frac{1}{3}u$ (9.78)	u energy density
Chandrasekhar mass[c]	$M_{Ch} \simeq 1.46 M_\odot$ (9.79)	M_{Ch} Chandrasekhar mass
Maximum black hole angular momentum	$J_m = \frac{GM^2}{c}$ (9.80)	J_m maximum angular momentum
Black hole evaporation time	$\tau_e \sim \frac{M^3}{M_\odot^3} \times 10^{66}$ yr (9.81)	τ_e evaporation time
Black hole temperature	$T = \frac{\hbar c^3}{8\pi GMk} \simeq 10^{-7}\frac{M_\odot}{M}$ K (9.82)	T temperature k Boltzmann constant

[a] From two bodies, m_1 and m_2, in circular orbits about their centre of mass. Note that the frequency of the radiation is twice the orbital frequency.
[b] Particle velocities $\sim c$.
[c] Upper limit to mass of a white dwarf.

9.6 Cosmology

Cosmological model parameters

Hubble law	$v_r = Hd$ (9.83)	v_r radial velocity H Hubble parameter d proper distance
Hubble parameter[a]	$H(t) = \frac{\dot{R}(t)}{R(t)}$ (9.84) $H(z) = H_0[\Omega_{m0}(1+z)^3 + \Omega_{\Lambda 0} + (1-\Omega_{m0}-\Omega_{\Lambda 0})(1+z)^2]^{1/2}$ (9.85)	$_0$ present epoch R cosmic scale factor t cosmic time z redshift
Redshift	$z = \frac{\lambda_{obs}-\lambda_{em}}{\lambda_{em}} = \frac{R_0}{R(t_{em})} - 1$ (9.86)	λ_{obs} observed wavelength λ_{em} emitted wavelength t_{em} epoch of emission
Robertson–Walker metric[b]	$\mathrm{d}s^2 = c^2\mathrm{d}t^2 - R^2(t)\left[\frac{\mathrm{d}r^2}{1-kr^2} + r^2(\mathrm{d}\theta^2 + \sin^2\theta\,\mathrm{d}\phi^2)\right]$ (9.87)	$\mathrm{d}s$ interval c speed of light r,θ,ϕ comoving spherical polar coordinates
Friedmann equations[c]	$\ddot{R} = -\frac{4\pi}{3}GR\left(\rho + 3\frac{p}{c^2}\right) + \frac{\Lambda R}{3}$ (9.88) $\dot{R}^2 = \frac{8\pi}{3}G\rho R^2 - kc^2 + \frac{\Lambda R^2}{3}$ (9.89)	k curvature parameter G constant of gravitation p pressure Λ cosmological constant
Critical density	$\rho_{crit} = \frac{3H^2}{8\pi G}$ (9.90)	ρ (mass) density ρ_{crit} critical density
Density parameters	$\Omega_m = \frac{\rho}{\rho_{crit}} = \frac{8\pi G\rho}{3H^2}$ (9.91) $\Omega_\Lambda = \frac{\Lambda}{3H^2}$ (9.92) $\Omega_k = -\frac{kc^2}{R^2H^2}$ (9.93) $\Omega_m + \Omega_\Lambda + \Omega_k = 1$ (9.94)	Ω_m matter density parameter Ω_Λ lambda density parameter Ω_k curvature density parameter
Deceleration parameter	$q_0 = -\frac{R_0\ddot{R}_0}{\dot{R}_0^2} = \frac{\Omega_{m0}}{2} - \Omega_{\Lambda 0}$ (9.95)	q_0 deceleration parameter

[a]Often called the Hubble "constant." At the present epoch, $60 \lesssim H_0 \lesssim 80\,\mathrm{km\,s^{-1}\,Mpc^{-1}} \equiv 100h\,\mathrm{km\,s^{-1}\,Mpc^{-1}}$, where h is a dimensionless scaling parameter. The Hubble time is $t_H = 1/H_0$. Equation (9.85) assumes a matter dominated universe and mass conservation.

[b]For a homogeneous, isotropic universe, using the $(-1,1,1,1)$ metric signature. r is scaled so that $k = 0, \pm 1$. Note that $\mathrm{d}s^2 \equiv (\mathrm{d}s)^2$ etc.

[c]$\Lambda = 0$ in a Friedmann universe. Note that the cosmological constant is sometimes defined as equalling the value used here divided by c^2.

Cosmological distance measures

Look-back time	$t_{\mathrm{lb}}(z)=t_0-t(z)$	(9.96)
Proper distance	$d_{\mathrm{p}}=R_0\int_0^r \frac{\mathrm{d}r}{(1-kr^2)^{1/2}}=cR_0\int_t^{t_0}\frac{\mathrm{d}t}{R(t)}$	(9.97)
Luminosity distance[a]	$d_{\mathrm{L}}=d_{\mathrm{p}}(1+z)=c(1+z)\int_0^z\frac{\mathrm{d}z}{H(z)}$	(9.98)
Flux density–redshift relation	$F(\nu)=\frac{L(\nu')}{4\pi d_{\mathrm{L}}^2(z)}$ where $\nu'=(1+z)\nu$	(9.99)
Angular diameter distance[d]	$d_{\mathrm{a}}=d_{\mathrm{L}}(1+z)^{-2}$	(9.100)

$t_{\mathrm{lb}}(z)$	light travel time from an object at redshift z
t_0	present cosmic time
$t(z)$	cosmic time at z
d_{p}	proper distance
R	cosmic scale factor
c	speed of light
$_0$	present epoch
d_{L}	luminosity distance
z	redshift
H	Hubble parameter[b]
F	spectral flux density
ν	frequency
$L(\nu)$	spectral luminosity[c]
d_{a}	angular diameter distance
k	curvature parameter

[a]Assuming a flat universe ($k=0$). The apparent flux density of a source varies as d_{L}^{-2}.
[b]See Equation (9.85).
[c]Defined as the output power of the body per unit frequency interval.
[d]True for all k. The angular diameter of a source varies as d_{a}^{-1}.

Cosmological models[a]

Einstein – de Sitter model ($\Omega_k=0$, $\Lambda=0$, $p=0$ and $\Omega_{\mathrm{m0}}=1$)	$d_{\mathrm{p}}=\frac{2c}{H_0}[1-(1+z)^{-1/2}]$	(9.101)
	$H(z)=H_0(1+z)^{3/2}$	(9.102)
	$q_0=1/2$	(9.103)
	$t(z)=\frac{2}{3H(z)}$	(9.104)
	$\rho=(6\pi Gt^2)^{-1}$	(9.105)
	$R(t)=R_0(t/t_0)^{2/3}$	(9.106)
Concordance model ($\Omega_k=0$, $\Lambda=3(1-\Omega_{\mathrm{m0}})H_0^2$, $p=0$ and $\Omega_{\mathrm{m0}}<1$)	$d_{\mathrm{p}}=\frac{c}{H_0}\int_0^z\frac{\Omega_{\mathrm{m0}}^{-1/2}\,\mathrm{d}z'}{[(1+z')^3-1+\Omega_{\mathrm{m0}}^{-1}]^{1/2}}$	(9.107)
	$H(z)=H_0[\Omega_{\mathrm{m0}}(1+z)^3+(1-\Omega_{\mathrm{m0}})]$	(9.108)
	$q_0=3\Omega_{\mathrm{m0}}/2-1$	(9.109)
	$t(z)=\frac{2}{3H_0}(1-\Omega_{\mathrm{m0}})^{-1/2}\operatorname{arsinh}\left[\frac{(1-\Omega_{\mathrm{m0}})^{1/2}}{(1+z)^{3/2}}\right]$	(9.110)

d_{p}	proper distance
H	Hubble parameter
$_0$	present epoch
z	redshift
c	speed of light
q	deceleration parameter
$t(z)$	time at redshift z
R	cosmic scale factor
Ω_{m0}	present mass density parameter
G	constant of gravitation
ρ	mass density

[a]Currently popular.

Index

Section headings are shown in boldface and panel labels in small caps. Equation numbers are contained within square brackets.

B

I

I

I

F

G

I

M

N

Q

S

T

I

Z